Aerospace Engineering on the Back of an Envelope

Irwin E. Alber

Aerospace Engineering on the Back of an Envelope

 Springer

Published in association with
Praxis Publishing
Chichester, UK

Dr. Irwin E. Alber
Irvine
California
U.S.A.

SPRINGER–PRAXIS BOOKS IN ASTRONAUTICAL ENGINEERING

ISBN 978-3-642-22536-9   e-ISBN 978-3-642-22537-6
DOI 10.1007/978-3-642-22537-6
Springer Heidelberg Dordrecht London New York

Library of Congress Control Number: 2011935856

© Springer-Verlag Berlin Heidelberg 2012
This work is subject to copyright. All rights are reserved, whether the whole or part of the material is concerned, specifically the rights of translation, reprinting, reuse of illustrations, recitation, broadcasting, reproduction on microfilm or in any other way, and storage in data banks. Duplication of this publication or parts thereof is permitted only under the provisions of the German Copyright Law of September 9, 1965, in its current version, and permission for use must always be obtained from Springer. Violations are liable to prosecution under the German Copyright Law.
The use of general descriptive names, registered names, trademarks, etc. in this publication does not imply, even in the absence of a specific statement, that such names are exempt from the relevant protective laws and regulations and therefore free for general use.

Cover design: Jim Wilkie
Project copy editor: David M. Harland
Typesetting: OPS Ltd., Gt. Yarmouth, Norfolk, U.K.

Printed on acid-free paper

Springer is part of Springer Science + Business Media (www.springer.com)

Contents

Preface xiii

Acknowledgments xv

List of figures xvii

List of tables xxi

1 Introduction 1
  1.1 Why Back-of-the-Envelope engineering? 1
      1.1.1 Back-of-the-Envelope engineering; an important adaptation and survival skill for students and practicing engineers 2
      1.1.2 Design of a high school science fair electro-mechanical robot 2
      1.1.3 Design of a new commercial rocket launch vehicle for a senior engineering student's design project 3
      1.1.4 Preliminary design of a new telescope system by an engineer transferred to a new optical project 3
      1.1.5 Examining the principles and ideas behind Back-of-the-Envelope estimation. 4
  1.2 What is a Back-of-the-Envelope engineering estimate?. 4
      1.2.1 Tradeoff between complexity and accuracy. 5
      1.2.2 Back-of-the-Envelope reasoning 6
      1.2.3 Fermi problems 7
      1.2.4 An engineering Fermi problem 8
  1.3 General guidelines for building a good engineering model 12
      1.3.1 Step by step towards estimation. 13
      1.3.2 Quick-Fire estimates 14

vi **Contents**

    1.4   Quick-Fire estimate of cargo mass delivered to orbit by the Space Shuttle 15
           1.4.1   Cargo mass problem definition 15
           1.4.2   Level-0 estimate: the empirical "rule of thumb" model. . 16
           1.4.3   Level-1 estimate: cargo mass using a single stage mathematical model based on the ideal rocket velocity equation 17
           1.4.4   Level-2 estimate: cargo mass using a two stage vehicle model based on the ideal rocket velocity equation..... 21
           1.4.5   Level-3 estimate: cargo mass delivered by a two stage vehicle; based on a revised estimate for second stage structural mass fraction 23
           1.4.6   Impact of added knowledge and degree of model complexity 25
           1.4.7   Moving from the Shuttle to the Hubble Space Telescope 27
    1.5   Estimating the size of the optical system for the Hubble Space Telescope... 28
           1.5.1   System requirements for the HST................. 29
           1.5.2   Shuttle constraint on HST size 29
           1.5.3   Estimating the length of the HST optical package..... 30
    1.6   Concluding remarks.................................... 35
    1.7   Outline of this book 35
    1.8   References... 38

2   **Design of a high school science-fair electro-mechanical robot** 41
    2.1   The Robot-Kicker Science Fair Project 41
    2.2   Back-of-the-Envelope model and analysis for a solenoid kicking device 41
           2.2.1   Defining basic dimensions and required soccer ball velocity...................................... 42
           2.2.2   Setting up a BotE model for the solenoid kicking soccer ball problem 43
           2.2.3   Model for solenoid kicker work and force 50
           2.2.4   Final design requirements for linear-actuator solenoid and supporting electrical system....................... 62
    2.3   Appendix: Modeling of the temperature rise produced by ohmic heating from single or multiple solenoid-actuator kicks 63
           2.3.1   Quick-Fire problem approach 63
           2.3.2   Problem definition and sketch 69
           2.3.3   The baseline mathematical model................. 69
           2.3.4   Physical parameters and data 71
           2.3.5   Numerical results 72
           2.3.6   Interpretation of results 74
    2.4   References... 75

Contents vii

3   Estimating Shuttle launch, orbit, and payload magnitudes 77
    3.1   Introduction 77
          3.1.1   Early Space Shuttle goals and the design phase 78
          3.1.2   The Shuttle testing philosophy and the need for modeling   79
          3.1.3   Back-of-the-Envelope analysis of Shuttle launch, orbit, and payload magnitudes 80
    3.2   Shuttle launch, orbit, and reentry basics 80
          3.2.1   The liftoff to orbit sequence 80
          3.2.2   Reentry 82
    3.3   Inventory of the Shuttle's mass and thrust as input to the calculation of burnout velocity 83
          3.3.1   Burnout velocity 83
          3.3.2   The velocity budget 84
          3.3.3   Mass inventory 84
          3.3.4   Thrust and specific impulse inventory 84
    3.4   Mass fraction rules of thumb 87
    3.5   Quick-Fire modeling of the takeoff mass components and takeoff thrust using SMAD rules of thumb 89
          3.5.1   Quick-Fire problem approach 89
          3.5.2   Problem definition and sketch 90
          3.5.3   Mathematical/"Rule of Thumb" empirical models 90
          3.5.4   Physical parameters and data 92
          3.5.5   Numerical calculation of total takeoff mass, cargo bay mass, and total takeoff thrust 95
          3.5.6   Interpretation of the Quick-Fire results 97
          3.5.7   From Quick-Fire estimates to Shuttle solutions using more accurate inputs 98
    3.6   Ideal velocity change Δv for each stage of an ideal rocket system   98
          3.6.1   Propellant mass versus time 99
          3.6.2   Time varying velocity change 100
          3.6.3   Effective burnout time and average flow rate 100
          3.6.4   Ideal altitude or height for each rocket stage 101
    3.7 Δv_{ideal} estimate for Shuttle first stage, without gravity loss 102
          3.7.1   Estimate of SSME propellant mass burned during first stage 103
          3.7.2   First stage mass ratio and average effective exhaust velocity 104
          3.7.3   Average specific impulse for the "parallel" (solid + liquid) first stage burn 104
          3.7.4 Δv_{ideal} estimate for Shuttle first stage 106
          3.7.5 Δv_{ideal} and altitude as functions of time, for the Shuttle first stage 106
    3.8   The effect of gravity on velocity during first stage flight 108
          3.8.1   Modeling the effects of gravity for a curved flight trajectory 108

viii Contents

		3.8.2	Time-varying pitch angle model	110
		3.8.3	Effect of gravity on rocket velocity during first stage flight	111
		3.8.4	Effect of gravity on rocket height during first stage flight	111
		3.8.5	Comparing model velocity and altitude with Shuttle data	112
		3.8.6	Gravity loss magnitudes for previously flown launch systems	113
		3.8.7	Model velocity, with gravity loss, compared with flight data	114
		3.8.8	Calculation of gravity-loss corrected velocity at first stage burnout	115
	3.9	The effect of drag on Shuttle velocity at end of first stage flight		116
		3.9.1	Modeling the effects of drag in the equation of motion .	116
		3.9.2	Estimating first stage drag loss	117
		3.9.3	Final drag and gravity-corrected velocity at first stage burnout; key elements of the overall "velocity budget" for the first stage	120
	3.10	Calculation of second stage velocities and gravity losses		120
		3.10.1	Pitch and gravity loss modeling for the second stage flight period	121
		3.10.2	Time-varying gravity loss solution, region 2a	122
		3.10.3	Time-varying velocity solution, region 2b	124
		3.10.4	Combined velocity solution for regions 1, 2a, and 2b and v(MECO)	125
	3.11	Summary of predicted Δv budget for the Shuttle		126
	3.12	Comparison of Back-of-the-Envelope modeled Shuttle velocity and altitude as a function of time to NASA's numerical prediction for all stages		128
		3.12.1	Comparison of model velocity with NASA's numerical prediction	128
		3.12.2	Comparison of model altitude with NASA's numerical prediction	129
		3.12.3	Modeled altitude sensitivity to pitch time scale	130
	3.13	Estimating mission orbital velocity requirements for the Shuttle .		132
		3.13.1	Part 1: circular orbital velocity	132
		3.13.2	Part 2: elliptical orbits and the Hohmann transfer Δv's . .	135
		3.13.3	Numerical values for transfer orbit Δv's	138
		3.13.4	Time of flight for a Hohmann transfer	139
		3.13.5	Direct insertion to a final orbital altitude (without using a parking orbit)	139
	3.14	A Back-of-the-Envelope model to determine Shuttle payload as a function of orbit altitude		140
		3.14.1	Analytic model for payload as a function of orbital altitude	141

          3.14.2   Approximate linearized solution for payload   145
          3.14.3   Reduction in useful cargo mass due to increases in OMS
                    propellant mass   146
          3.14.4   OMS models for correcting cargo or payload mass   148
          3.14.5   Model for rate of change of "useful cargo" with altitude   150
          3.14.6   Approximate analytic model for useful cargo   152
          3.14.7   Modeling missions to the International Space Station. . .   154
   3.15   Tabulated summary of Back-of-the-Envelope equations and nu-
          merical results.   155
   3.16   References   166

4 **Columbia Shuttle accident analysis with Back-of-the-Envelope methods** .   169
   4.1   The Columbia accident and Back-of-the-Envelope analysis.   169
          4.1.1   BotE modeling goals for the Columbia accident   171
          4.1.2   Quick estimation vs accurate estimation   172
   4.2   Quick-Fire modeling of the impact velocity of a piece of foam
          striking the Orbiter wing.   172
          4.2.1   Interpretation of Quick-Fire results.   176
          4.2.2   The bridge to more accurate BotE results   176
   4.3   Modeling the impact velocity of a piece of foam debris relative to
          the Orbiter wing; estimations beyond the Quick-Fire time results   177
          4.3.1   Looking at the collision from an earth-fixed or moving
                    Shuttle coordinate system   178
          4.3.2   The constant drag approximation.   181
          4.3.3   Analytically solving for the impact velocity and mass,
                    given the time to impact   182
          4.3.4   Summary of results for constant acceleration model
                    compared to data   189
          4.3.5   The non-constant acceleration solution   189
          4.3.6   An estimate of impact velocity and particle mass, taking
                    the time to impact as given (the "inverse" problem) . . .   194
          4.3.7   Comparing Osheroff's "inverse" calculations to our
                    "direct" estimate results   195
          4.3.8   Concluding thoughts on the impact velocity estimate . . .   197
   4.4   Modeling the impact pressure and load caused by impact of foam
          debris with an RCC wing panel   197
          4.4.1   The impact load   197
          4.4.2   Impact overview   198
          4.4.3   Impact load mathematical modeling   199
          4.4.4   Elastic model for the impact stress   200
          4.4.5   Elastic–plastic impact of a one-dimensional rod against a
                    rigid-wall.   203

x Contents

    4.4.6 The elastic–plastic model 204
    4.4.7 Numerical results and plotted trends. 205
    4.4.8 Impact area estimate 207
    4.4.9 Load estimate 208
    4.4.10 Impact loading time scale (BotE) 209
    4.4.11 Loading time histories, numerical simulations 210
  4.5 Develop a Back-of-the-Envelope engineering stress equation for the maximum stress in the RCC panel face for a given panel load 211
    4.5.1 BotE panel stress model 211
    4.5.2 Estimates for the allowable maximum stress or critical load parameters for failure 213
    4.5.3 Final comments on the prediction of possible wing damage or failure 215
  4.6 Summary of results for Sections 4.2, 4.3, and 4.4   216
  4.7 References 220

5 Estimating the Orbiter reentry trajectory and the associated peak heating rates 221
  5.1 Introduction 221
  5.2 The deorbit and reentry sequence 222
  5.3 Using Quick-Fire methods to crudely estimate peak heating rate and total heat loads from the initial Orbiter kinetic energy 223
    5.3.1 Quick-Fire problem definition and sketch 224
    5.3.2 The Quick-Fire baseline mathematical model, initial results, and interpretation 224
  5.4 A look at heat flux prediction levels based on an analytical model for blunt-body heating 231
    5.4.1 Numerical estimates of Stanton number using the Sutton–Graves constant 236
  5.5 Simple flight trajectory model 237
    5.5.1 A simple BotE model for the initial entry period; the entry solution 239
    5.5.2 The equilibrium glide model 249
  5.6 Calculating heat transfer rates in the peak heating region 263
    5.6.1 Selecting the nose radius 265
    5.6.2 Comparing the model maximum rate of heat transfer, $\dot{q}_{w_{max}}$, with data 266
    5.6.3 Model estimate for nose radiation equilibrium temperature, T_{max} 267
    5.6.4 Model calculations of $\dot{q}_w$ as a function of time 268
    5.6.5 Model calculations for total heat load at the stagnation point 271
  5.7 Appendix: BotE modeling of non-Orbiter entry problems 274
  5.8 References 275

Contents xi

6 Estimating the dimensions and performance of the Hubble Space Telescope... 277
- 6.1 The Hubble Space Telescope 277
  - 6.1.1 HST system requirements 277
  - 6.1.2 HST engineering systems 278
  - 6.1.3 Requirements for fitting the HST into the Orbiter..... 279
- 6.2 The HST Optical Telescope design 280
  - 6.2.1 The equivalent system focal length 281
  - 6.2.2 How do designers determine the required system focal ratio, F_{eq}?................................ 284
  - 6.2.3 Telescope plate scale 288
  - 6.2.4 Selection of HST's primary mirror focal ratio, $F_1 = |f_1|/D$ 289
  - 6.2.5 Calculating the magnification m and exact constructional length L................................. 290
  - 6.2.6 Estimating the secondary mirror diameter 291
  - 6.2.7 Estimating the radius of curvature of the HST secondary mirror 292
- 6.3 Modeling the HST length 295
  - 6.3.1 The light-shield baffle extension................. 296
  - 6.3.2 Modeling the length of the light shield 297
  - 6.3.3 The length of the instrument section.............. 298
  - 6.3.4 Calculating the total HST telescope length.......... 298
- 6.4 Summary of calculated HST dimensions 299
- 6.5 Estimating HST mass.............................. 300
  - 6.5.1 Primary mirror design 300
  - 6.5.2 Estimating primary mirror mass.................. 301
  - 6.5.3 The estimated total HST system mass and areal density . 303
  - 6.5.4 Some final words on the HST mass estimation exercise . 305
  - 6.5.5 Onward to an estimate of HST's sensitivity 306
- 6.6 Back-of-the-Envelope modeling of the HST's sensitivity or signal to noise ratio 306
  - 6.6.1 Defining signal to noise ratio 307
  - 6.6.2 Modeling the mean signal, S................... 307
  - 6.6.3 Modeling the noise 310
  - 6.6.4 Final equation for signal to noise ratio 316
  - 6.6.5 Final thoughts on BotE estimates for HST sensitivity .. 318
- 6.7 References 319

Index.. 321

For my father, Louis Alber, a good and caring man, who taught me to keep true to my dreams.

For my mother, Charlotte Alber, still going strong at 97, whose great belief in my potential helped me to achieve success in school.

For my wife, Lana, the backbone of our family, whose love, guidance, and perseverance throughout my college and aerospace engineering years gave me the strength to take on the many professional challenges that I often faced. She always made sure that I took time off from research and writing to spend with our family and friends.

For my daughters, Deborah, Barbara, and Elizabeth, who have grown up to be strong, intelligent women in their own right.

For my wonderful grandchildren: Zachary, Samantha, and Leah Levine; and Thomas, Michael, and Anne Kenney. They make us proud every day. They are our hope for the future!

Preface

In my more than forty years working as an engineer on a variety of increasingly difficult problems in the aerospace industry, I have come to appreciate the strong computer-assisted design and analysis skills of many of my younger colleagues. Many recent engineering graduates are now quite adept at using, or modifying existing complex computer codes to design and to analyze engineering systems in those particular disciplines where they have previously obtained substantial education and hands-on experience.

To stay fully employed in the aerospace engineering business, working on challenging and rewarding problems, an engineer must always be able to learn and to utilize the technical skills needed to support the latest engineering contract goals.

The ability to quickly adapt to and master a new engineering challenge or design problem is a very important survival skill for today's engineers. Most importantly, one must revive certain basic technical skills or concepts that one learned long ago and subsequently forgot in the stressful day-to-day working environment.

A critical survival skill is the ability to conceptually work out in one's own mind, and on paper, the basic technical underpinnings (or the physics) of a complex real-world problem that may involve several new analysis and design elements. To be able to quickly comprehend a new engineering problem or goal and to set forth on a new design effort, an engineer needs the skills to conceptualize and develop *simplified* mathematical and/or empirical models for a range of complex physical processes using a number of different engineering approximations; and the ability to quickly calculate *quantitative* estimates of the problem's key design parameters and to interpret their meaning, even if there is little or no supporting data at hand.

Another way of restating these points is:

An engineer must have strong "Back-of-the-Envelope" engineering estimation skills.

It is said that a good system designer is usually skilled in the *art of estimation*. However, I have observed that many of today's aerospace engineers are *unable* to estimate key engineering dimensions and requirements without the aid of an already existing engineering model or software code. Because they do not know the basics of estimation, engineers often cannot quickly select the most important physical scales and parameters to characterize real aerospace engineering systems. It follows that they have difficulty identifying those design variables that can be safely neglected, or bypassed, when tackling a new problem. The ability to estimate can be quite useful in limiting the extent of a major analysis effort or the range of testing needed to advance a new design.

To be able to develop quick design concepts and to estimate the important geometric, performance, and design dimensions for a new problem, an engineer needs to learn how to effectively utilize Back-of-the-Envelope (BotE) estimation skills. These provide important rough quantitative guidelines and/or answers to real-world problems, particularly for projects with enormous complexity and very limited resources.

In the six chapters of this book, we illustrate the importance and use of BotE techniques by primarily applying them to two major real-world engineering case studies, the Space Shuttle and the Hubble Space Telescope. We also examine a few less-well-known engineering systems.

The goal of this book is not to try to cover all of the major design issues for these case studies, but rather to illustrate how an engineer might use BotE skills to obtain a first-order quantitative "feel" for the magnitudes and sensitivities of several of the key system variables driving the design and performance metrics that characterize these two complex aerospace systems. We show, by example, how an engineer can quickly use BotE techniques (and perhaps a #2 pencil) to obtain initial *quantitative* estimates of the sizes and performance measures of key components for the Space Shuttle and Hubble Space Telescope projects.

Such engineering estimates can often be used to guide the development of more complex computer modeling studies and initial test programs so that engineers can confidently advance to the next step in a project's system design.

Irwin Alber
Irvine, California

Acknowledgments

I wish to express a deep measure of gratitude to my mentors, Professors Lester Lees and Toshi Kubota for their unique approaches to teaching, and for their personal guidance when I was a graduate student in Aeronautics at Caltech. The regular "Back-of-the-Napkin" coffee breaks in the cafeteria by our Hypersonics Group led me many years later to the basic theme for this book.

A number of professors, colleagues, and friends reviewed key portions of this book and made quite a few valuable suggestions for improvement.

Dr. Thomas Tyson (a former member of the Caltech Hypersonics Group) and Angelo Vassos (high school physics teacher, retired) helped greatly by reviewing Chapter 4, where I address the high-speed foam impact problem associated with the Space Shuttle Columbia accident. This was the first topic that I addressed using Back-of-the-Envelope techniques.

Jake Jacobs (an electrical engineer) reviewed Chapter 2 on the design of a solenoid robot kicking device. He electronically redrew a number of important figures to help me clarify some of the chapter's electromagnetic concepts and recommended circuitry.

Dr. Ian Clark, Georgia Tech visiting aerospace engineering professor, and two of his top graduate students performed a thorough and insightful review of Chapter 5, which focused on the application of the Back of the Envelope methods developed in this book to the Space Shuttle reentry problem. Based on their detailed comments, an appendix was added to the chapter that discusses how these basic methods can be more broadly applied to a number of *non-Shuttle* entry systems with differing aerodynamic and thermal characteristics.

I want to thank Dr. Donald Rapp (Jet Propulsion Laboratory, retired) who has written six technical books. In a casual lunchtime conversation on a boat cruise to Alaska, we discussed our mutual interests and I gave him a brief description of my ongoing book effort. As a result he recommended my project to his publisher. He later reviewed Chapter 3 on the launch, orbit, and payload magnitudes of the

Shuttle, and recommended that I add a number of Quick-Fire estimation examples to the text.

In addition, I received substantial encouragement from Caltech Professor emeritus, Anatol Roshko; Drs. Susan and James Wu (Professors emeritus at the University of Tennessee Space Institute); and Dr. Wilhelm Behrens (Technical Fellow, Northrop Grumman Aerospace Systems). Finally, I want to especially thank my wife, Lana, for her full support and for bearing with me throughout the long periods of research and the writing of this book.

Figures

1.1	Relative cost/time vs resolution in problem solving	5
1.2	Back of the Envelope estimation of 747 engine thrust	9
1.3	Modeling methodology flow diagram	13
1.4	Space Shuttle on the launch pad	16
1.5	Shuttle cargo mass estimates under different assumptions	26
1.6	Single stage cargo mass vs specific impulse, Isp	27
1.7	Configuration of Hubble Space Telescope (HST) system	28
1.8	HST in Orbiter cargo bay prior to launch	30
1.9	BotE estimate for HST optical package length L	31
2.1	Schematic of model soccer field with player	43
2.2	Forces acting on a linearly translating spherical ball	44
2.3	Ratio of natural roll ball velocity to initial plunger velocity	49
2.4	Ratio of plunger velocity to natural roll velocity vs kicking height	49
2.5	Solenoid and plunger schematic	51
2.6	Push solenoid schematic	51
2.7	Push-pin solenoid showing time varying gap length	52
2.8	Plunger velocity vs distance from initial plunger position	56
2.9	Solenoid force as a function of gap distance	58
2.10	Maximum final velocity and maximum work vs ampere-turns	61
2.11	Circuit schematic for solenoid driver	63
2.12	Thermal balance "lumped capacitance" models for solenoid	70
3.1	Shuttle Orbiter with main engines and solid boosters on launch	78
3.2	Shuttle launch, orbit, and reentry mission elements	81
3.3	Shuttle drawing with mass components annotated	85
3.3a	Shuttle drawing with mass components to be determined	90
3.4a	Ideal Shuttle velocity as a function of flight time	107
3.4b	Ideal Shuttle altitude as a function of flight time	107
3.5	Force balance on point mass rocket on a curved trajectory	109
3.6	Shuttle velocity vs flight time with gravity loss	111
3.7	Shuttle altitude vs flight time with gravity loss	112

xviii **Figures**

3.8	Shuttle altitude with gravity loss compared to STS-1 data, 1st stage	113
3.9	Shuttle velocity with gravity loss compared to STS-1 data, 1st stage	114
3.10	Modeled flight dynamic pressure, q, vs flight time	117
3.11	Nose on view of Shuttle	119
3.12	Shuttle velocity vs time, 1st stage, 2nd stage: (2a), 2nd stage: (2b)	124
3.13	Comparison of model and NASA predicted flight velocities	129
3.14	Shuttle altitude vs time; gravity-loss model, 1st stage, 2nd stage: (2a)	130
3.15	Comparison of model and NASA predicted flight altitudes	131
3.16	Comparison of two versions of three region model for altitude	131
3.17	Point mass in circular orbit	133
3.18	Hohmann elliptical transfer orbit	135
3.19	Shuttle payload vs orbital altitude compared to mission data	145
3.20	Shuttle payload prediction with OMS propellant correction	147
4.1	Insulating foam hitting the left wing of the orbiter; graphical image	170
4.2	Foam motion viewed in earth-fixed and Shuttle fixed coordinates	179
4.3	Model foam triangular-pyramid shape	184
4.4	Predicted impacted velocity vs Ballistic Number, BN	187
4.5	Predicted impacted time vs Ballistic Number, BN	188
4.6	Comparison of exact velocity solution to approximation model	191
4.7	Model impact velocity vs BN; exact relative to approximate	192
4.8	Model impact time vs BN; exact relative to approximate	193
4.9	Stress vs strain for foam; elastic and plastic regions	199
4.10	Moving 1D rod impacting a rigid wall generating elastic wave	200
4.11	Elastic–plastic rod impacting rigid wall generating plastic shock	203
4.12	Modeled foam impact stress vs normal velocity	205
4.13	Impact stress dependence on impact speed and incidence angle	207
4.14	Footprint of impacts against RCC panels	208
4.15	Impact load vs impact speed and incidence angle	209
4.16	Impact loading curve vs time; schematic	209
4.17	Impact load curve vs time; finite element simulation results	210
4.18	Impact of foam cylinder on flat plate; plate bending under pressure	212
4.19	Damage/No-damage model transition curve	215
5.1	Orbiter reentry mission elements	222
5.2	Sketch of reentry length scales and drag force	226
5.3	Orbiter stagnation region shock layer and boundary layer	227
5.4	Force balance on reentry vehicle on a curved trajectory	238
5.5	Flight path angle solution during the early entry period	245
5.6	Altitude vs time for the early entry period; comparisons to data	245
5.7	Velocity vs time for the early entry period; comparisons to data	246
5.8	Flight path angle during the early entry period; compared to data	249
5.9	Altitude vs time for early entry through equilibrium glide periods	257
5.10	Velocity vs time for early entry through equilibrium glide periods	257
5.11	Flight path angle vs time for early entry and equilibrium glide	258
5.12	Altitude and velocity vs t; entry interface to touchdown (1820s)	259
5.13	Typical Orbit entry trajectory data including bank angle vs time	261
5.14	Altitude and velocity vs t; impact of bank angle correction	262
5.15	Altitude and velocity vs t; impact of change in angle of attack	263
5.16	Modeled heat flux vs time; $275 < t < 1{,}200$ s for glide regime	269
5.17	Heat flux vs time; $0 < t < 1{,}200$ s for initial entry and glide regime	269

5.18	Heat flux model vs time; $0 < t < 1,200$ s compared to STS-5 data	270
5.19	Heat flux and total heat load; model vs data from several sources	273
6.1	Configuration of Hubble Space Telescope System	278
6.2	Two-mirror Cassegrain optics with dimensions and mirrors specified	280
6.3	Effective or equivalent focal length defined graphically	282
6.4	Ray geometry for telescope based on calculated dimensions	283
6.5	Airy diffraction pattern	285
6.6	Diffraction patterns; two stars separated by Sparrow resolution scale	285
6.7	Relating position on focal plane to measured sky angle; ray sketch	286
6.8	Reflection of rays from spherical mirror	292
6.9	Stray light blocked by HST light shield schematic	297
6.10	HST primary mirror with honeycomb core	301
6.11	Structural components of support systems module	304
6.12	Poisson distribution	311
6.13	SNR vs star magnitude m for several exposure times	317
6.14	Exposure time vs star magnitude m for fixed SNR	318
6.15	Exposure time vs m; background sky limited approximation	319

Tables

1.1	747 thrust levels for three jet engine manufacturers	12
2.1	Summary of BotE calculations; dynamics of rolling soccer ball	64
2.2	Summary of BotE calculations; basic solenoid actuator design	66
3.1	Shuttle propellant, structure and payload masses; 1st and 2nd stages	86
3.2	Thrust, specific impulse, and mass flow rates; 1st and 2nd stages	87
3.3	Mass fractions for typical spacecraft compared to Space Shuttle	88
3.4	Shuttle velocity budget by stage and flight region	127
3.5	OMS propellant mass data for selected missions	148
3.6	Summary of BotE calculations, 1st stage burnout; no gravity model	156
3.7	Summary of BotE calculations, 1st stage burnout; with gravity and drag	158
3.8	BotE calculations, 2nd stage burnout; with gravity and drag loss	161
3.9	BotE calculations, required orbital velocity and payload mass estimate	164
4.1	STS-107 flight data; velocity, altitude and time for foam break off	183
4.2	Direct and indirect calculations of impact velocity and impact time	196
4.3	Data for properties of foam: crush stress, elastic wave speed	206
4.4	Summary of calculations for impact velocity, time, and accelerations	217
4.5	Summary of calculations for impact stress, load, loading time scale	218
4.6	Summary of calculations for RCC bending stress and failure criterion	219
5.1	Equivalent nose radius for Orbiter heat transfer calculations	266
6.1	Calculated dimensions/scales for Hubble Space Telescope	299

1
Introduction

1.1 WHY BACK-OF-THE-ENVELOPE ENGINEERING?

I was first made aware of the value of Back-of-the-Envelope (BotE) estimation techniques when I was an aeronautics graduate student at Caltech in the early 1960s.

My thesis advisor and mentor, Professor Lester Lees, taught us, by example, his approach to building basic engineering models for complicated real-world problems. He showed us how simple models could be used to estimate the solutions to problems that not only served the needs of our particular graduate research projects, but often had broader import in a number of related aerospace engineering endeavors. We didn't learn these techniques in the classroom, but in the informal setting at the coffee table in the Caltech student cafeteria. Professor Lees and his colleague, Professor Toshi Kubota, would sit with us grad students and discuss one or more of our latest research roadblocks. At some key point in the discussion, Professor Lees would grab whatever was handy—a napkin or a paper envelope—and begin to sketch out some particular aerodynamic flow field while discussing the underlying physics of the problem with us. He would add a few applicable fluid or thermodynamic equations and perhaps make an estimate of the magnitude of a particular parameter that was central to that problem (e.g. Mach number or Reynolds number). Professor Lees would then hand the napkin or "envelope" to one of us and ask that particular student to give some thought to the assumptions and physics of the model that he had just jotted down. He would give us a week or two to ponder the meaning of his "Back-of-the-Napkin" scrawls, and then invite one or more students to his office to discuss what we had learned and to recommend the next step along the research road to solving our particular thesis problem.

These teacher/student give and takes, assisted by a series of simple Back-of-the-Envelope estimates of the problem at hand, helped many of us to get to the heart of the difficult research issues that faced us. These coffee seminars had a profound influence on all of Professor Lees' students, and we carried forward into our later

---

I. E. Alber, *Aerospace Engineering on the Back of an Envelope*.
© Springer-Verlag Berlin Heidelberg 2012.

professional lives this process of sketching out and thinking through a problem in order to quickly arrive at an initial solution.

1.1.1 Back-of-the-Envelope engineering; an important adaptation and survival skill for students and practicing engineers

Before embarking on the case studies that demonstrate the use of BotE estimation methods for complex aerospace systems, the following important question should be considered: Who should read this book?

In the following section we suggest that the readership should include: (a) physics and mathematics students, beginning as early as the high school level; (b) college undergraduate and graduate engineering students; and (c) practicing aerospace engineers. It is our contention that a wide range of students and professionals can benefit from a thorough understanding of the fundamentals and case-study applications of Back-of-the-Envelope engineering estimation techniques. We illustrate this proposition by considering three basic scenarios.

Let's start by considering a high school physics student setting out to design and build his first science fair project.

1.1.2 Design of a high school science fair electro-mechanical robot

You are a junior in high school and have just completed your first AP physics class. You want to enter a project in the regional science fair based on your interest in designing, building, and operating a moving electro-mechanical robot that will be a team "player" in a robot soccer tournament. Specifically, your aim is to design a robot that has the ability to grab and trap a loose soccer ball that is in play, and then to rapidly kick the ball past a defending goalie and into the net of the opposing team. With help from your physics instructor you are encouraged to work out the design of a basic robot kicker based on a simple physics model for such a device. The initial design concept is sketched out by you and your teacher literally on the "back of an envelope" and then on the blackboard.

Your initial assignment is to mathematically model—using mechanical and electro-magnetic principles learned in your high school physics classes—the performance of a robotic kicker powered by a rapidly-acting linear solenoid actuator. Your initial Back-of-the-Envelope mathematical analysis for the robot kicker, along with several graphs of the derived solutions for a range of design parameters (e.g. solenoid current and actuator plunger size), is presented in Chapter 2 of this book. After using the results of this simplified model analysis to define your design, you then proceed to locate and purchase (with the financial support of your family) the supplies and tools to construct your kicker, which you go on to enter in a state-wide robot soccer competition.

Sec. 1.1]   Why Back-of-the-Envelope engineering?   3

1.1.3 Design of a new commercial rocket launch vehicle for a senior engineering student's design project

You are a senior mechanical/aerospace engineering student about to graduate from a major university, and you have just started to interview for your first job.

Your senior year research project requires you to join with a group of your fellow classmates in designing a new rocket launch vehicle which must be low cost and capable of delivering a space capsule containing seven astronauts into low earth orbit to perform a number of different satellite maintenance missions for NASA. Your commercial launcher should also be capable of delivering astronaut crews and supplies to the International Space Station on a regular basis at a cost well below that of the Space Shuttle. Your professor and an aerospace engineer friend encourage you to first acquaint yourself with the early design and analysis studies undertaken for the Shuttle, because they consider that this engineering case study will serve as a very important real-world reference for your senior research project.

You are given only a month to complete your effort, so you decide to use simple Back-of-the-Envelope estimation techniques to develop a quick design. In order to gain some practical experience using these modeling techniques you first tackle the problem of determining the cargo mass that the Shuttle is able to deliver to different low earth orbits. The basic Back-of-the-Envelope launch and propulsion analysis that you carry out for the Shuttle, together with several graphs demonstrating the estimated reduction in cargo mass with increasing orbital altitude, is presented in Chapter 3 of this book. You find that your initial Shuttle modeling estimates compare favorably to NASA data. This gives you confidence that the new commercial rocket design that your team is tasked with developing can be quickly analyzed with some degree of confidence.

You then estimate the major propulsion and structural parameters for your new commercial vehicle design using the Back-of-the-Envelope techniques that you've just mastered. You perform a rapid parametric analysis of the key design components, select the best of the candidate solutions, and present your design recommendations in a final summary report.

You not only earn high marks from your aerospace professor, you also use your senior study as an example of the valuable technical capabilities that you possess when applying for an engineering staff position with several recently established commercial aerospace companies.

1.1.4 Preliminary design of a new telescope system by an engineer transferred to a new optical project

You are an engineer working for a large aerospace engineering company. You have 10 years of experience in launch systems design. However, your firm's launch system department has just lost its bid to extend its long-term contract with the government. Your only hope of remaining employed with the firm is to transfer into a newly established department that has a new contract to mathematically model, design, and test a new telescope system. This new telescope will be a central part of a remote-sensing satellite that is to be launched in three year's time.

You know very little about telescope design and it has been 15 years since you took an optics course at your engineering *alma mater*. You need to get up to speed quickly in telescope optics and to comprehend the supporting technology. You seek to convince your assigned optics design manager that you can handle the project's proposed engineering work. You read a few technical reports and several texts in this area, but you simply don't understand the magnitude and functionality of the basic design parameters, nor how all the optical elements work together. You decide to construct a basic analytical model of the optical system in order to better understand the key elements driving the new design. You first carry out a basic Back-of-the-Envelope mathematical analysis for the reflective mirror components of a similar telescope system, the Hubble Space Telescope (HST). Your model equations, plus your estimate of the HST's focal plane signal to noise ratio for weak light sources, is given in Chapter 6 of this book. After comparing your model calculations to published HST optical design dimensions and signal to noise ratios for a range of star magnitudes, you feel sufficiently confident to begin a transition to the new optical engineering project.

1.1.5 Examining the principles and ideas behind Back-of-the-Envelope estimation

Before proceeding with these and other BotE case studies (including a BotE study of the Shuttle Columbia accident of 2003, and a separate study to determine the peak heating rates for the vehicle during reentry), we turn our attention to examining the fundamental principles and ideas underpinning the estimation process. We cite a number of important papers and books on the subjects of modeling and estimation below, and highlight the famous "Fermi problem" as a classical estimation problem for which little *a-priori* information is available.

We start by examining several key definitions of what we mean by BotE reasoning and estimation.

1.2 WHAT IS A BACK-OF-THE-ENVELOPE ENGINEERING ESTIMATE?

Using the concepts outlined in Section 1.1, we can define a BotE engineering estimate as a quick, sometimes rough, quantitative estimate of the key design and performance parameters for real-world engineering systems. The calculations are based on simplified mathematical and/or empirical models that approximate the physics of complex engineering processes. Often there is little or no credible data available to serve as input to these calculations.

In outlining the University of California at Davis course entitled "Principles of Environmental Science", Peter Richerson observes that environmental scientists often use simple Back-of-the-Envelope methods to model complex environmental processes. He characterizes BotE model calculations as follows [1]:

"The idea is to boil a complex physical problem to a simplicity that lies at its heart. You try to make some conservative simplifying assumptions, perhaps heroic simplifying assumptions. This often requires a willingness to ignore all sorts of possible objections ... Then you turn the crank and see what pops out. Then you ask if the answer is reasonable given the science you know and whatever data you have at hand. If not, you take another cut ... No surefire recipe exists to pick the right simple model to apply to a complex problem."

This statement captures for us the meaning and spirit of BotE model building and estimation.

1.2.1 Tradeoff between complexity and accuracy

When using BotE techniques, one should provide the recipient of your work with a measure of the approximations made at every step of model development. It is also important to note that there is always a tradeoff between model complexity (i.e. the cost and time) and the certainty (the statistical error) of your engineering estimate. Figure 1.1 shows the tradeoff between "problem cost" and "problem certainty" (or the error in estimating the parameters of a given problem) for four different estimation techniques. These estimation classes range in complexity from crude best-guess

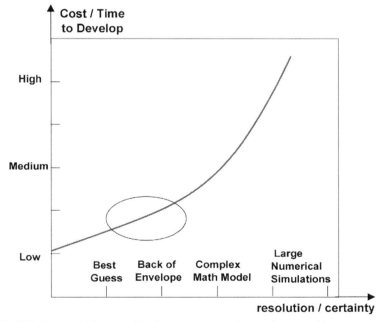

Figure 1.1. Relative cost/time to develop an answer to a given problem vs resolution or certainty of the resulting solution. The BotE techniques presented in this book fall within the ranges of certainty and cost denoted by the "oval".

methods to large complex numerical simulations. Each estimation method would be symbolically assigned a level of certainty. Lower costs are obviously associated with those methods with the lowest degree of certainty, and increasingly higher costs are associated with those methods with increasingly higher levels of certainty.

The methods and costs range in order of certainty from: (1) an educated best guess which has the lowest certainty (or the highest statistical error) but has the lowest cost; to (2a) a crude BotE order of magnitude estimate; to (2b) a better BotE estimate based on a simple mathematical model; to (3) an even more complex mathematical model requiring sophisticated analytical and/or numerical methods; to (4) very large, often multi-dimensional numerical code simulations which have the highest potential certainty (or the least statistical error) but cost the most. One should not think of each such method as pertaining to a discrete range of certainty and cost, rather they represent a continuum of estimation techniques with one blending into another. The BotE methods employed in this book are represented by the oval in Figure 1.1, and they range in development difficulty from "best guess" to "detailed", but not "overly complex" mathematical models.

We reiterate the quote from Richerson [1]: "No surefire recipe exists to pick the right simple model to apply to a complex problem."

In Sections 1.4 and 1.5 we present several estimation examples and demonstrate by example how a given engineering problem can be tackled using estimation techniques constructed with different model complexities driven by key input parameters derived from different knowledge databases. We will show how the *certainties* of our results can be altered by the amount of *a-priori* data and model knowledge that is available to the engineer when taking on a specific estimation problem.

1.2.2 Back-of-the-Envelope reasoning

There are a number of key papers in the literature that emphasize one or more aspects of the BotE estimation problem. We cite one such definition by Paritosh and Forbus [2]. In "Common Sense on the Envelope", the authors present the following definition of BotE reasoning:

> "Back of the envelope reasoning involves the estimation of rough but quantitative answers to questions. Most of the questions are real-world problems where usually one does not have complete or accurate models or model parameters. This type of reasoning is particularly common in engineering practice and experimental sciences, including activities like evaluating the feasibility of an idea, planning experiments, sizing components, and setting up and double-checking detailed analyses."

They also state that BotE estimation can be decomposed into two distinct (but not independent) processes:

- *Direct parameter estimation*—This involves directly estimating a parameter based on previous experience or domain knowledge. The methods here are either direct retrieval or statistical estimation based on a set of available data.

- *Building an engineering estimation model*—When the parameter to be estimated is not directly stored or encountered, one has to build a model that relates the parameter in question to other parameters that have been estimated by direct parameter estimation.

Stated in our own terminology, in Process 1 an engineer can try to look up in a textbook or recall from memory a specific magnitude for a parameter. He also, for example, can calculate the mean value of that particular parameter from one or more available data sets. In Process 2, an engineer starts out by developing a mathematical or empirical model of the physical processes under study in order to calculate the parameter of interest. One uses other related parameters (from Process 1) as input to the model calculations (e.g. the approximate mass and diameter of a soccer-ball for the robot kicker problem considered in this book).

In some estimation problems we cannot create a well-defined mathematical or empirical model for the problem at hand, we can only guess at certain loose relationships between the problem variables. To compound the difficulty, there may be almost no reliable supporting data. Here, a simple solution path must be conceptualized and rough approximations made so as to generate an estimate. Fortunately, the primary goal in such problems is often to estimate only the *order of magnitude* of the solution. We cite here a famous example of such a problem.

1.2.3 Fermi problems

The eminent physicist Professor Enrico Fermi used to give unusual estimation problems to his graduate students that appeared to be impossible. In his honor, they were called Fermi problems [3]. He posed such difficult problems as a way of teaching his students how to: (1) make use of commonly available knowledge regarding the scale of a problem; (2) make reasonable assumptions regarding the connections between problem variables; (3) make reasonable estimates of unknown data; and (4) use all available information and assumptions to perform simple order of magnitude calculations to generate the results. In particular, he emphasized that it is very important to be explicit about each approximation so that one can later recheck each step in order to find out why one's *final* answer might be "way out of the ball park".

Perhaps the most famous Fermi problem posed more than 50 years ago and often quoted is: *How many piano tuners are there in Chicago?*

Fermi taught his students that many different approaches, often using non-traditional methods, can be exploited in pursuit of a solution. Here is one such approach that has been cited in many textbooks.

To initiate the piano tuner problem, one might guess the number of people in Chicago [4]. You might only know this to an order of magnitude, usually to within a factor of two. For example, let's say there are 5 million people today in greater Chicago. (As a point of reference, the latest census says that there are nearly 3 million within the city boundaries, and close to 10 million in the greater Chicago area.) An estimate is now needed of the fraction of people in the city that might own or rent a piano. So let's assume that the number of families in the city equals 1/5th of the population, and that 1 family in 20 has a piano that might be in need of a piano tuning. Hence there are $(5 \times 10^6)(1/5)(1/20) = 50{,}000$ pianos in the city. If we further assume that each piano gets tuned every two years (and here we are merely guessing) then there are approximately 25,000 tunings per year. Now, if a piano tuner can perform his tuning job at the rate of one piano per hour and he works for 2,000 hours per year, he is able to tune 2,000 pianos per year. The approximate number of piano tuners in the city is then calculated as follows:

$$\# \text{ piano tuners} = (\# \text{ tunings/year})/(\# \text{ tunings/piano tuner/year})$$
$$= (25 \times 10^3)/(2 \times 10^3)$$
$$\approx 13$$

How good is the result?

This particular estimate of 13 piano tuners is probably correct to within one order of magnitude, but it is possible to make a crude check of reality by looking up the number of advertisements for piano tuners in Chicago on the internet, or in the Chicago yellow pages. Note that in the absence of hard data we have guessed a number of the intermediate estimation parameters (e.g. that there are 5 people per family). We often just use common sense and our own personal experience.

1.2.4 An engineering Fermi problem

We will follow the spirit of the Fermi problem approach when we make BotE estimates for the engineering case study problems presented in this book. In recognizing that there is often little readily-available engineering data, we try to guess or estimate particular values using our "engineering" experience in order to try to obtain a realistic final estimate.

For our real-world engineering case study problems the goal is to generate BotE estimates that differ from engineering "truth" by less than a factor of two.

To get a feel for the possible estimation approaches let's try our techniques on the following estimation question; let's call it an *engineering* Fermi problem: *What is the thrust of a Boeing 747 jet engine?*

This aerospace engineering Fermi question was one of many posed by Benjamin Linder in his 1995 PhD thesis entitled "Understanding estimation and its relation to engineering education" [5]. We note that the question as posed is not explicit when it comes to defining the thrust. Linder commented that the person solving this problem does not know if "thrust" refers to the thrust required for takeoff, or the thrust for the nominal cruising speed, or indeed to the maximum engine thrust. It is up to the engineer, or perhaps his manager, to further refine the problem. For our BotE estimate, detailed below, we assume that the quantity of interest is "takeoff thrust", which as propulsion engineers know is close to the maximum thrust that a jet engine generates.

The objective of Linder's thesis was a survey of MIT mechanical engineering students (and of their counterparts at five other schools) to try to understand any difficulties that students might have in making simple estimates of engineering quantities such as force and energy.

In a brief test set by Linder, students were given 5 minutes to estimate the energy stored in a common 9 volt chemical battery, and their answers varied by about 9 orders of magnitude! The students were also asked to estimate the drag on a bicycle rider and the power output of a small motor. Linder found that engineering graduates—even from the best engineering schools—are not very good at estimation primarily because they lack the fundamental knowledge and skills required to make reasonable estimates. He also found that students were unable to perform BotE calculations because often they could not specify the correct engineering units for the quantities under study [6].

We will now present an example set of approximations, assumptions, and methods to calculate a BotE estimate for the 747 thrust problem.

1.2.4.1 Problem definition for the 747 thrust estimation problem

- *Quickly estimate the jet engine thrust for a Boeing 747 using Back-of-the-Envelope techniques.*
- Assume that the quantity of interest is *takeoff* thrust.

For this prototype problem, the estimation steps have been written out on the back of a standard-size envelope (see Figure 1.2).

The following is a detailed explanation of the assumptions and calculations I made. This explanation of the particular steps taken here uses typical BotE solution procedures as a *guide*. You might formulate a different alternative solution path. No one path is deemed optimal.

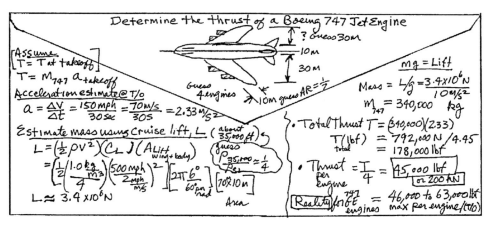

Figure 1.2. My hand-drawn sketch and calculations, estimating the thrust of a Boeing 747 engine directly on the back of an envelope using BotE techniques. The final single engine thrust estimate is compared with published results for those jet engines built primarily by General Electric.

10 Introduction [Ch. 1

1.2.4.2 The sketch

At the top of the "envelope" shown in Figure 1.2, I've drawn a sketch of a crude 747 aircraft with some relevant dimensions. These dimensions are based on my personal visual recollection of the 747 jet aircraft that I flew on as a passenger to Australia about 10 years ago.

1.2.4.3 Gathering the input data needed for the calculation

I recall the 747 had four engines and some pretty large wings. I estimate that the wing, on each side of the fuselage, is close to 100 feet in length, or about 30 m. The width of each wing is perhaps approximately 1/3rd of its length, call it 10 m.

When the engines are fully revved up, the 747 moves down the runway and accelerates to full takeoff speed. I estimate the takeoff speed at about 150 mph or 75 m/s since it seemed to me that just prior to leaving the ground the speed of the aircraft, in relative terms, was about twice the speed of a car on the nearby freeway. I remember looking at my watch to ascertain the length of time it took the plane to become airborne, and it was about 30 seconds. After reaching cruising altitude on most long flights the pilot usually announces "we're leveling off at 35,000 ft". I recall that most jet aircraft travel at a near constant speed of about 500 mph or about 250 m/s at this altitude.

1.2.4.4 The approach: formulating the physical model

From Newton's second law of motion we can write down the following well known relationship

Total takeoff thrust ≈ (mass of 747) × (average acceleration; start to takeoff).

If I can estimate both the average acceleration and the mass of the 747, then I can readily calculate an estimate for the total takeoff thrust generated by all four engines. Dividing that result by four gives the estimated thrust per engine.

How do I calculate the estimated acceleration at takeoff?

The average acceleration, "a" (units of m/s^2), is approximately equal to a best guess of the takeoff velocity (at wheels up) divided by an estimate of the total takeoff time required to reach that velocity.

How do I calculate the estimated mass, m, of the 747?

Here is where I utilize a little of my engineering knowledge of aerodynamics by recalling the equation for steady-state wing lift during cruise conditions. Lift is proportional to the dynamic pressure of the oncoming air multiplied by the wing area. The constant of proportionality is the lift coefficient, C_{lift}. We can estimate the total 747 lift, L, at cruise conditions (velocity $v \approx 250$ m/s and altitude = 35,000 ft) based on an estimate for dynamic pressure, ($\frac{1}{2}\rho v^2$), wing area, and a very crude

approximation for C_{lift} using the following equation
$$L = (\tfrac{1}{2}\rho v^2)(\text{Lift coefficient})(\text{wing area}) = (\tfrac{1}{2}\rho v^2)(C_l)(A_{\text{lift}})$$
where ρ = air density at cruise altitude of 35,000 ft (or 10 km) $\approx \tfrac{1}{4}\rho_{\text{sea level}} \approx \tfrac{1}{4}(1.0 \text{ kg/m}^3)$.

From my college aerodynamics course, I recall
$$C_l \approx 2\pi \cdot (\text{effective wing angle of attack in radians})$$
based on thin airfoil theory.

My best guess for effective angle of attack is 6° or approximately 0.1 radians. For those with an aerodynamics background, the approximation is effective angle of attack $\approx$ (true angle of attack + effective camber angle).

From Figure 1.2, let's assume that the effective lift area A_{lift} (wing area + the approximately 10 m wide section of the fuselage where the wings are attached) is $\approx (2 \times 30 \text{ m} + 10 \text{ m}) \times (10 \text{ m width}) = 700 \text{ m}^2$.

Since the plane is flying at a fixed cruise altitude, the total aerodynamic lift must equal the weight of the aircraft, i.e. Lift = $m_{\text{aircraft}} \cdot g$ where g is the acceleration of gravity; about 10 m/s².

Dividing the estimated Lift by g yields the approximate mass m of the 747.

Estimating takeoff thrust

From Newton's second law
$$\text{Total takeoff thrust} = (\text{mass of 747}) \times (\text{acceleration at takeoff})$$

1.2.4.5 *The calculations*

Based on the model defined in Section 1.2.4.4, takeoff acceleration is estimated to be
$$a = \frac{\Delta v}{\Delta t} = \frac{70 \text{ m/s}}{30 \text{ s}} = 2.33 \text{ m/s}^2$$

Lift at cruise conditions
$$L = (\tfrac{1}{2}\rho v^2)(\text{Lift coefficient})(\text{wing area})$$
$$= \left(\frac{1}{2}\left(\frac{1.0 \text{ kg/m}^3}{4.0}\right)(250 \text{ m/s})^2\right)(2\pi \cdot 0.1 \text{ radians})(70 \text{ m} \times 10 \text{ m})$$
$$= 3.4 \times 10^6 \text{ N} \approx 764{,}000 \text{ lbf}$$

Takeoff mass estimate is then
$$m = \frac{L}{g} \approx \frac{3.4 \times 10^6 \text{ N}}{10 \text{ m/s}^2} = 3.4 \times 10^5 \text{ kg}$$

Takeoff thrust (4 engines)
$$T_{4\,\text{engines}} = m_{\text{aircraft}} \cdot a_{\text{average}} \approx (3.4 \times 10^5 \text{ kg})(2.33 \text{ m/s}^2) = 792{,}000 \text{ N} = 178{,}000 \text{ lbf}$$

12 Introduction   [Ch. 1

and so the thrust per engine is

$$T_{1\,engine} \approx 198{,}000 \text{ N} = 44{,}500 \text{ lbf}$$

1.2.4.6 Comparing results with reality

Table 1.1 lists tabulated values of takeoff thrust, units of lbf, for several different versions of the 747.

Note that the tabulated thrusts range from 46,500 lbf to 63,000 lbf depending on the engine manufacturer and the particular version of the aircraft. Our BotE estimate of 44,500 lbf (or nearly 200 kN) is close to the lower end of the range of measured thrusts. Note that 46,500 lbf is the thrust level produced by the earliest Pratt & Whitney versions of the 747's jet engines.

Our thrust estimate is $\approx 96\%$ of the *lowest* tabulated thrust. Our thrust estimate is $\approx 71\%$ of the *highest* thrust listed in Table 1.1. Our estimate of aircraft mass of 3.4×10^5 kg is about 20% lower than the nominal aircraft mass in the literature; the published mass of the 747-400 is approximately 4×10^5 kg at takeoff.

Our predicted takeoff thrust is well within the factor of two uncertainty for a reasonably "good" Back-of-the-Envelope estimate. The estimated magnitudes of many of the individual parameters utilized in our calculation also fall within this factor of two level of uncertainty.

Table 1.1. Tabulated 747 thrust levels for three jet engine manufacturers. Thrusts are listed by aircraft version [7].

Measurement	747-100B	747-200B	747-300	747-400
Thrust per engine, lbf Pratt Whitney	46,500	54,750	54,750	63,000
Thrust per engine, lbf Rolls Royce	50,100	53,000	53,000	59,500/ 60,600
Thrust per engine, lbf General Electric		52,500	55,640	62,100

1.3 GENERAL GUIDELINES FOR BUILDING A GOOD ENGINEERING MODEL

The flow diagram in Figure 1.3 outlines a basic "step by step" roadmap that should be used in the development and evaluation of approximate solutions for any engineering design or analysis problem. It is derived from the charts given in [8, 9]. It is a useful guide when employing BotE techniques or any other estimation methodology, because it is essentially based on the principles of the "scientific method".

Sec. 1.3]   General guidelines for building a good engineering model   13

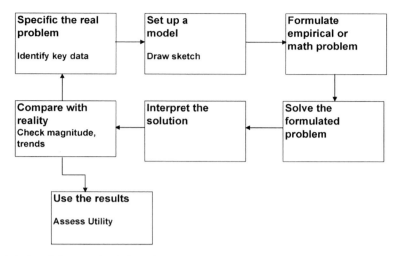

Figure 1.3. Modeling methodology based on Penrose in 1978 [8] and Clements in 1982 [9].

1.3.1 Step by step towards estimation

Using Figure 1.3 as our estimation roadmap, we see that the modeling process follows these eight sequential steps:

(1) Define the "real" problem and gather all the available data needed to define key input variables and parameters.
(2) Make a simple sketch of the principal physical problem geometry and use it to identify key input and output variables. Organize your approach. Develop your solution path.
(3) Formulate empirical or mathematical models that attempt to get to the heart of the governing physics. Sometimes an empirical rule of thumb is sufficient to estimate the order of magnitude of a solution.
(4) Solve the proposed model (or set of models) by analytical or numerical methods and calculate the magnitude of the desired engineering quantity. It is very, very important to keep track of all physical units (meters, seconds, watts, etc) at each step in the solution chain.
(5) Interpret the solution in both qualitative and quantitative terms. Evaluate, as simply as possible, the sensitivity of the solution to uncertainties or changes in the primary system parameters. If the solution varies wildly with slight changes in the input parameters, then you should attempt to revise your physical and/or mathematical model to make it more robust.
(6) Finally compare your results with both experiment and with other related information—theoretical or empirical—published in the printed or electronic literature. Determine if the estimated magnitude is "in the ball park".
(7) Check the predicted trends of the derived results for the *expected* trends or behaviors of the system. If you pass this reality check, then present your final

estimates in an oral briefing or in a short written report. It is important to assess how these results may alter any preliminary project design. If you are starting from scratch, you can often employ your estimates to develop a crude *first-cut* design.

(8) If the solution falls *out of bounds* with respect to either magnitude or trend, redefine your problem and repeat the estimation process. At this point you hopefully have a clearer understanding of: (a) what must be corrected in your initial physical model, and/or (b) what new or revised data is required as input to your model.

1.3.2 Quick-Fire estimates

In this book, when we are asked to produce an engineering estimate very quickly, say, in less than an hour, we call that estimate a "Quick-Fire" problem. We still follow the "scientific method" steps listed above, but will take judicious short-cuts depending on the resources and on the time available to complete our task.

The name, "Quick-Fire" comes from the specific short-term tasks assigned to competing chefs in a popular television cooking show *Top Chef*, currently still in production on the Bravo television network [10]. The competitors are judged on the quality of a single dish that they have to prepare in a very short period of time, typically 30 minutes to an hour. The challenge is that they are "given" very few ingredients and a limited number of cooking utensils.

For all the Quick-Fire problems in this book, we will try to simplify model complexity to minimize the time required to complete the eight steps listed above. Here are the recommended *Quick-Fire* steps that we have adopted:

(a) Define and/or conceptualize the problem using a sketch or brief mathematical description.
(b) Select a model or approach, either mathematical or empirical that describes the basic physics of the problem.
(c) Determine the input data parameters and their magnitudes as required to solve the problem, either from data sources or by scaling values by analogy, or simply by using an educated guess.
(d) Substitute the input data into the model and compute a value or range of values for the estimate.
(e) Present the results in a simple form and provide a brief interpretation of their meaning.

In the next section, we apply BotE Quick-Fire methods to a very important question that challenged engineers in the preliminary design phase of the Space Shuttle project: *Can the Shuttle deliver the largest Air Force satellites into low earth orbit?* To obtain an approximate answer to this question we will solve the related estimation problem: *Determine the maximum "cargo or satellite" mass that the Shuttle design can deliver into low earth orbit.* We have little detailed knowledge of key Shuttle propulsion and structure parameters, but need these "ingredients" in order to estimate the cargo mass delivered to orbit. In addition, since we have only a limited time for engineering analysis, our basic estimation "cooking utensil", which is our physics model, must be simple to build and easy to use or solve.

1.4 QUICK-FIRE ESTIMATE OF CARGO MASS DELIVERED TO ORBIT BY THE SPACE SHUTTLE

Using Quick-Fire BotE techniques, we will develop a number of simple estimates of cargo mass that the Shuttle can deliver to low earth orbit. Each of the different estimates is based on increasingly improved levels of knowledge regarding the magnitudes and the physics of the key system parameters.

We use simple models for the Shuttle structures and for the thrust generating performance of the Shuttle's rocket engines. On obtaining our first crude solution we will examine the sensitivity of our results to different input data and different model refinements. Although it may be a time-consuming process to obtain more detailed propulsion or structural information, it will improve the robustness of our physics model. With time and effort our BotE cargo-mass estimate generally (but not always) becomes more credible.

In Chapter 3 of this book we will revisit this whole Shuttle launch problem, but with the benefit of considerably more information and physical understanding of the overall problem at hand. In that more detailed treatment (done on the back of a somewhat "larger" envelope, but still with pencil and paper) we will be better able to accurately model both solid and liquid rocket engine thrust magnitudes. Using more complex flight dynamics models it will be shown that we can calculate time-dependent vehicle velocities for each Shuttle stage while still keeping our results relatively simple and easy to interpret. We will compare all our static and time-dependent predictions to corresponding NASA data sets, both in order to test the accuracy of our improved models and to assess the gains in accuracy attainable by using better quality input data. The results will demonstrate that it is possible to estimate, with considerable confidence, the cargo mass that the Shuttle can deliver to orbit.

First, however, we use our Quick-Fire techniques to construct the simplest mathematical models based on well-known rocket engineering principles. We also compare our analytical estimations to the predicted cargo mass calculated from a simple empirical rule of thumb established by the "space systems" community. We designate this rule of thumb estimate as our "level-0" approximation.

So let's get started!

1.4.1 Cargo mass problem definition

- *Perform a preliminary estimate of the Shuttle's payload and cargo mass for a low earth orbit mission. Compare the estimates from several simple models with input data based on a variety of assumptions and knowledge sources.*

16 Introduction [Ch. 1

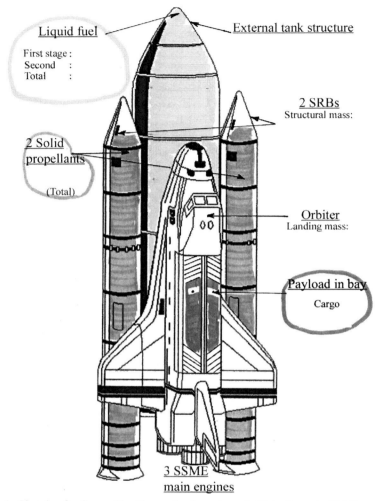

Figure 1.4. Sketch of a Space Shuttle on the launch pad. Note items in Orbiter cargo bay.

- Assume that the initial Shuttle takeoff mass $= 2 \times 10^6$ kg and the approximate mass of the manned Orbiter $= 1 \times 10^5$ kg.

1.4.2 Level-0 estimate: the empirical "rule of thumb" model

In the book "Space Mission Analysis and Design" [11] (also referred to as SMAD) the typical useful payload mass delivered to orbit, referred to here as cargo mass, m_{cargo}, is a very small fraction of the overall total system mass, or initial launch mass, m_i. This rule of thumb fraction of cargo mass divided by total takeoff mass, is typically in the 1% range based on cargo data listed in SMAD for a number of flown launch vehicles.

Given an initial estimated value for Shuttle mass of $m_i \approx 2$ million kg, we can readily calculate a crude but useful estimate for the cargo mass.

Inversely, given the cargo mass that is to be delivered to orbit we can crudely estimate the initial Shuttle takeoff mass, m_i.

We will show later in Chapter 3 how one can also determine m_i based on a second rule of thumb in which the total launch structural mass (mass that is not associated with expendable propellants or delivered cargo) is approximately 14% of the initial Shuttle mass, or conversely that m_i is about 7 times larger than the structural mass which can be estimated independently.

Using the above 1% rule of thumb for *cargo mass* as a fraction of initial rocket mass, the Shuttle cargo mass delivered to orbit is estimated to be

$$m_{cargo} \approx 0.01(m_i)$$

i.e.

$$m_{cargo} = 0.01(2{,}000{,}000) = 20{,}000 \text{ kg} \qquad \text{SMAD rule of thumb} \qquad (1.1)$$

This is actually a pretty good estimate, as the values of Shuttle cargo mass published by NASA range from approximately 25,000 to 30,000 kg for mission orbits of 110 to 150 km altitude.

1.4.3 Level-1 estimate: cargo mass using a single stage mathematical model based on the ideal rocket velocity equation

How accurate is our estimate of the cargo mass when we rely on a bit of physics, namely the equation for the ideal velocity change of a *single stage* rocket, Δv_{ideal}? The governing equation is based on Newton's second law of motion. We start by defining our model and its main parameters, and apply a little algebra to set up the calculation for the resulting cargo mass.

1.4.3.1 Single stage mathematical model

The well-known *ideal* expression for the incremental velocity increase, Δv_{ideal}, achieved by a rocket after a certain mass of propellant has been burned or consumed is

$$\Delta v_{ideal} = -C \ln\left(\frac{m_{final}}{m_{initial}}\right) = -C \ln\left(\frac{m_{initial} - m_{propellant}}{m_{initial}}\right) \qquad (1.2)$$

This equation is sometimes referred to as the "Tsiolkovsky rocket equation", named after the scientist-mathematician who derived it in 1903. For those new to the subject, this equation is derived in Chapter 3 based on Newton's second law of motion applied to a rocket system undergoing zero gravitational forces.

1.4.3.2 Parameter definitions

In Equation 1.2, m_{initial} is the initial mass at launch for the rocket and m_{final} is the "final" rocket mass. Let's assume that all of the propellant is totally consumed at the end of the single stage flight. Then the final rocket mass is simply equal to the initial mass minus the initial propellant mass, $m_{\text{propellant}}$. In this equation, C is the effective rocket exhaust velocity defined by

$$C \equiv \frac{\text{engine thrust}}{\text{engine mass flow rate}}$$

C can also be expressed in terms of the *specific impulse*, Isp, which is the thrust per unit weight flow rate (units of seconds). It is a basic measure of the chemical energy released during combustion of a given mix of fuel and oxidizer. The heated gaseous combustion products when accelerated through a rocket nozzle produce a given exhaust velocity C, which is approximately equal to the engine thrust, T, divided by the engine mass flow rate, $\dot{m}$. Thus C can be written as

$$C = \frac{T}{\dot{m}} \equiv \text{Isp} \cdot g \quad \text{(or specific impulse)}$$

where g is the acceleration of gravity = $9.81 \, \text{m/s}^2$.

For the liquid hydrogen/liquid oxygen propellants used by the main engines of the Orbiter, the specific impulse is approximately Isp ≈ 450 s in vacuum. For the solid rocket motors, which are packed with a mixture of ammonium perchlorate, aluminum, and iron oxide, the Isp ranges from 242 s at sea level to 268 s in near-vacuum conditions. As we will see, the total payload (i.e. the combined cargo and Orbiter spacecraft mass) that can be delivered to orbit by a given launch vehicle is strongly influenced by the magnitude of the average specific impulse for each stage and for the overall mission.

Let's first model the complex Shuttle system as an "equivalent" single-stage-to-orbit rocket.

We define the final payload as the sum of the mass of the Orbiter plus its cargo.

Note that the Orbiter has a crew of seven astronauts and that it carries all its supporting equipment into space plus the additional propellants needed for maneuvering. The estimated mass of the Orbiter without cargo, as given in Section 1.4.1, is approximately 100,000 kg.

We now define all of the relevant Shuttle system mass components and their interdependent relationships. In particular the initial takeoff mass is given by

$$m_i = m_{\text{propellant}} + m_{\text{structure}} + m_{\text{orbiter}} + m_{\text{cargo}} \quad (1.3)$$

where

$$m_{\text{propellant}} = m_{\text{p/solid}} + m_{\text{p/liquid}}$$

and

$$m_{\text{structure}} \approx m_{\text{structure/solid motors}} + m_{\text{structure/external liquid fuel tank}}$$

We also define the payload mass, $m_{p/l}$, to be the sum of the Orbiter mass and the cargo mass

$$m_{payload} \equiv m_{p/l} = m_{orbiter} + m_{cargo} \qquad (1.4)$$

In some studies, the Orbiter mass is defined as part of the overall structural mass. For that particular definition, the cargo mass is equal to the payload mass. As long as we are consistent in its use, either definition of payload and structural mass is acceptable.

1.4.3.3 Empirical model for structural mass

Vehicle structural mass (without the Orbiter) is often empirically assumed to be a fixed fraction, f_s, of the initial takeoff mass less the payload mass

$$m_s = f_s \cdot (m_i - m_{p/l}) \qquad (1.5)$$

For many launch systems, the structural fraction $f_s \approx 0.08$ or 8% as cited by Humble, et al. in "Space Propulsion Analysis and Design" [12].

The final rocket mass for a given stage, after all propellant is consumed, can be written as

$$m_{final} = m_s + m_{p/l} = f_s m_i + (1 - f_s) m_{p/l} \qquad (1.6)$$

1.4.3.4 The mathematical solution

Taking the logarithm of both sides in Equation 1.2, the final to initial mass ratio is

$$\frac{m_{final}}{m_i} = \exp(-\Delta v / C) \qquad (1.7)$$

Substituting Equation 1.6 into Equation 1.7 yields the following simple equation for the ratio of payload to initial mass

$$\frac{m_{p/l}}{m_i} = \frac{\exp(-\Delta v / C) - f_s}{(1 - f_s)} \quad \text{(single stage rocket equation)} \qquad (1.8)$$

The final equation for delivered Shuttle cargo mass is

$$\boxed{m_{cargo} = m_i \left(\frac{\exp(-\Delta v / (\text{Isp} \cdot g)) - f_s}{(1 - f_s)} \right) - m_{orbiter}} \qquad (1.9)$$

To calculate cargo mass for a single-stage-to-orbit vehicle we simply specify the value of each of the parameters on the right side of Equation 1.9. Here are our first-order estimates or guesses for these values.

1.4.3.5 Input data

$$m_i = 2 \times 10^6 \text{ kg}$$

$$\Delta v \approx 9.0 \text{ km/s} = 9{,}000 \text{ m/s}$$

20 Introduction [Ch. 1

To reach orbit

$$\Delta v = \text{circular LEO orbital velocity} + (\text{gravity loss} + \text{drag loss} - \text{earth rotation})_{\text{launch}}$$

To better understand and to obtain a more precise estimate of Δv the reader should review the orbital mechanics and launch-related velocity loss models in Chapter 3 of this book.

To estimate a single value for Isp for the performance of the combined solid propulsion and liquid propulsion systems, we note that for the first 2 minutes of flight (stage 1), the thrust from the solid rocket motors are dominant (Isp ≈ 250 s) while for the remaining 6 minutes of the of flight (stage 2) the thrust comes solely from the main engines of the Orbiter (Isp ≈ 450 s) [13].

Let's *weight* these respective specific impulses by the fraction of time during the total 8 minute ascent-to-orbit period where each is dominant, and then add the weighted solid and liquid Isp values together. This scaling provides us with a crude estimate for an equivalent single-stage-average specific impulse, $(\text{Isp})_{\text{avg}}$.

Assuming that 25% of the ascent-to-orbit period the thrust performance is dominated by the solid rocket motors and that 75% of the flight time the Orbiter's liquid engines constitute the sole form of propulsion, we calculate the average Isp to be

$$\text{Isp}_{\text{avg}} \approx (0.25)(250) + (0.75)(450) = 400 \text{ s}$$

We also assume the following values for mass fraction and Orbiter mass

$$f_s \approx 0.08 = \text{structural mass fraction (as defined by Equation 1.5)}$$

and

$$m_{\text{orbiter}} \approx 100{,}000 \text{ kg}$$

1.4.3.6 Calculation of cargo bay mass for single stage model

We now calculate the total payload (Orbiter mass + cargo mass) inserting the input data into our single stage model using Equation 1.8

$$m_{p/l} = m_i \left(\frac{\exp(-\Delta v/C) - f_s}{(1 - f_s)} \right) = (2 \times 10^6) \left(\frac{\exp(-(9{,}000)/(400)(9.81)) - .08}{(1 - .08)} \right)$$
$$= 45{,}400 \text{ kg}$$

Hence

Cargo mass = payload mass − Orbiter mass = 45,400 kg − 100,000 kg

i.e.

$$\boxed{\text{Estimate cargo mass} = -54{,}600 \text{ kg}}$$

1.4.3.7 Interpretation of the Level-1 estimate

This result, −54,600 kg, of course, makes no physical sense!

Our single stage model does predict that the total payload mass which can be lifted into orbit by this hypothetical single stage vehicle is 45,400 kg. This low

Sec. 1.4] Quick-Fire estimate of cargo mass delivered to orbit 21

payload mass indicates that a single stage Shuttle vehicle, with a 2 million kg initial mass, driven by an engine with an average specific impulse of 400 s and supporting a 100,000 kg Orbiter spacecraft, could *never* reach orbit, no matter how small the cargo mass might be.

That is not to say that a single stage Shuttle with an average Isp as high as 470 s couldn't reach orbit with an Orbiter and a substantial payload. It could. But current engine and fuel technology cannot achieve this level of performance. If we were able to construct a much larger single stage vehicle, say with an initial mass nearly triple that of the Shuttle, then theoretically we could do the job with a lower specific impulse. Unfortunately, overall cost and construction constraints make this a very unlikely scenario as well.

So our model is not adequate since it's performance falls considerably short of reality. Let's rework the problem using a *two stage* model of the Shuttle launch system.

1.4.4 Level-2 estimate: cargo mass using a two stage vehicle model based on the ideal rocket velocity equation

A two stage analytical model can be constructed by applying our simple payload mathematical solution (Equation 1.8) to each stage of the launch vehicle.

1.4.4.1 Two stage mathematical model

We note that the payload from the first stage becomes the initial mass of the second stage, since at first stage burnout the solid motors and their supporting structures are dropped from the Shuttle into the ocean. Thus, in equation form

$$m_{\text{1st stage p/l}} = m_{\text{i 2nd stage}}$$

Using this identity, the equation for the final payload at the end of the second stage, $m_{\text{2nd stage p/l}} \equiv m_{\text{payload}}$ is equal to the initial mass m_i multiplied by two nearly identical terms for each stage based on Equation 1.8

$$m_{\text{2nd stage p/l}} = (m_i) \left(\frac{\exp(-\Delta v_1/(\text{Isp}_1 \cdot g)) - f_{s,1}}{(1 - f_{s,1})} \right) \left(\frac{\exp(-\Delta v_2/(\text{Isp}_2 \cdot g)) - f_{s,2}}{(1 - f_{s,2})} \right)$$

(1.10)

1.4.4.2 Two stage input data

The most challenging task, prior to any calculation, is to select (or determine by analysis) the incremental velocity changes achieved by each of the two stages, Δv_1 and Δv_2. The sum of these velocity changes must equal $\Delta v_{\text{required}}$, which is the loss-corrected orbital velocity for the mission. Taking the same total required orbital velocity that we used in our single stage model calculations, we set the following condition

$$\Delta v_1 + \Delta v_2 = \Delta v_{\text{required}} \approx 9.0 \text{ km/s}$$

Next, we decide on a rule for partitioning the two Δv components. Notice that the first stage solid rocket motors operate for approximately 2 minutes and are then jettisoned, then the Orbiter's liquid rocket motors operate for approximately 6 minutes. There is of course a thrust contribution from the liquid engines in the first stage of flight, but we assume for now that it is small compared to the solid rocket motor contribution at takeoff. Using these two time scales, 2 minutes and 6 minutes, as a crude guide, we assume that the velocity change for stage one is $2/8\,\Delta v_{required}$, and the velocity change for stage two is $6/8\,\Delta v_{required}$. Hence, we set the following values for Δv_1 and Δv_2

$$\Delta v_1 = 1/4(\Delta v_{required}) \approx 2.25 \text{ km/s}$$

and

$$\Delta v_2 = 3/4(\Delta v_{required}) \approx 6.75 \text{ km/s}$$

Using the specific impulses already cited for the solid and liquid propellants of the Shuttle, the values of Isp_1 and Isp_2 are fixed at

$$\text{Isp}_1 = 250 \text{ s} \quad \text{and} \quad \text{Isp}_2 = 450 \text{ s}.$$

The liquid rocket engine contribution to the first stage thrust is included in the more accurate treatment of this problem in Chapter 3.

As to the structural mass fractions assigned for each stage, a reasonable first guess is that they are both equal at 8%

$$f_{s_1} = f_{s_2} = 0.08$$

1.4.4.3 Calculation of two stage model cargo bay mass; Level-2

We now calculate the payload at the end of the second stage of flight by inserting the estimated input data for both stages into Equation 1.10

$$m_{\text{2nd stage p/l}} = (2 \times 10^6) \left(\frac{\exp(-2{,}250/(250 \cdot 9.81)) - .08}{(1 - .08)} \right)$$

$$\times \left(\frac{\exp(-6{,}750/(450 \cdot 9.81)) - .08}{(1 - .08)} \right)$$

$$= 103{,}200 \text{ kg}$$

Using this estimated total payload mass for the two stages, the corresponding cargo mass, carried in the Orbiter's cargo bay is

Cargo mass = payload mass − Orbiter mass = 103,200 − 100,000

$$\boxed{\text{Estimated cargo mass} = 3{,}200 \text{ kg}}$$

1.4.4.4 Interpretation of the Level-2 estimate

This two stage model result, 3,200 kg, is positive and therefore makes some physical sense (unlike our previous single stage model). However, this is far too small a cargo magnitude for a viable rocket system the size of the Shuttle. It would mean that the Shuttle could send into orbit only the smallest of satellites. The Shuttle was initially designed to be capable of orbiting the largest Air Force satellites, typically $\geq 20{,}000$ kg. The mass of the Hubble Space Telescope that was delivered to an orbit of over 600 km in 1990 is about 11,000 kg [14]. In some cases a single mission deployed several satellites. The Shuttle also was given the job of carrying large equipment modules into space to construct the International Space Station.

So our payload calculations appear to be low, but physically on the right track. We will need additional information to get into the "right-sized" payload ball park. Let's speculate what that added information might be in order to improve our next BotE estimate for payload mass.

1.4.5 Level-3 estimate: cargo mass delivered by a two stage vehicle; based on a revised estimate for second stage structural mass fraction

During the second stage of the ascent, the single biggest structural component is the external tank that supplies the propellants to the Orbiter's engines. It can be thought of simply as a very large thin-walled pressurized vessel.

1.4.5.1 Questioning the magnitude of the Level-2 structural mass fraction

Our model calculations indicate that once the first stage solid propellants have been consumed in the first two minutes of flight, the initial mass for the second stage of flight is about 700,000 kg. Assuming an 8% structural mass fraction and the Level-2 payload mass of 103,200 kg, we estimate, from Equation 1.5, the second stage structural mass to be $\approx 0.08(700{,}000 - 103{,}200) \approx 48{,}000$ kg.

We are then led to ask the following two questions:

- Is the structural mass fraction of the second stage significantly less than the 8% that we assumed for the first stage?
- Is the mass of the external tank (i.e. without fuel) approximately equal to 48,000 kg (ignoring the mass of the Orbiter's engines)?

At this point in our Quick-Fire analysis we might not have time and resources to answer these questions, so we would report our preliminary findings for the estimated cargo mass as equal to approximately 3,400 kg and state the caveat that we do not know the actual second stage structural mass; nor do we have a good guess as to what it might be other than about 48,000 kg. We also realize that for every 1 kg *reduction* in structural mass there is available an additional 1 kg in payload or cargo mass. So clearly structural designers are tasked to make the rocket structure as light as possible, subject to the proviso that they also ensure that all stages of the Shuttle meet or surpass the pressure and bending load requirements for the vehicle at takeoff and in flight.

1.4.5.2 If we have more time to build a better model for the external tank mass

With enough research time (it could be minutes, hours, or days) an engineer can make his or her way to a body of data that lists all the masses of all the structural elements on the Shuttle. For more obscure engineering systems (not the Space Shuttle) the reader should know that he cannot simply use Google to search out a magnitude for each and every parameter of interest!

For our Shuttle problem, another approach to estimating the key structural mass fraction parameter for the second stage would be to make a very simple mathematical model for the external tank based on what we know of the functionality of that tank, given the sketch in Figure 1.4. In fact, we will develop such an estimate in the Quick-Fire portion of Chapter 3.

1.4.5.3 A simple model of the Shuttle's external tank

From the sketch of Figure 1.4 and our knowledge of how the Shuttle operates we might conclude that the external tank is for the most part a pressurized cylindrical vessel carrying the Orbiter's liquid oxygen and liquid hydrogen propellants. It has an approximately conical dome.

We first estimate the volume of material in the cylindrical section of the tank by making educated guesses as to its length, its diameter, and its wall thickness. Assuming that the tank material is aluminum we can quickly look up the material density of aluminum and find it to be about $2,700 \, kg/m^3$. The hardest part of the calculation is choosing the wall thickness, "t". We know it is small, but how small? Well, we could calculate it by assuming that the walls of the tank are designed to keep the pressure-induced circumferential hoop stresses in the wall below the yield strength of aluminum. However, that would in turn require some knowledge of the tank pressures. You can see that this estimation process can get rather complicated. Nevertheless, it can be done this way if we assume that the internal over-pressure is of order of one atmosphere.

1.4.5.4 Estimation by analogy

Another approach is to estimate the wall thickness by considering some type of engineering analogy. In this case we choose (as shown in Chapter 3) to estimate "t" by *scaling-up* the wall thickness of a similar pressurized vessel, namely an aluminum soda-pop can. A typical pressurized "soda can" has a diameter of 2.6 inches (66 mm) and a wall thickness of only 0.005 inches (0.127 mm). Scaling up the diameter to the size of the Shuttle's external tank (to approximately 8.4 m or 8,400 mm) requires a diameter scale increase of $8,400/66 = 127$. If we "scale up" the soda can wall thickness "t" by the same factor we obtain a wall thickness "t" = $(127)(0.127 \, mm) = 16 \, mm = 1.6 \, cm$ [15].

It turns out that if we assume a wall thickness of order of 1 cm, the tank mass for an aluminum cylinder will be 30,000 kg; which is pretty close to the value for the external tank of the Shuttle published by NASA. (See Chapter 3 for further details.)

1.4.5.5 Revising the structural mass input parameter

With this new estimate of 30,000 kg (38% less structural mass than assumed in our Level-2 calculation) we are able to revise our second stage structural mass fraction, $f_{s,2}$, as follows

$$f_{s,2} \equiv \frac{m_{s,2}}{(m_{i,2} - m_{\text{payload},2})} \approx \frac{30{,}000}{(700{,}000 - 103{,}200)} = 0.05 \text{ or } 5\%$$

We see that this revised structural mass fraction is approximately 5%, not 8% as originally assumed.

1.4.5.6 Calculation of two stage model cargo bay mass with the revised input parameter; Level-3

Let's repeat our payload and cargo mass calculations using this new structural fraction parameter. With 0.05 as the second stage mass fraction in Equation 1.10, the revised cargo mass estimate increases to

> Estimated cargo mass = 22,000 kg   (Level-3 estimate)

A delivered cargo mass of approximately 22,000 kg is significantly higher than the previous estimate of 3,200 kg. This increase is due solely to a lowering of our estimate for the structural mass for the external tank. In essence, we applied the same two stage model as in the Level-2 estimate but significantly improved our knowledge of one of the key parameters, namely second stage structural mass fraction, which strongly altered the predicted cargo mass.

1.4.6 Impact of added knowledge and degree of model complexity

Added knowledge generally improves the resolution or level of certainty of our engineering estimates, but almost always at the cost of additional research and analysis time. Figure 1.1 is a graphical depiction of this tradeoff of analysis cost vs problem resolution or certainty.

From our Level-3 example it is clear that knowledge gained by the additional modeling of other key system parameters (such as the Shuttle's average specific impulse representing the parallel thrusting of the solid and liquid rocket motors during the first stage of flight, Isp_1) will most likely improve our overall predictions. In Chapter 3 we will develop a better BotE model of the dynamics of this complex first stage of Shuttle flight.

One way of illustrating the gains from added system knowledge on the level of certainty of the estimates for a given problem is to examine the evolution of a problem's quantitative predictions over the full range of model complexity and assumptions. Figure 1.5 summarizes our cargo mass predictions for the various Shuttle payload model and parameter assumptions that we examined, namely: Level-0 (rule of thumb), Level-1 (single stage model), Level-2 (initial two stage model), and Level-3 (two stage model with additional second stage structural

26 Introduction [Ch. 1

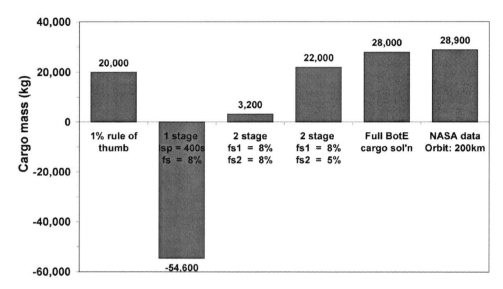

Figure 1.5. Comparison of Shuttle cargo mass model predictions with different model assumptions against NASA Shuttle data for typical deployment missions to a nominal 200 km parking orbit.

knowledge). We also compare these four preliminary results with the more detailed BotE analysis cargo estimates developed in Chapter 3. In that "full BotE" analysis, all major structural and propellant masses are assumed known *a-priori* from NASA reports. Finally all our modeling is compared with NASA cargo payload data for a nominal 200 km altitude orbit, Shuttle cargo mass ≈ 29,000 kg [16].

It is clear that a very important element for reducing model error (improving model confidence) is accurate knowledge of the key parameters that are input to our rather simple dynamics models. We ought not to conclude from this BotE Shuttle model comparison that in every instance all additional improvements in knowledge (such as improving the estimate of a single modeling parameter) brings you closer to the *physical truth* of a problem as determined by experiment. An improvement in only *one* of the model parameters can sometimes upset a delicate balance of terms in the baseline model. Given that there may be other parts of the problem's physics that have not been accounted for in the model, a single parameter change governing a sensitive variable may (in certain cases) dramatically pull the estimated solution away from the truth. This is not often the case, but the sensitivity of a model to the system parameters and overall solution robustness are important. These sensitivities should be taken into consideration when estimating the error bars that should accompany your reported estimation results.

For example, it is apparent that our cargo calculations are quite sensitive to the average specific impulse that we select for each stage. This is dramatically shown in Figure 1.6, calculated from Equation 1.9 for a *single stage* model of the Shuttle. The

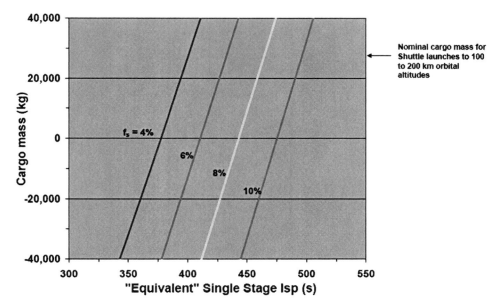

Figure 1.6. Equivalent single stage Shuttle cargo mass modeled as a function of specific impulse (Isp) for equivalent single stage structure mass fractions, f_s (Equation 1.9). Initial rocket mass = 2 million kg.

estimated cargo mass is plotted as a function of the equivalent single stage Isp for a range of structure mass fractions. The slope of these curves is quite steep; the derivative of cargo mass with respect to Isp is around 1,300 kg/s. A 10 second error in an estimate of the average specific impulse can yield a 13,000 kg difference in the estimated cargo mass. Note also that for a structural mass fraction $f_s = 0.08$, the predicted cargo mass is only positive for an average Isp > 443 seconds.

1.4.7 Moving from the Shuttle to the Hubble Space Telescope

We now transition our discussion from estimating Shuttle cargo mass to a related problem: estimating the dimensions of a very important item of Shuttle cargo that was launched into orbit in 1990—the Hubble Space Telescope (HST) which has revolutionized our understanding of the universe.

The HST was specifically designed to be capable of being tended in orbit by the Shuttle's astronaut crews every few years. This meant that the HST not only had to be carried into space initially in the Orbiter's cargo bay, but it also had to be snatched from its orbit, pulled back into the cargo bay, updated, and returned to its orbital path numerous times to maintain its optical performance. Because of this important linkage between the HST and the Shuttle, an early constraint on the HST design was for the HST to be able to *fit* into the Shuttle's cargo bay. In particular, HST system engineers needed to make sure that the satellite's length and width did not exceed the dimensions of the Orbiter's cargo bay. This design requirement for a

28 Introduction [Ch. 1

good fit leads us to test how well we can estimate overall HST dimensions using only our basic Quick-Fire approach to estimation.

In order to keep the Quick-Fire estimation time short, we focus solely on the important problem of obtaining a first-order estimate of the length of the *optical* system of the HST. The length of its two-mirror optical package is an important portion of the overall HST length. However, there are other contributors to the overall HST length problem, and consideration of these issues will be deferred until Chapter 6.

1.5 ESTIMATING THE SIZE OF THE OPTICAL SYSTEM FOR THE HUBBLE SPACE TELESCOPE

In this section we turn our attention to the second major real-world engineering project that is addressed in this book. Figure 1.7 is a NASA contractor drawing of the major components of the Hubble Space Telescope. Note the callouts that specify the primary and secondary mirrors as well as the light shield tube which precedes the optics section.

As in our treatment of the Shuttle problem, we seek to use BotE techniques to estimate key sizing and performance data that are unique to the HST system.

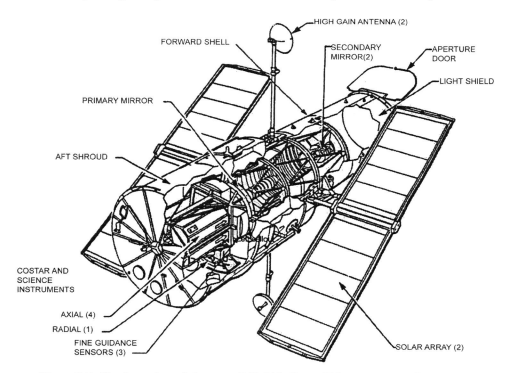

Figure 1.7. Configuration of the overall Hubble Space Telescope system [14, p. 14].

1.5.1 System requirements for the HST

Our primary estimation goals are to calculate the basic optical telescope scales and the optical performance necessary for this unique space telescope to meet the following major resolution and sensitivity system requirements:

- High angular resolution—the ability to image fine detail.
- High sensitivity—the ability to detect very faint objects.

With a primary mirror diameter of only 2.4 m, the HST would be considered a medium sized telescope on the ground. But its high angular resolution and high sensitivity in space, and its unprecedented pointing and control, make it a unique astronomical observatory.

1.5.2 Shuttle constraint on HST size

The HST was designed to be launched into a low earth orbit (600 km altitude) and subsequently serviced using the Shuttle. It was launched aboard Discovery on April 27, 1990 as STS-31. Design tradeoffs had to be made in order to keep the HST within the weight and cost constraints imposed by Shuttle payload limits. It was designed with maintenance in mind as the strategy for maintaining its full scientific potential over an operational lifetime of several decades.

Once the decision was made to design, develop, build, and maintain the HST in conjunction with the Shuttle, this defined several specific design requirements. In particular the HST couldn't exceed the Orbiter's cargo bay dimensions of 18.3 m in length and 4.6 m in width. Considering the presence of other equipment in the cargo bay, the HST package should typically have a length no greater than about 16 m (or 52 feet). Also, the HST system mass had to be below the maximum cargo mass capability of the Shuttle when placing a satellite in an orbit of about 600 km, limiting it to about 15,000 kg.

We are therefore led to ask whether the basic optical design for the HST, a Ritchey–Chrétien Cassegrain reflecting telescope with a 2.4 m diameter concave primary mirror and a smaller convex secondary that feeds an array of supporting instruments, meets the cargo bay size and mass limitations of the Shuttle design. The main advantage of a two-mirror Cassegrain telescope is that the telephoto optics produce a relatively short total optical package.

The upper limit on the cargo mass that can be delivered to orbit by the Shuttle was estimated in Section 1.4 using Quick-Fire BotE methods.

Here we address the companion problem of determining the maximum length for the HST optical system in order to partially answer the important engineering question: *Will the HST fit into the Shuttle's Orbiter cargo bay?*

As you can see in Figure 1.8, a photo of the HST installed in the Orbiter's cargo bay, taken prior to its deployment mission in 1990, the full HST system just fits into the bay, although it appears there is more available cargo space longitudinally than there is laterally.

30 Introduction [Ch. 1

Figure 1.8. Hubble Space Telescope in Orbiter Cargo Bay prior to launch of STS-31 in April, 1990, a public domain photo from NASA [17].

1.5.3 Estimating the length of the HST optical package

Let's first generate a quick estimate of the length of the HST optics using only the basic parameters driving the optical design of its Cassegrain reflecting telescope with a 2.4 m diameter primary mirror.

For this initial estimate we define *optical length* to be the length of the optical package, L. This length, also referred to as the "constructional length", is defined to be the distance from the Cassegrain secondary mirror to the focal plane of the telescope. On the Back-of-the-Envelope sketched in Figure 1.9 this is the distance L. While this is a critical dimension for any telescope system (it is a measure of the telephoto *compactness* of the optical system due to the magnification introduced by the secondary mirror), it constitutes only about 50% of the total length of the full HST system. To estimate the total HST length it is necessary to add to L the following additional components: (a) the length of the light shield or hood in front of the telescope optical package that shields the optical system from stray light, ($L_{\text{light-shield}}$), and (b) the length of the scientific instrument package that is located behind the focal plane, ($L_{\text{instrument}}$). In Chapter 6 we will address the full

Sec. 1.5] **Estimating the size of the optical system for the Hubble Space Telescope** 31

scaling problem in greater detail and work out many of the significant dimensions for the HST, including a simple BotE estimate for $L_{\text{light-shield}}$.

Here we take a crude, first-order cut at this overall sizing problem using our Back-of-the-Envelope Quick-Fire methodology to estimate the key optical package length L. We start with the definition of the problem.

1.5.3.1 Problem definition for the HST optical problem

- *Estimate the optical package length L for the HST two-mirror Cassegrain telescope.*

In order to model this simplified problem, we will use the geometrical paraxial ray optics relationships for a two-mirror system, derive an equation for L and approximate it for a Cassegrain telescope for a given "back" distance b, the distance of the point of focus behind the primary. We will also assume that the optical magnification due to the secondary mirror is very large.

Figure 1.9 shows the basic estimation steps in the form of a sketch and hand-written calculations on the back of an envelope. A detailed explanation of the assumptions and calculations follows.

1.5.3.2 The sketch

At the top of the envelope, is my hand drawn sketch of the incoming light rays from a distant light source (e.g. a star). They are assumed to be parallel, having come from an object at infinity. The incoming rays reflect off the concave primary mirror and are directed by its curvature towards the on-axis focal point in accordance with the simple reflectance law in which the angle of incidence equals the angle of reflection. But just short of the primary focus the rays reflect from the convex secondary mirror

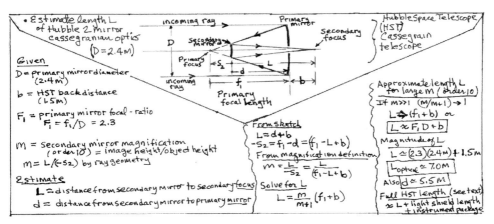

Figure 1.9. What a simple BotE estimate for the HST length L looks like. Estimating the length of the Hubble Space Telescope's two-mirror Cassegrain optics using simple BotE estimation techniques.

and pass through the hole in the primary mirror. All rays then coalesce at the secondary focal point (the system focal plane). The secondary focus is located at an axial distance, b, behind the primary mirror. The key to the solution is to relate all the lengths in the sketch to the magnitude of the principal optical scales: the diameter D of the primary mirror, the primary mirror focal ratio F_1, and the secondary magnification, m_2. The magnification m_2 is written here as "m". Simple ray geometry shows that "m" is equal to the ratio of the secondary mirror's image distance L to $-s_2$, the object distance. We use the small angle approximation applicable to any light ray that is close to and nearly parallel to the optical axis. This is the paraxial approximation. It is central to the derivation of the optical equations presented in all first-year optics courses.

In the paraxial optical approximation it can be shown that "m" is the equivalent focal length of the entire optical system, f_{eq}, divided by f_1, the focal length of the primary mirror. The system focal length, f_{eq}, is determined by the final resolution requirements for the system. For the HST we require that the telescope be diffraction-limited, and this requires a very large system focal length. When "m" is large (as it is when f_{eq} is very large) it can be shown that the actual magnitude of "m" disappears from the key length scale estimate; thereby simplifying the calculation of optical package length.

1.5.3.3 Gathering the input data

The principal dimensions $D \equiv D_{\text{primary}} = 2.4\,\text{m}$, $F_1 = 2.3$, and $b = 1.5\,\text{m}$ are given in Figure 1.9. Each of these parameters was chosen by the optical engineers at the Johns Hopkins Space Telescope Institute after extensive design tradeoff studies. The overall weight and cost of the system depends on D since the linear scale of all of the optical elements is proportional to this key parameter. Clearly mass (or weight) and cost are additional design constraints.

1.5.3.4 The approach: formulating the physical model

From the geometry of Figure 1.9, we see that L is defined equal to the distance from the secondary mirror to the secondary focus. Note that L is the sum of two distances, d (the distance from the secondary mirror to the primary) plus b (the back distance from the primary to the secondary focus)

$$L = d + b \qquad (1.11)$$

The distance $-s_2$ from the secondary mirror to the primary focal point (a distance measured in the negative) is also equal to the distance to the virtual *object* for the secondary mirror. The distance to the *image* for the secondary mirror is defined to be $= L$. Thus the secondary magnification, m, is calculated as follows

$$m = \frac{\text{image distance}}{(-\text{object distance})} = \frac{L}{-s_2} = \frac{L}{(f_1 - d)} \qquad (1.12)$$

Note from the axial positions in our drawing, we observe that $-s_2$ is given by
$$(-s_2) = \text{(primary focal length} - d) \equiv (f_1 - d)$$
Since $d = L - b$ from Equation 1.11, the expression for m becomes
$$m = \frac{L}{(f_1 - L + b)} \tag{1.13}$$
Solving Equation 1.13 for L
$$L = \frac{m}{(m+1)}(f_1 + b) \tag{1.14}$$

If we had a separate equation for the secondary magnification, "m" (which will be derived later in Chapter 6 as equal to the ratio of the system focal length to the primary focal ratio F_1) then we could calculate L directly from Equation 1.14.

In general, we know that a good Cassegrain has a large degree of telephoto compression (meaning that the optical package is small) due to the large secondary mirror magnification. This means that "m" is usually a large number, typically of order 10. If $m \gg 1$, then the ratio $m/(m+1)$ approaches unity. As an approximation for large magnification, we therefore set
$$m/(m+1) \approx 1$$

Then from Equation 1.14, we see that L is simply the sum of the primary focal length and the back distance b
$$L \approx (f_1 + b) \tag{1.15}$$
Since the primary-mirror focal ratio is by definition $F_1 \equiv \frac{f_1}{D}$ (we assume that we know that specific focal ratio from our given inputs), Equation 1.15 can also be written
$$L \approx (F_1 D + b) \tag{1.16}$$

1.5.3.5 The calculations

Substituting the input data for F_1, D, and b into Equation 1.16 gives the approximate magnitude of the optical package length, L
$$L \approx (2.3)(2.4 \text{ m}) + 1.5 \text{ m} = 7.0 \text{ m} \tag{1.17}$$

From Equation 1.11 we also have that $d = L - b$. Hence the approximate value for the distance d (between the secondary and primary mirrors) is
$$d \approx 5.5 \text{ m} \tag{1.18}$$
which is, of course, the focal length of the primary, $F_1 D$.

When Cassegrain telescope systems are constructed, the separation distance d between the two mirrors must be set very accurately to prevent distortions of the focal plane image [18].

1.5.3.6 Comparing results with reality

The published value of the secondary-to-primary distance for the HST, listed in "Characteristics of the HST" [19], is

$$d_{HST} = 4.90 \text{ m}$$

Our crude estimate for d provided by Equation 1.18 is therefore greater than d_{HST} by 0.6 m or 12%.

In addition, the published value for the optical package length, L is

$$L_{HST\,optical\,package} = 6.4 \text{ m}$$

Our estimate for L provided by Equation 1.17 is therefore greater than L_{HST} by only 0.6 m or 9%.

So at 7 m our very simple optical model has given us a pretty good estimate for the optical package length, L.

As previously stated, in order to obtain an estimate of the total HST length it is necessary to add to L both the length of the light shield which projects in front of the telescope optical package to shield the optical system from stray light, ($L_{light\text{-}shield}$), and the length of the scientific instrument package behind the focal plane that contains the cameras and fine guidance sensors ($L_{instrument}$). In Chapter 6 we will work out our own simple BotE estimate for $L_{light\text{-}shield}$. Typical values for these quantities published in the literature [19] are

$$L_{light\text{-}shield} \approx 4.0 \text{ m}$$

and

$$L_{instrument} \approx 3.0 \text{ m}$$

If we add the published $L_{light\text{-}shield}$ and $L_{instrument}$ to our estimate for optical package length $L = 7$ m, we find that our total HST system module length is estimated to be

$$L_{estimated\,total\,system} \approx 7.0 + 4.0 + 3.0 = 14.0 \text{ m}$$

A typical value published in the literature for the total HST length [20], is

$$L_{HST\,total\,system} \approx 13.2 \text{ m}$$

Based on our BotE estimate of the HST total system length, it indeed looks like the HST will fit lengthwise into the 18.3 m long Orbiter cargo bay with a little more than 2 m to spare on both ends!

Additional analysis will be necessary to model the diameter of the HST structural support frame in order to estimate (with BotE techniques) the outer diameter for the entire HST system. Typical diameter or width values listed by a number of HST sources give $width_{HST\,total\,system} \approx 4.3$ m [19]. This is the maximum diameter of the cylindrical aft shroud and bulkhead that encloses the scientific instruments, the subsystems, and the telescope's focal plane structure.

1.6 CONCLUDING REMARKS

In summary, we have shown by our Quick-Fire case study examples that BotE methods allow us to quickly estimate the dimensions and performance for real engineering systems like the Shuttle and the Hubble Space Telescope. Our calculations show that most of our estimates for key engineering quantities, such as Shuttle cargo mass and HST optical length, agree with available data to better than a factor of two, with error levels sometimes as low as 20%.

However, there are times when the key assumptions or input information are so inaccurate that the final estimated output of our effort makes little or no physical sense—for example our single stage Shuttle launch model predicted a negative cargo mass, when clearly only a positive cargo mass has meaning.

Additional BotE analyses utilizing better information for the systems under study are presented in Chapters 3 to 6, where we add complexity to our BotE models and base our calculations on more accurate parameter inputs. In some cases this added knowledge is required in order to significantly drive down the uncertainties of rather sensitive calculations. As pointed out at the beginning of this chapter, reduction in uncertainty is not cost free. The additional effort needed to drive down estimation error costs the engineer additional time, resources, and money. Project management and program pressures often dictate how long an engineer has for such analysis. But as the following chapters will demonstrate, *detailed* BotE analyses and parameter estimates for the Shuttle, Hubble Space Telescope, and "solenoid robot kickers" can be formulated and carried out simply with just pencil and paper.

In many real-world instances, preliminary BotE models and estimates can guide a subsequent full scale design effort that is often supported by complex numerical calculations and detailed engineering trade studies.

To develop any meaningful engineering model, no matter how simple, an engineer must, however, have some basic knowledge and understanding of the governing physics and engineering principles that drive the process under study, for example Newtonian mechanics, classical electromagnetism, and Gaussian ray optics. The following chapters should help to reacquaint the reader with the physics and mathematics that engineers learned in freshman (or high school) physics and calculus classes. These chapters demonstrate how these fundamentals help to guide our physical intuition and serve to underpin the conceptual development of the basic BotE models that we put together when first tackling a number of complex engineering problems like the Shuttle and the Hubble Space Telescope.

1.7 OUTLINE OF THIS BOOK

In this book we present a wide range of Back-of-the-Envelope model calculations for a number of real-world systems and problems.

In this introductory chapter, Chapter 1, we defined BotE engineering estimates as "Quick and sometimes rough, quantitative estimates of key geometric, performance and design parameters for real-world engineering systems". To illustrate the process, we developed Quick-Fire case study models for the 747 engine thrust problem, the Space Shuttle launch and cargo calculation problem, and the Hubble Space Telescope sizing problem. We discussed the tradeoff between model complexity and estimation accuracy, and illustrated this issue by working the Shuttle cargo problem at four levels of complexity ranging from a purely empirical rule of thumb estimate to estimates based on two-stage models for the Shuttle launch to orbit derived from the classical Tsiolkovsky rocket equation. These preliminary examples, and the rather more detailed ones presented in the subsequent chapters of this book are developed using basically the same basic BotE techniques.

Chapter 2 illustrates the use of mechanical and electromagnetic principles to model a simple solenoid actuator device for a student robot-soccer competition. Based on an estimate of the required mechanical energy and design time scales, we develop a BotE model for a solenoid actuator device that provides the kinetic energy required to successfully kick a soccer ball past a robot goalie. We apply Quick-Fire BotE techniques to estimate the required values for solenoid current and voltage and examine the related solenoid Ohmic heating problem. Finally we recommend a simple electrical circuit design to power the robot kicking device that we plan to enter in our hypothetical high school science fair.

Chapter 3 provides a more detailed examination and modeling of the Shuttle cargo mass problem first touched on in Chapter 1. We consider the Shuttle as a complex rocket launch system where the early testing philosophy showed a need for accurate launch performance models. In Chapter 3 we construct, using BotE methods, simple mathematical models for the Shuttle launch to orbit. We start by examining the available mass and propulsion data for the Shuttle and illustrate, by Quick-Fire estimation techniques, how an engineer can quickly estimate the mass of the principal Shuttle components prior to launch (e.g. structural and propellant mass for both the solid and liquid fuel systems). After deriving the Tsiolkovsky rocket equation, we calculate the ideal altitude and velocity changes (without gravity) produced by the propulsion system for each rocket stage. We increase the complexity of our model by including the reduction in rocket velocity due to gravity. To accomplish this we create a simple model for Shuttle pitch angle as a function of time during the curved flight trajectory. We also compare our calculated time-dependent velocities and altitudes with NASA data for the first stage of flight. We summarize our calculations by compiling a "Δv budget" for all stages. Our time-dependent ascent velocity estimates are then compared with NASA's numerical launch code predictions. After reviewing the principles of orbital mechanics, we estimate the required Shuttle burnout velocity for "direct insertion" to the final orbit. We then calculate predicted Shuttle cargo mass as a function of orbital altitude using our BotE equations and show that our results compare well with NASA cargo data to orbits less than about 300 km. We modify our initial results to model the reduction in cargo mass due to the increase in the propellant mass carried on the Orbiter to power its Orbital Maneuvering System. The Orbiter needs this added propellant in order to reach higher orbital altitudes such as the 600 km orbit for the Hubble Space Telescope mission.

Chapter 4 considers the Shuttle accident of 2003 in which Columbia and its seven member crew perished during reentry into the earth's atmosphere. This loss was determined to have been triggered by an impact-induced breach of the leading edge of the Orbiter's wing when a piece of insulation foam separated from the external tank at the start of the mission. We start by developing a crude BotE model for the estimated foam impact velocity. The requirement for more accurate impact velocity estimates leads us to develop a set of detailed BotE models for the physics of the foam impact. We look at the overall estimation problem from several points of view and different initial assumptions. We model both the "direct" problem (in which we are supplied the foam mass) and the "indirect" problem (using a separate estimate of the time to impact). From these models we calculate the impact velocity at the wing leading edge and determine either the time to impact or the foam mass using a simple aerodynamic model for the foam drag. We then develop elastic and elastic-plastic BotE models for the short-lived impact pressures and wing loads created as the foam collides with the wing. Finally we construct a BotE model for the resulting bending stress in the leading edge wing panel due to the impact. On incorporating our results into a parametric structural failure model, we show how this model predicts the possibility of wing damage and failure. Then we demonstrate that the failure threshold depends strongly on the incidence angle of the foam impact relative to the wing.

Chapter 5 investigates the thermally stressing reentry problem. A crude BotE estimate of the average aerodynamic heat transfer to the Orbiter is developed using a model for the strong dissipation of vehicle kinetic energy produced during reentry. We then develop a more specific fluid-dynamics-based heat transfer model and apply it to an "equilibrium glide" reentry trajectory. We follow a simplified approach in modeling the high temperature convective boundary layer heating of the Orbiter's nose or wing leading edge. The maximum heat flux to these surfaces is estimated, as well as the total heat load transferred to the Orbiter throughout the reentry period from atmospheric entry to landing. Our model heat flux predictions are then compared with full scale Orbiter flight data and NASA predictions.

In Chapter 6 we continue the analysis of the HST optical length scale problem started in Chapter 1. We estimate the system focal ratio for a Cassegrain two-mirror optical system that is capable of diffraction-limited resolution. We discuss the importance of the primary mirror focal ratio, which is constrained by the ability to manufacture highly curved large reflecting mirrors with short focal lengths. Using the law of reflection for spherical mirrors we show that the magnification of the secondary mirror is equal to the system focal ratio divided by the primary focal ratio. We then estimate the length of the optical package. The length of the light-shield in front of the optics package is calculated based on a simple angular criterion for limiting off-axis stray light. We combine these results to estimate the total HST telescope length. In addition, we also develop a BotE estimate for the sensitivity performance of the telescope, i.e. its ability to detect very faint stars. We model the focal plane signal to noise ratio (SNR) and deduce the exposure time to image a weak star with a prescribed SNR magnitude. Finally we generate a crude estimate of total HST mass, to assess whether it is within the Shuttle's cargo capability for a 600 km orbit.

1.8 REFERENCES

[1] Richerson, Peter, "Principles of Environmental Science", course outline UC Davis, Environmental Science and Policy 110, see part IV Approach, winter 2006. http://www.des.ucdavis.edu/faculty/Richerson/esp%20110%20syllabus%2006.htm
[2] Paritosh, Praveen and K. Forbus, "Common Sense on the Envelope," Proceedings of the 15th International Workshop on Qualitative Reasoning, San Antonio, Texas, 2001.
[3] Morrison, P. "Fermi Questions," American Journal of Physics, Vol. 31, No. 8, pp. 626–627, 1963.
[4] Swartz, Clifford, *Back of the Envelope Physics*, The Johns Hopkins University Press, 2003.
[5] Linder, Benjamin, "Understanding Estimation and its Relation to Engineering Education," Ph.D. thesis, Massachusetts Institute of Technology, September, 1999.
[6] Shakerin, Said, "The Art of Estimation", *International Journal of Engineering Education*, Vol. 22, No. 2, pp. 273–278, 2006.
[7] Boeing 747 article, online, Wikipedia website. http://en.wikipedia.org/wiki/Boeing_747—747-100
[8] Penrose, O., Journal for Mathematical Modeling for Teachers, Vol. 1, page 31, 1978. Also cited in Houston, Ken, "Assessing the Phases of Mathematical Modeling" published in *Modeling and Applications in Mathematics Education*, New ICMI Study Series, Volume 10, Part 3, Section 3.3, pp. 249–256, 2007.
[9] Clements, Dick, *Mathematical Modeling: a case study approach*, Cambridge University Press, 1989.
[10] *Top Chef*, Television show, Bravo Network, Season 1, March 2006.
[11] Wertz, J.R. and W.J. Larson, *Space Mission Analysis and Design, 3rd edition*, published by Space Technology Library, 1999.
[12] Humble, R. and G. Henry, W. Larson, "Space Propulsion Analysis and Design" Space Technology Series, McGraw-Hill Companies, Inc, 1995.
[13] Braeunig, R.A., "Rocket and Space Technology" web site: see Space Shuttle Specifications for dimensions, mass, thrust of various components, http://www.braeunig.us/space/
[14] "STS-31 Press Information", Rockwell International, Space Transportation Systems Division, Publication 3546-V, Rev. 4-90, see p. 13, April 1990.
[15] The External Tank: educational reading materials and graphics developed by NASA KSC and funded by NASA SOMD http://www.alivetek.com/klass/Curriculum/Get_Training%201/ET/RDG_ET.pdf
[16] http://shuttlepayloads.jsc.nasa.gov/flying/capabilities/capabilities.htm is the original hyperlink listing values for nominal Shuttle cargo mass or payloads (accessed 9/28/06) (This link is no longer viable). Currently published Shuttle payloads are given at the following hyperlink: http://science.ksc.nasa.gov/shuttle/technology/sts-newsref/stsover-prep.html
[17] Photo of Hubble Space Telescope (HST) installed in the cargo bay of the Space Shuttle Orbiter Discovery for the STS-31 Mission prior to launch on April 24, 1990. Photos

compiled at the Marshall Space Flight Center Image Exchange website, MIX, http://mix.msfc.nasa.gov/ A search on MIX for "Hubble Space Telescope" yields the image. http://mix.msfc.nasa.gov/IMAGES/MEDIUM/9009375.jpg

[18] Bely, Pierre, *The Design and Construction of Large Optical Telescopes*, Section 4.5.2 "Selection of f-ratio", Springer-Verlag New York, 2003.

[19] Lockheed Martin Servicing mission 3A media reference guide, Section 5, "Hubble Space Telescope Systems", Goddard Space Flight Center, December 1999. http://sm3a.gsfc.nasa.gov/downloads/sm3a_media_guide/HST-systems.pdf

[20] Soulihac, D. and D. Billeray "Comparison of Angular Resolution Limit and SNR of HST and large ground based telescopes", (see Table 1: Characteristics of the Hubble Space Telescope, p. 507), published in "Space Astronomical Telescopes and Instruments; Proceedings of the Meeting, Orlando, FL, Apr. 1-4, 1991", edited by Bely, Pierre, and J. Breckinridge, SPIE Proceedings. Vol. 1494, 1991.

2

Design of a high school science-fair electro-mechanical robot

2.1 THE ROBOT-KICKER SCIENCE FAIR PROJECT

A student's project for the high school science fair is to design, build, and operate a moving electro-mechanical robot that will be an active "player" in a local robot soccer tournament. His primary goal is to design the device so that it can trap a close-by soccer ball, and then rapidly kick it past a defending goalie into the net of the defending team. His physics teacher suggests that as a first step he should consider mathematically modeling the performance of a light-weight kicking device powered by a rapidly-acting linear solenoid actuator. The initial goal is to determine the basic design dimensions and the required kicking speed/force and stored magnetic energy required for the robot kicker. His teacher recommends that he use simple physics-based Back-of-the-Envelope methods to calculate the dependence of the kicked-soccer-ball speed on such design parameters as the mass and cross-sectional area of the solenoid plunger, and the electric current required to operate the solenoid actuator. His advisor points out that the analysis needs to determine the height above the ground where the "solenoid toe" of the kicker should strike the soccer ball in order to have the best chance of scoring a goal.

Sections 2.2.1 and 2.2.2 present the basic BotE mathematical analysis for the robot kicking device, along with several plots of the derived solutions for a range of design parameters.

2.2 BACK-OF-THE-ENVELOPE MODEL AND ANALYSIS FOR A SOLENOID KICKING DEVICE

To initiate the modeling and analysis effort we (as observers representing the student) are lead to ask the following question: Can physics-based Back-of-the-Envelope modeling provide a simple credible estimation of: (a) the required initial

---

I. E. Alber, *Aerospace Engineering on the Back of an Envelope*.
© Springer-Verlag Berlin Heidelberg 2012.

ball velocity after the impact kick, (b) the total kinetic energy required for the ball, taking into account both its linear and rolling motion, (c) the dependence of the speed of the kicked ball on such important system parameters as the electric current supplied to the solenoid actuator and the mass of the steel solenoid plunger, and (d) the vertical height on the soccer ball where the "solenoid toe" should impulsively strike in order to have the best chance of scoring a goal? As we will demonstrate, the answer to each of these questions is: yes!

To calculate such detailed performance measures, one would require as input, to the baseline mathematical model, certain quantitative measures that uniquely characterize the soccer ball and the basic characteristics of a linear-solenoid actuator. Specifically, the analysis that follows shows the need for the following inputs: (1) the mass (m) and radius (R) of the soccer ball, (2) the initial solenoid plunger position (or gap) and the mass and diameter of the plunger within the solenoid, (3) the nominal maximum work output required by the actuator, as well as the "force vs stroke" characteristics of the mathematical model emulating the performance of a similar industrial solenoid actuator, and (4) the required time scale for the actuator to complete a single kick and the duty cycle for its operation (in other words, the number of kicks per second required for it to be competitive). In addition, we require a basic sketch of the solenoid actuator device required to initiate the BotE analysis.

In Section 2.2.1 we define the basic dimensions of the soccer field, the size and mass of the soccer ball, and then determine the "natural roll" velocity along the ground required for the ball to successfully get past the goalie. In fact, this "goal-scoring" velocity will be a key design requirement for the solenoid-kicker system.

2.2.1 Defining basic dimensions and required soccer ball velocity

We first define the dimensions of the robot playing field (e.g. 12 m wide by 18 m long for a "Robocup" middle-size robot league), the ball's diameter (0.111 m) and the mass of a competition-sized soccer ball (0.45 kg) [1]. The soccer ball is modeled as a thin-walled spherical shell.

2.2.1.1 Estimating the required soccer ball velocity

Assume that the ball is positioned in front of the robot at a distance $x = 3$ m from the goal line with a goal net width of 2 m. If the robot goalie is positioned halfway between the edges of the net and is able to move laterally at a speed of 2.0 m/s, it will take him about 0.5 s to reach the edge of the 2 m wide net and block our robot's kick to the edge of the net. If the ball is kicked from the penalty spot; $x = 3$ m and $y = 1$ m (measured from the far edge of the net), as shown in Figure 2.1, then the ball must travel a distance of 3.16 m.

The required average ball velocity must be $\geq 3.16 \text{ m}/(0.5 \text{ s})$. Therefore we set our minimum velocity requirement as

$$V_{\text{required}} = 6.32 \text{ m/s (or 22.7 km/h)}$$

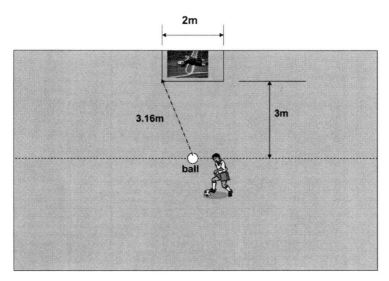

Figure 2.1. Schematic of model soccer field with player attempting to score a goal. Goal net is 2 m in width.

We assume, following the initial actuator kick, that $V_{required}$ is the velocity after the ball of radius R achieves a "natural roll" for which the horizontal velocity is related to the angular rate of rotation of the ball, ω, by the expression, $V_{natural\text{-}roll} = \omega R$.

2.2.2 Setting up a BotE model for the solenoid kicking soccer ball problem

When the ball is kicked, the imparted momentum produces a ball velocity at ground level with the direction of the velocity vector set by the direction of the applied force vector generated by the kicker's toe. Only horizontal impacts along the vertical plane of symmetry of the ball are considered. There is no side-spin or lift of the ball in our simple scenario. Think of this problem as a horizontal pool cue (or stick) hitting a cue-ball along the ball's vertical plane of symmetry. If the soccer ball is kicked at a height above or below the centerline of the ball, rotation or "spin" of the ball will accompany the forward motion (See the analysis below). A skidding ball, with a low spin rate, generates a frictional force at the ground contact point that acts to increase the spin until the "natural roll" (or zero friction) condition that we noted in Section 2.2.1 is obtained. This ground-contact-induced frictional force also slows down the forward motion of the ball. The solution to the post-kick spin and roll problem is presented in the following section.

2.2.2.1 Model for dynamics of a rolling ball struck by a thin plunger

As depicted in Figure 2.2, a hollow thin-shell spherical soccer ball at rest is struck by a thin "red" plunger moving at speed v_0 at a height h above the centerline of the ball,

44 Design of a high school science-fair electro-mechanical robot [Ch. 2

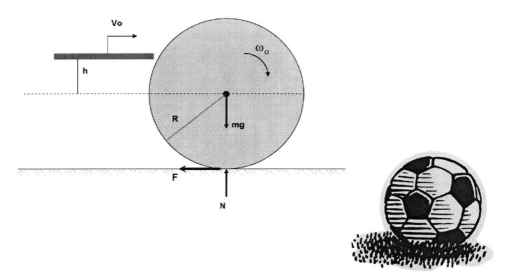

Figure 2.2. Forces acting on a linearly translating spherical ball with velocity V_b that is spinning at angular velocity ω_0, following impact by a thin red plunger moving initially at velocity v_0. F represents the backwards-pointing frictional force (positive in the negative x direction). N is the normal force at the point of contact needed to balance the weight of the ball.

imparting a horizontal impulse, I, due to the collision force X caused by the plunger acting for a short period of time on the ball.

This initial modeling problem requires us to solve the governing equations of motion to determine the plunger velocity required for the ball to achieve a particular final natural roll velocity V_{NR}.

We first develop a closed-form solution for the required plunger velocity as a function of both normalized impact height (h/R), and the ratio of the mass of the ball to the mass of the plunger (m_b/m_p). The conservation of momentum and energy equations are then used to calculate the specific plunger velocity, v_0, that yields a required natural-roll velocity of 6.32 m/s (This is the velocity required to kick the ball past the goalie in our simulated *Robocup* match; as calculated in Section 2.2.1.1).

In solving for the initial spin on the ball immediately after being struck by the plunger, we note that the law of conservation of angular momentum is *not* valid here since there is a net impulsive torque applied to the ball by the plunger. This impulse abruptly changes the ball's angular momentum. We ignore the effects of sliding friction during the short duration plunger-induced impulse. We follow the derivation presented in the solution to a comparable MIT physics class problem entitled "Rotation and Translation" [2].

The impact imparts a horizontal momentum, or Impulse, arising from the collision force X acting on the soccer ball that accelerates the center of mass m_b. So now let's define the velocity of the ball after impact $\equiv V_b$. The mass of the ball is

$\equiv m_b$. The resulting expression for the applied impulse is

$$\text{Impulse} = \int X \, dt = m_b V_b \tag{2.1}$$

Note that in Equation 2.1 we ignore the effects of ground-contact friction during the short impulse time.

The short-duration collision force X, applied to the ball by the plunger at the height h above the center-line also exerts an *external torque*, $\tau = X \cdot h$, on the ball. The corresponding angular impulse A imparted to the ball is

$$\text{A} = \int \tau \, dt = h \cdot (\text{Impulse}) = h m_b V_b \tag{2.2}$$

Newton's second law of motion for a rigid body rotating about a fixed axis subject to a net torque τ (or sum of torques) is given by

$$\tau = I_{cm} \frac{d\omega}{dt} \tag{2.3}$$

where I_{cm} is the moment of inertia of the body (our spherical ball) about its center of mass and $\omega =$ the ball's angular velocity (radians/s).

For the thin hollow sphere modeled here, the moment of inertia is given by

$$I_{cm} = \tfrac{2}{3} m_b R^2$$

Integrating Equation 2.3 over a short time interval Δt, yields the *angular form* of the impulse relationship given by Equation 2.1

$$\text{A} = \int \tau \, dt = I_{cm}(\Delta \omega) = I_{cm}(\omega_{\text{final}} - \omega_{\text{initial}}) \tag{2.4}$$

We assume that there is no initial spin on the soccer ball before it is hit (i.e. $\omega_{\text{initial}} = 0$). We also define the final angular velocity after impact to be $\omega_{\text{final}} \equiv \omega_b$. Substituting the expression for the torque impulse, Equation 2.2, into Equation 2.4 yields

$$h m_b V_b = I_{cm} \omega_b \tag{2.5}$$

Solving Equation 2.5 for the final angular velocity after impact

$$\omega_b = \frac{h m_b V_b}{I_{cm}} = \frac{h m_b V_b}{2/3 m_b R^2} = \frac{3}{2}\left(\frac{h}{R}\right)\left(\frac{V_b}{R}\right) \tag{2.6}$$

According to Equation 2.6, if the plunger strikes the ball at its centerline ($h = 0$), there is no imparted spin after impact; i.e. $\omega_b = 0$. As a result the ball initially skids or slides along the playing surface.

The torque produced by the frictional force F acting at a distance R relative to the center of the ball, as shown in Figure 2.2, immediately produces a positive angular acceleration that leads to an increase in the ball's angular velocity (for $h > 0$). The subsequent steady state or natural roll value of ω is proportional to the final natural roll velocity, i.e. $\omega_{\text{steady state}} = V_{\text{Natural Roll}}/R$.

From Equation 2.6, if the plunger rod hits the ball at a relative height $h/R = 2/3$ then $\omega_b = \left(\dfrac{V_b}{R}\right)$, which is the equation for the natural roll of a ball moving at a linear velocity $= V_b$. As we will see for this natural roll case, the angular velocity remains constant for all times after impact, i.e. $\omega(t) = $ constant $= \omega_b$, as does the translational velocity $V = $ constant $= V_b$.

Let's consider the case where the ball is hit at some arbitrary height h. We ask: What is the velocity of the ball after impact, V_b, given the velocity of the piston before impact?

To solve for V_b we first assume that there is no energy lost during the impact (an assumption that is not strictly valid for the soccer ball materials considered here). This allows us to use both the equation for the conservation of kinetic energy and the equation for the conservation of linear momentum. These two equations express the fact that the linear momentum and the linear and rotational energies of the plunger and ball system *after* impact are equal to the momentum and energy of the plunger and ball *before* impact.

The following expressions are used to calculate the ball velocity, V_b after impact, as well as the velocity of the plunger after impact V_p [3]. Note that the ball has zero initial velocity and the plunger, prior to impact, is moving at velocity v_0 (Figure 2.2).

Conservation of momentum (in the x direction)

$$\left. \begin{array}{c} \text{initial momentum} = \text{final momentum} \\ m_p v_0 = m_p V_p + m_b V_b \end{array} \right\} \quad (2.7)$$

Conservation of energy (kinetic energies of linear and rotational motion)

$$\left. \begin{array}{c} \text{initial energy} = \text{final energy} \\ \tfrac{1}{2} m_p (v_0)^2 = \tfrac{1}{2} m_p (V_p)^2 + \tfrac{1}{2} m_b (V_b)^2 + \tfrac{1}{2} I_{cm} (\omega_b)^2 \end{array} \right\} \quad (2.8)$$

where $I_{cm} = \tfrac{2}{3} m_b R^2$.

We first solve Equation 2.7 for V_p and substitute this expression into Equation 2.8. The expression derived for the rotational velocity of the ball after impact, namely $\omega_b = \dfrac{3}{2}\left(\dfrac{h}{R}\right)\left(\dfrac{V_b}{R}\right)$ from Equation 2.6, is then substituted into Equation 2.8. After a little algebra we obtain the following equation for the ball velocity after impact V_b

$$V_b = \dfrac{2v_0}{\left[1 + \dfrac{m_b}{m_p} + \dfrac{3}{2}\left(\dfrac{h}{R}\right)^2\right]} \quad (2.9)$$

Substituting Equation 2.9 into Equation 2.7 yields Equation 2.10, the companion equation for the piston velocity, V_p after impact [3], is

$$V_p = v_0 \frac{\left[1 - \frac{m_b}{m_p} + \frac{3}{2}\left(\frac{h}{R}\right)^2\right]}{\left[1 + \frac{m_b}{m_p} + \frac{3}{2}\left(\frac{h}{R}\right)^2\right]} \qquad (2.10)$$

After the initial skid and speedup of the ball's rotational velocity due to the frictional torque at the contact point, the ball subsequently develops a "natural roll" where $\omega_{\text{steady state}} = V_{\text{natural roll}}/R$. To determine the actual magnitude of $V_{\text{natural roll}}$, we can use an equation for the conservation of angular momentum, measured relative to the contact point S between the ball and the ground.

Conservation of angular momentum

In freshman physics we learned the following definition of angular momentum. The angular momentum vector $\vec{L}_s$ associated with a particle of mass m translating at velocity $\vec{v}$ relative to a given reference point S is defined as

$$\vec{L}_s = \vec{r} \times m\vec{v} \qquad (2.11)$$

where $\vec{r}$ is the position vector from S to the particle. The symbol "×" denotes the cross product.

If a body made up of a collection of particles both translates and spins, then the total angular momentum of that body is simply the sum of the translational motion of its center of mass with respect to the point S ($\vec{L}_{\text{trans}} = \vec{r}_{S,\text{cm}} \times m\vec{v}_{\text{cm}}$) and the spin of the body about its center of mass ($\vec{L}_{\text{spin}} = I_{\text{cm}}\vec{\omega}$).

For the kicked ball problem shown in Figure 2.1, our soccer ball (of radius R) moves in the x direction with scalar velocity V_b and spins with angular velocity ω_b just after impact by the plunger. If there is no net torque on a body, then the total angular momentum is conserved. After impact, the condition of zero torque will apply if we select the point S at which the ball makes contact with the ground as our reference location for evaluating the total angular momentum. There is no moment caused by the frictional force F since the moment arm is zero when it is measured from S to the contact point where the sliding frictional force is applied. The resulting conservation of angular momentum relationship for our ball, valid at any later time after impact, is then given by the following expression, Equation 2.12 [2]

$$\left. \begin{array}{c} \text{angular momentum just after impact} = \text{angular momentum at a later time} \\ m_b R V_b + I_{\text{cm}} \omega_b = m_b R V + I_{\text{cm}} \omega \end{array} \right\} \quad (2.12)$$

Solving Equation 2.12 for the "later" velocity V, using $I_{\text{cm}} = \frac{2}{3} m_b R^2$, yields

$$V = V_b - \tfrac{2}{3} R(\omega - \omega_b) \qquad (2.13)$$

When the ball achieves a natural roll, $V = V_{NR}$, and $\omega = V_{NR}/R$. Substituting these natural roll expressions into Equation 2.13 we obtain

$$V_{NR} = \tfrac{3}{5} V_b + \tfrac{2}{5}(\omega_b R) \qquad (2.14)$$

A similar expression, but with different constants for a solid (as opposed to a hollow) sphere was obtained by Shepard [4].

Using the derived expression for ω_b from Equation 2.6 for a ball struck at a height h above its centerline, we obtain

$$V_{NR} = \tfrac{3}{5} V_b \left(1 + \frac{h}{R}\right) \qquad (2.15)$$

Note from Equation 2.15 that if $h = 0$, the subsequent natural roll velocity is reduced to 60% of the ball's velocity after the plunger impact, i.e. $V_{NR} = \tfrac{3}{5} V_b$. However, if $h/R = 2/3$, then $V_{NR} = V_b$, which is a significant velocity increase over the $h = 0$ result. Clearly a faster natural roll velocity is obtained if the plunger hits higher up on the ball. If we consider the vertical distance measured from the ground plane, the natural roll height is $(5/6)d$, where d = diameter of the ball.

Substituting Equation 2.9 for V_b into Equation 2.15, gives the following equation for the natural roll velocity as a function of the initial plunger velocity, v_0, and the height ratio h/R for a fixed ratio of ball to plunger mass. (It should be noted that Shepard's problem 3.11 is the corresponding solution for a solid sphere [3].)

$$V_{NR} = \frac{6}{5} \left\{ \frac{\left(1 + \dfrac{h}{R}\right)}{\left(1 + \dfrac{m_b}{m_p} + \dfrac{3}{2}\left(\dfrac{h}{R}\right)^2\right)} \right\} v_0 \qquad (2.16)$$

This equation for the ratio of natural roll velocity to plunger velocity, V_{NR}/v_0, is plotted vs h/R in Figure 2.3 for several values of m_b/m_p.

The plunger strike position that maximizes V_{NR}/v_0 can be determined either from Figure 2.3 or by differentiating Equation 2.16 and then setting the derivative to zero. This value of h/R also is the one that minimizes plunger velocity for a given value of the natural roll velocity. As we will see later on, this minimum v_0 solution reduces the amount of solenoid energy (and current) required by our activated solenoid kicker.

The reciprocal of the ordinate in Figure 2.3, v_0/V_{NR}, is plotted in Figure 2.4 for positive values of h/R. We use this plot to readily estimate v_0 for a given required natural roll velocity.

Let's first assume, as an example, that the mass ratio for the soccer ball and plunger $m_b/m_p = 0.75$. We can see from Figure 2.4 that for this ratio the required plunger velocity for a given V_{NR} is a minimum at $h/R \approx 0.50$. At this value of h/R, the velocity ratio $v_0/V_{NR} \approx 1.18$, i.e. the required "plunger" velocity must be about

Sec. 2.2]   Back-of-the-Envelope model and analysis for a solenoid kicking device   49

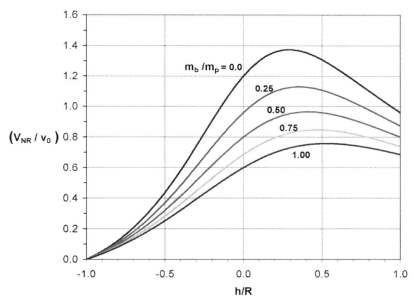

Figure 2.3. The ratio of natural-roll soccer ball velocity to initial plunger velocity, V_{NR}/v_0 as a function of the vertical height to ball radius ratio, h/R, for selected ball/plunger mass ratios, m_b/m_p.

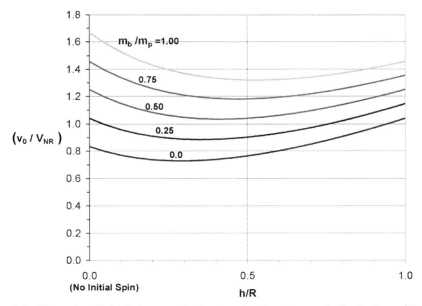

Figure 2.4. The ratio of initial plunger velocity to natural-roll soccer ball velocity v_0/V_{NR} as a function of the vertical height to ball radius ratio, h/R, for selected ball/plunger mass ratios, m_b/m_p.

20% greater than the natural soccer ball roll velocity obtained by our estimate. If the required natural roll velocity to score a goal is $V_{NR} = 6.32\,\text{m/s}$ (as calculated in Section 2.2.1.1) then the corresponding plunger velocity must be

$$v_0 = 1.18 \times 6.32\,\text{m/s} = 7.46\,\text{m/s}$$

Rounding this to the nearest 0.1 m/s, we set the *required* maximum plunger velocity to be delivered by our linear-actuator solenoid kicker design to be

$$v_{0_{required}} = 7.5\,\text{m/s}$$

Interestingly, the minimum plunger energy solution for an impact height $h/R \approx 0.50$ is slightly *below* the ideal impact height necessary for a "natural roll"; $h/R = 2/3$.

For our minimum velocity impact solution, the solenoid plunger must strike a standard soccer ball at a height of about 3 cm (or approximately 1 inch) above the centerline of any captured soccer ball. This strike-point determines the "best height" for positioning the solenoid actuator on the moving robot platform.

2.2.3 Model for solenoid kicker work and force

2.2.3.1 *Why a solenoid kicker?*

The challenge set for the student is to design a simple robot kicking device that is able to kick a soccer ball with sufficient speed to send it past a defender and into the goal. We can choose from several different mechanical or electro-mechanical drive mechanisms to power the moving kicking device: elastic springs or rubber bands, pneumatic or pressure reservoirs, electric current driven linear solenoids, or a range of electric servo-motors (e.g. sim-motors). All of these mechanisms convert some form of stored potential energy (e.g. elastic spring energy, stored pressure, or stored electro-magnetic energy) into kinetic energy. The magnitude of this kinetic energy, KE (or equivalent mechanical work), is used to calculate the effective "plunger" speed $= \sqrt{2(KE/m_p)}$ needed to initiate the subsequent kicking motion.

While a compressed spring is the simplest device, it takes a considerable amount of time to reload the spring. One typically uses an auxiliary dc motor driven by a small battery to do the reloading. The basic limitation of this device for our application is that it typically takes many seconds to reload the spring. When reviewing the performance of the motor designed several years ago for Robocup competitions, the builders found that it took of the order of 5 seconds to make such a reload [5], which they considered too long a time between kicks for a fast-moving game.

Simple commercial linear-solenoids have reload times of order 0.1 seconds and can be powered by compact dc battery supplies driving a simple circuit. They are readily available and fairly cheap. Our design challenge is to determine the solenoid size that will provide the required kicking speed, recycle time, and momentum necessary for "game" conditions.

2.2.3.2 Linear-solenoid fundamentals

A solenoid linear-actuator is a long solenoid coil wound in a helical pattern, with a steel (or other ferromagnetic metal) plunger core housed within the winding. The plunger is pulled into the center of the coil when energized by an electric current. The linear solenoid actuator has many applications including: locks, doorbells, switches, and relays. When the current is passed through the coil a magnetic field is set up, with the magnetic field inside the coil much stronger than that outside the coil. When the steel plunger is placed near or within one end of the energized coil, the magnetic field causes the cylindrical plunger to become a temporary magnet of opposite polarity to that of the coil. As a result the steel plunger is virtually sucked into the center of the coil by this magnetic force and travels freely along the centerline of the coil, towards the ferromagnetic back stop. For steel or other ferromagnetic plungers the direction of the intake force on the plunger is *independent* of the direction of the current in the coil. Most solenoids include a fixed cylindrical stop that extends part way into the center bore, which improves performance. After being halted the plunger is returned to its initial gap position, usually by a modest spring, when the current is cut off. A sketch of a generic solenoid and plunger is shown in Figure 2.5.

A drawing of a typical solenoid and plunger (with a "push pin" attached, as needed for a robot kicker) is shown in both Figure 2.6 and Figure 2.7. We might add

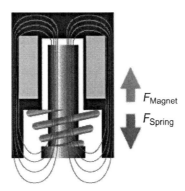

Figure 2.5. Solenoid and plunger schematic with return spring [6].

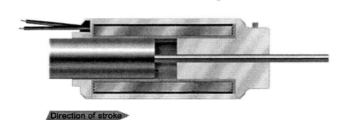

Figure 2.6. Push solenoid schematic. The push rod or plunger moves to the right when the solenoid is energized. [7, 8].

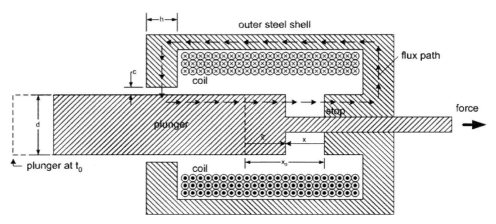

Figure 2.7. Push-pin solenoid schematic. $x(t)$ is the time-varying air gap length between the plunger face and the stop; x_0 = initial air gap at $t = 0$. Note the small annular gap, of length c, between the sliding plunger and the outer steel shell. The plunger diameter = d. $\tilde{x}$ is the distance moved by the plunger (positive to the right). [Illustration courtesy of J. Jacobs]

a small disk perpendicular to the end of the push pin to assist in the transfer of the plunger force to the soccer ball—in effect, it is the kicker's "boot".

2.2.3.3 Linear-solenoid theory and calculation of magnetic field energy

In this section we derive an expression for the energy or work, U_B, stored in the magnetic field of a solenoid. The simple mathematical model is based on the standard freshman physics lectures that introduce the basic laws governing magnetic fields, namely Ampere's law and Faraday's law of induction.

Using the derived equation for the available magnetic energy produced by the solenoid, we then assume that this energy is ideally converted into mechanical work associated with the attractive force F. This force moves a steel plunger of mass m_p inside our ideal solenoid linear actuator over a distance Δx according to

$$U_B = U_{\text{mechanical}} = -\int_0^{\Delta x} F(\mathbf{x})\, dx \qquad (2.17)$$

where in the absence of friction, $U_{\text{mechanical}} = \frac{1}{2} m_p V_p^2$.

Note that the minus sign in Equation 2.17 indicates that the direction of the solenoid force is opposite to the direction of increasing gap size, x, as depicted in Figure 2.7.

From Equation 2.17, we see that the force on the plunger, F, is simply the first derivative of $U_B(x)$ with respect to the variable gap distance x

$$F(x) = -\frac{dU_B}{dx} \qquad (2.18)$$

The mechanical work is converted directly into a corresponding amount of kinetic energy for the accelerating steel plunger, $\tfrac{1}{2}m_p V_p^2$. The final push-rod or plunger velocity, V_p, is thus calculated from the released stored magnetic energy, U_B.

2.2.3.3.1 The magnetic field of a solenoid

Ampere's law states that the line integral of the magnetic field vector **B** around a closed loop of incremental vector length $d\vec{\ell}$ is equal to the total current enclosed within, or

$$\oint \mathbf{B} \cdot d\vec{\ell} = \mu_0 N I \tag{2.19}$$

where N = number of turns of the solenoid coil
I = electric current carried in the coil (the current that pierces the closed loop). Unit is ampere or A.
μ_0 = permeability of free space = $4\pi \times 10^{-7}$ henry/m
   = $4\pi \times 10^{-7}$ newton/(ampere)^2

The units for **B** are tesla or newton/ampere-m.

Assuming a uniform magnetic field inside a solenoid with a centerline air gap length ℓ we obtain from Equation 2.19 the well known solution for a very long solenoid,

$$B = \mu_0 N I / \ell \tag{2.20}$$

2.2.3.3.2 The magnetic energy of a solenoid actuator

Using classical electromagnetic theory one can show that the magnetic energy, U_B, stored in a volume of space occupied by a magnetic field (in a vacuum or in a non-magnetic substance like air) is proportional to the square of the magnetic field magnitude B integrated over the free space volume [9]

$$U_B = \frac{1}{2\mu_0} \int B^2 d(\text{Vol}) = \frac{1}{2\mu_0} \int_0^{\ell_{max}} B^2 A_p \, dl \tag{2.21}$$

For our cylindrical solenoid, the incremental volume, $d(\text{Vol})$, is equal to the plunger face cross-sectional area A_p multiplied by the incremental gap length $d\ell$. Integrating Equation 2.21, with B given by Equation 2.20, over a cylindrical volume with gap length varying between 0 and ℓ_{max}, yields

$$U_B = \frac{\mu_0 (NI)^2}{2}(A_p) \int_0^{\ell_{max}} \ell^{-2} d\ell = -\frac{\mu_0 (NI)^2}{2}(A_p)\left[\frac{1}{\ell}\right]_0^{\ell_{max}} \tag{2.22}$$

Note that if we evaluate Equation 2.22 at the lower limit, where the gap distance $\ell \to 0$, the total stored energy becomes unbounded, i.e. $U_B \to \infty$.

In real linear-actuator systems, there is always some small *additional* air gap besides the primary one along the main axis of the solenoid between the plunger and backstop. One such *additional* gap is the annular air gap between the outer steel wall (or pole) of the solenoid and the sliding plunger. This second gap in Figure 2.7 with width "c" makes it possible for the plunger to smoothly move in and out of the solenoid. This gap adds an additional amount of *reluctance* that reduces the level of the magnetic field B carried by the plunger and solenoid body. You can think of reluctance as a kind of "resistance" in an ideal magnetic circuit, e.g. the flux path in Figure 2.7 is analogous to the corresponding path for the current in a standard battery-driven electrical circuit. The effect of this added *reluctance* is particularly important when the centerline solenoid gap is small. Note that in this treatment the equations governing the magnetic field in a general magnetic circuit with a number of air gaps are not presented, nor utilized, thereby enabling us to use simplified equations in the solenoid kicker problem.

Following our BotE approach, we approximate this additional gap effect by adding a constant length "a" to the basic plunger/stop gap length ℓ in Equation 2.22. In an ideal actuator, "a" is proportional to the width of the clearance "c" as shown in Figure 2.7. Hence, as an *approximation*, we amend Equation 2.22 for stored energy to include an additional *empirical gap* length "a" as follows

$$U_B \approx \frac{\mu_0(NI)^2}{2}(A_p)\int_0^{\ell_{max}}(\ell+a)^{-2}\,d\ell = -\frac{\mu_0(NI)^2}{2}(A_p)\left[\frac{1}{\ell+a}\right]_0^{\ell_{max}} \quad (2.23)$$

Evaluating Equation 2.23, noting that the previous singularity disappears for non-zero values of "a", we obtain a *bounded* expression for the maximum energy stored in our solenoid actuator

$$U_{B_{max}} = C\left[\frac{1}{a} - \frac{1}{(\ell_{max}+a)}\right] \quad (2.24)$$

where

$$C = \frac{\mu_0(NI)^2}{2}(A_p).$$

Using an arbitrary gap length x (the coordinate shown in Figure 2.7) instead of a fixed gap ℓ_{max}, the energy solution can be written as the following function of x

$$U_B(x) = -C\left(\frac{1}{x+a}\right) + \frac{C}{a} \quad (2.25)$$

Combining the two fractions in Equation 2.24 produces a compact equation for maximum energy stored in a solenoid actuator with a given initial gap length ℓ_{max}

$$U_{B_{max}} = \frac{C\ell_{max}}{a(\ell_{max}+a)} \quad (2.26)$$

2.2.3.3.3 The force on the plunger

Employing Equation 2.18, we see that the solenoid-driven force on the plunger $F(x)$ is given by the derivative with respect to x of $U_B(x)$ in Equation 2.25

$$F(x) = -\frac{dU_B}{dx} = -C\left(\frac{1}{x+a}\right)^2 \quad (2.27)$$

The sign of the force is negative because the force $F(x)$ on the plunger acts opposite to the direction of a positive increase in the gap dimension; the plunger is pulled into (not out of) the solenoid cavity. The force decreases with increasing gap width x. The maximum force on the plunger, obtained at zero gap width, is obtained from Equation 2.27

$$F_{\max}(x=0) = -\frac{C}{a^2} \qquad (2.28)$$

We observe that the magnitude of $F_{\max}$ is strongly sensitive to the numerical value of the auxiliary gap length parameter "a". As "a" gets smaller, the absolute value of $F_{\max}$ grows as $1/a^2$. There is an approximate way to determine the empirical value for "a" to be used in our subsequent analysis. We find "a" by setting the magnitude of the stored energy, given by Equation 2.26, equal to the measured amount of work, $U_{B_{\max}}$, that is determined from plunger force measurements (as a function of gap distance) for a particular commercial linear actuator solenoid device. We select a commercial actuator device with an amount of energy sufficient to *potentially* meet the needs of our robot kicker.

For a particular solenoid, driven by a current I, the estimated work output of the actuator is found by numerically integrating the measured "force vs gap" curve for that device. A typical force vs stroke measurement curve is shown later in Figure 2.9. The required energy for our robot kicker problem is ≈ 17 joules which is equal to the kinetic energy of a 0.6 kg plunger moving at the required speed of 7.5 m/s, as calculated in Section 2.2.2.1.

2.2.3.3.4 The plunger velocity as a function of distance traveled

The kinetic energy imparted to the plunger as a function of the instantaneous gap spacing, x, can readily be calculated using the equations already derived for the energy, namely Equation 2.25. Let's first consider the change in kinetic energy for a push-rod plunger accelerated from zero velocity (at an initial gap distance x_0) to a final zero-gap position at the solenoid back-stop, as shown in Figure 2.7. We write the following expression for the change in kinetic energy, ΔKE, based on the change in stored energy for the plunger at two gap positions (x_0 and x). ΔKE is chosen so that the kinetic energy $= 0$ when the plunger is at the initial position x_0.

$$\Delta KE \equiv KE(x) = U_B(x_0) - U_B(x) \qquad (2.29)$$

Using Equation 2.25 for the stored energy, we obtain

$$KE(x) = -C\left(\frac{1}{x_0 + a}\right) + C\left(\frac{1}{x + a}\right) = \frac{C(x_0 - x)}{[(x - x_0) + (x_0 + a)][(x_0 + a)]} \qquad (2.30)$$

In order to more easily interpret Equation 2.30, let's define (from Figure 2.7) a plunger-based distance coordinate, $\tilde{x}$ with its origin at the initial gap position of the plunger ($x = x_0$), that is positive in value for all plunger positions short of the solenoid backstop,

$$\tilde{x} = x_0 - x \qquad (2.31)$$

The kinetic energy as a function of $\tilde{x}$ is then given simply by

$$KE(\tilde{x}) = \tfrac{1}{2}m_p V_p^2 = \frac{C\tilde{x}}{(b - \tilde{x})b} \qquad (2.32)$$

where $b \equiv x_0 + a$.

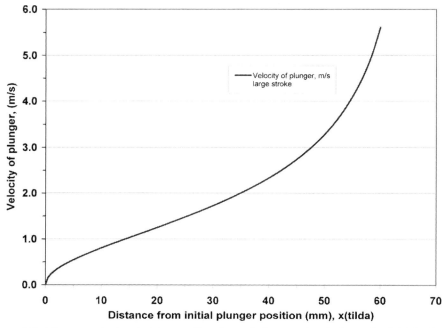

Figure 2.8. Plunger velocity (m/s) as a function of distance (mm) from the initial plunger position 60 mm from the solenoid stop. ($NI = 9{,}080$ ampere-turns) and plunger mass $= 0.6\,\text{kg}$. Auxiliary gap length parameter "a" $= 7\,\text{mm}$.

Solving Equation 2.32 for the corresponding plunger velocity as a function of the plunger distance $\tilde{x}$

$$V_p = \left[\frac{2C}{m_p b}\right]^{1/2} \left[\frac{\tilde{x}}{(b-\tilde{x})}\right]^{1/2} \tag{2.33}$$

where

$$C = \frac{\mu_0 (NI)^2}{2}(A_p)$$

A sample plot of this function is depicted in Figure 2.8 for a set of baseline parameters.

The maximum plunger velocity, at the zero gap stop position, $\tilde{x} = x_0$ is given by the simple equation

$$V_{p_{\max}} = \left[\frac{2C}{m_p a}\right]^{1/2} \left[\frac{x_0}{(x_0 + a)}\right]^{1/2} \tag{2.34}$$

For the set of parameters used in Figure 2.8, the maximum plunger velocity is 5.61 m/s, which is the maximum velocity shown at $\tilde{x} = 60$ mm. This velocity falls below the required plunger maximum velocity of 7.5 m/s required to produce a 6.3 m/s natural roll velocity for our penalty-kicked soccer ball.

As we will show, a 7.5 m/s plunger velocity can be obtained by increasing the magnitude of NI (or ampere-turns) for the baseline solenoid in Figure 2.8. Finally, we note that for large initial gap distances x_0 compared to "a", i.e. $\frac{x_0}{a} \gg 1$, the maximum velocity is given by the following expression independent of x_0

$$V_{p_{max}} \rightarrow \left[\frac{2C}{m_p a}\right]^{1/2} \qquad (2.35)$$

2.2.3.3.5 Estimated total plunger travel time

With the equation for plunger velocity as a function of plunger distance $\tilde{x}$ given by Equation 2.33, we can readily calculate the time required for the plunger to move from its initial gap position to the final zero gap position. We designate this time as t_{max}. Since the incremental travel time dt is equal to the incremental distance $d\tilde{x}$ divided by the instantaneous velocity, $V_p(\tilde{x})$, t_{max} is obtained from the following integral

$$t_{max} = \int_0^{x_0} \frac{d\tilde{x}}{V_p(\tilde{x})} \qquad (2.36)$$

We leave it as an exercise for the reader to show that the approximate value of the integral, valid for $a/x_0 \ll 1$, is

$$t_{max} \approx \frac{\pi}{2}\left(\frac{m_p x_0^3}{2C}\right)^{1/2} \qquad (2.37)$$

Equation 2.37 indicates that the total plunger travel time is proportional to the initial gap distance to the 3/2 power and inversely proportional to NI (ampere-turns) since $C = \frac{\mu_0 (NI)^2}{2}(A_p)$. Therefore the higher the current I, the shorter the travel time. This plunger travel time is important in our student's soccer tournament, since it helps to determine how fast the robot device can repeat a kick, assuming that another soccer ball is available (The ball must be *trapped* just in front of the plunger).

2.2.3.3.6 Power-up time scales for an R-L circuit

We have not included in our repeat-time estimate: the time for the return spring on the plunger to reset the plunger gap, or the time for a resistance-inductance (R-L) circuit to bring the solenoid to full current or power once the "kick" button is hit. The power-up time scale for this circuit is of order $5L/R$ where the solenoid inductance $L \approx \frac{\mu_0 N^2 A_p}{\ell_{solenoid}} \approx 0.01$ henry for the solenoid considered below. This formula for the inductance L of a long solenoid is derived in Section 33-2 of the classic physics textbook by Halliday and Resnick [9]. The calculated R-L power-up time scale ($5L/R$) is of order 25 milliseconds for a coil with a 2 ohm resistance. This interval can be reduced by decreasing the number of turns N in the solenoid design, because that reduces the inductance L. Equation 2.34 shows that for a fixed plunger velocity requirement, NI must be set at a particular level. This constant level implies that the current I must be increased if N is decreased. The increased current generates greater I^2R heating losses which have to be taken into account when determining the number of kicks in a game.

2.2.3.3.7 Calculated levels of maximum force, energy, and velocity

The following parameters are typical of a large push-type tubular solenoid such as produced by Magnetic Sensor Systems [10] with a 3.0 inch diameter and a 4.13 inch outer cylindrical shell length. The output work of this particular device, at its *nominal NI* settings, is lower, but it is still within a factor of two of our maximum energy requirement of about 17 joules.

- Maximum plunger stroke, $x_0 = 60\,\text{mm} = 0.06\,\text{m}$
- Ampere turns $= NI = 9{,}080$ for a 10% duty cycle for our reference case of a 3×4.13 inch push-type solenoid actuator
- Plunger diameter $= 1.68$ inches $= 0.0426\,\text{m}$
- Plunger Area, $A_\text{p} = 1.425 \times 10^{-3}\,\text{m}^2$
- Plunger mass, $m_\text{p} = 0.6\,\text{kg}$
- $\mu_0 = 4\pi \times 10^{-7} = 1.257 \times 10^{-6}$ newton/(ampere)^2
- From a numerical integration of the measured force vs stroke or from the displacement curve of Figure 2.9, the estimated measured work

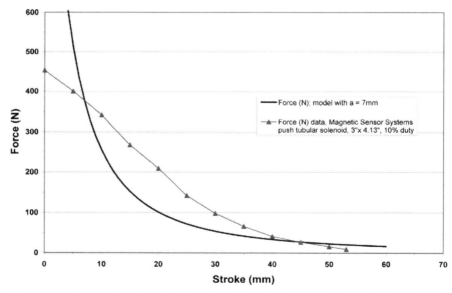

Figure 2.9. Solenoid force (newtons) as a function of stroke or gap distance (mm). Comparison of analytical force model (dark blue curve with length parameter $a = 7$ mm) against the measured force vs stroke data (red triangles) published for the Magnetic Sensor System push tubular solenoid operating at a 10% duty cycle with $NI = 9{,}080$ [11].

$$= \int_0^{\Delta x_0} F(\mathbf{x})\, dx \approx 9.5\, \text{J}$$ (for the 3×4.13 inch push-type solenoid actuator with $NI = 9{,}080$ ampere-turns). From Equation 2.26 we calculate "a" ≈ 7 mm $= 0.007$ m.

- Wire size (American wire gauge, AWG $= 17$); $d = 1.15$ mm
- Resistance $= 1.9$ ohm for $N = 713$ for 115 m length; resistance/wire-length $= 0.166$ ohms/m
- Estimated current $= 12.7$ amperes for a 10% duty cycle and $NI = 9{,}080$ ampere-turns as recommended by Magnetic Sensor Systems
- Voltage $= 24.2$ volts; or two 12 volt batteries

We can now calculate typical values for the maximum force and maximum energy (or delivered work) for our model solenoid given the solenoid/plunger parameters for the reference 3×4.13 inch push-type solenoid actuator at a 10% duty cycle

$$C = \frac{\mu_0 (NI)^2}{2}(A_p) = \frac{1.257 \times 10^{-6}(9{,}080)^2}{2}(1.425 \times 10^{-3})$$

$$= 7.38 \times 10^{-2}\ \text{newton-m}^2$$

$$F_{\max}(x=0) = -\frac{C}{a^2} = -\frac{7.38 \times 10^{-2}}{(0.007)^2} = 1{,}506\ \text{N}$$

$$U_{B_{\max}} = \frac{C\ell_{\max}}{a(\ell_{\max} + a)} = \frac{7.38 \times 10^{-2}(0.06)}{(0.007)(0.067)} = 9.44\ \text{J}$$

The maximum velocity is calculated by directly setting the kinetic energy to the maximum stored magnetic energy of 9.44 J

$$V_{P_{\max}} = \sqrt{\frac{2 U_{B_{\max}}}{m_p}} = \sqrt{\frac{2(9.44)}{0.60}} = 5.61\ \text{m/s}$$

Our simple asymptotic approximation for $V_{P_{\max}}$ (from Equation 2.35) yields about a 6% higher plunger velocity

$$V_{P_{\max}} \rightarrow \left[\frac{2C}{m_p a}\right]^{1/2} = \left[\frac{2(7.38 \times 10^{-2})}{(0.6)(0.007)}\right]^{1/2} = 5.93\ \text{m/s}$$

The total plunger travel time from Equation 2.37 is estimated to be

$$t_{\max} \approx \frac{\pi}{2}\left(\frac{m_p x_0^3}{2C}\right)^{1/2} = \frac{\pi}{2}\left(\frac{(0.60)(0.06)^3}{2(7.38 \times 10^{-2})}\right)^{1/2} = 0.047\ \text{s} = 47\ \text{millisec}$$

If 47 millisec was the only time to consider in our problem, the robot-kicker could theoretically kick a soccer ball over 20 times per second. If the power-up time for the solenoid circuit is of order 25 millisec then the soccer ball could be kicked every 72 millisec or about 14 times per second.

An additional important time scale must be considered for our solenoid. It is the time needed to cool the solenoid device to prevent damage to the electrical wires or insulation. This is the reason for the recommended low 10% duty cycle for the high current industrial solenoid considered here (i.e. 10% of the time the solenoid is energized, 90% of the time the current is turned off). If we adopt a 10% duty cycle, the ball can be kicked on average every 720 millisec, which is about every $\frac{3}{4}$ of a second, thus allowing about 4 kicks for every 3 seconds. This might be a little slow for robot soccer. However it is possible to make a significant number of kicks rapidly within a few tenths of a second, as needed. As we show in the Appendix at the end of this chapter, it is possible to use a *100% duty cycle* if the total number of kicks made in a game is limited to a few hundred.

An additional BotE analysis needs to be undertaken in order to generate an estimate for the maximum temperatures, the heating and "cooling" time scales, and the maximum recommended number of kicks in a game for our solenoid. In the Appendix, we perform a Quick-Fire analysis of the solenoid heating problem to estimate: (1) the temperature rise produced during a single solenoid-actuator initiated kick, and (2) the increased solenoid temperatures caused by a number of rapid follow-up kicks. We conclude that heating analysis by estimating the allowable number of kicks that can be performed without causing thermal damage to the electrical insulation of the wires of our solenoid coil.

2.2.3.3.8 Comparison of estimates with solenoid actuator data

In this section we compare our modeled force vs gap (or stroke) performance curve to actual data for the industrial solenoid actuator of interest. We also estimate the level of current (or ampere-turns) needed to satisfy the required solenoid plunger speed of 7.5 m/s. Figure 2.9 shows modeled and measured solenoid force as a function of stroke or gap distance (mm).

We see that our model force curve as a function of gap has more curvature than does the measured force vs stroke data for the Magnet Sensor Systems solenoid. But we have chosen the empirical gap parameter "a" for our calculations to be 7 mm in order to make our theoretical work output (in Joules) equal to the measured work output for the data given in Section 2.2.3.3.7. Figure 2.9 shows that our modeled force, F, falls below the data for stroke or gap lengths in the range of 10 to 40 mm. On the other hand, Equation 2.27 produces a higher force than that of the data when the plunger approaches zero stroke.

Industrial-built solenoids are usually designed to reduce the maximum load on the "stop face" at the zero gap point by using design modifications to "flatten" the overall force vs stroke curve while still maintaining maximum power output. One of the design changes involves changing the shape of the plunger face away from a simple flat configuration to that of a truncated cone as well as shaping the receiving fixed center pin [12]. Such detailed design modifications are beyond the scope of our simple BotE estimates.

Figure 2.10 shows the dependence of plunger maximum velocity and the maximum work output from our modeled solenoid actuator as a function of *NI*,

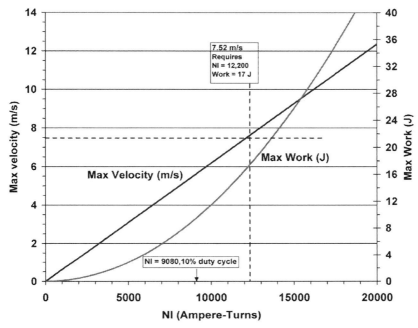

Figure 2.10. Theoretical maximum final velocity and maximum work as a function of NI, the product of applied current and the number of solenoid turns for a given solenoid plunger mass of 0.6 kg. Initial plunger position (or initial gap = 60 mm).

the ampere-turns for the activated solenoid, as derived from Equations 2.26 and 2.33.

Remember that our solenoid actuator must be capable of kicking a soccer ball to a required final natural roll velocity of 6.3 m/s to score a competitive goal. We previously determined that the maximum plunger velocity required to deliver a 6.3 m/s natural roll of a soccer ball is approximately equal to 7.5 m/s (as given by Equation 2.16 and plotted in non-dimensional form in Figure 2.3). For a plunger velocity of 7.5 m/s, Figure 2.10 shows that the required level of NI needed to produce this velocity is 12,200 ampere-turns. The solenoid actuator output work for this NI is $U_{B\text{-max}} = 17$ Joules. From Section 2.2.3.3.7 and the data for the Magnetic Sensor System solenoid being considered, the voltage for a 1.9 ohm solenoid operating at a 10% duty cycle is 24.2 volts. Ohm's law gives a current of 12.7 amperes to operate the reference solenoid with a 1.9 ohm resistance. If $NI = 9,080$ ampere-turns, as recommended for our reference commercial solenoid, then N must be 713 turns.

Note that for the plunger velocity requirement of 7.5 m/s, our solution of $NI = 12,200$ ampere-turns results in a higher nominal current of $I = NI/713 = 17.1$ A if we maintain N at 713 turns

$$(I_{\text{required}} \approx 17 \text{ A for } N = 713 \text{ turns})$$

The corresponding required voltage is then $E = I \cdot R = (17.1)(1.9) = 32.5$ V

($E_{\text{required}} \approx 33$ volts; or 3 twelve volt batteries)

This higher current requirement 17 A leads to an 80% higher Ohmic heating rate for our solenoid system compared to that of the reference push-tube solenoid if we keep the number of solenoid-turns the same as for our reference solenoid at 713 and also use 17 gauge solenoid wire.

For the required number of ampere-turns estimated above at 12,200, the calculated total time to carry out a single kick is approximately 60 millisec. This is based on estimates for both the travel time of the plunger and the solenoid power-up time. (For calculation details see Table 2.2, Section 3.3 below). For a 10% duty cycle, the time between kicks will then be about 600 millisec, corresponding to a little under 2 kicks per second based on industrial *near continuous* operational heat load and cooling considerations.

In real solenoid actuator systems the packing volume, namely how close the electrical components are packed in a small space, is also a design consideration, particularly when the solenoid is subjected to high heat loads [12].

When all the above issues are taken into consideration, some alteration of the design parameters may be needed in order to satisfy all the competing system constraints for our robot system. However, our estimate appears to be a good *first-cut* at the problem.

2.2.4 Final design requirements for linear-actuator solenoid and supporting electrical system

The following tables summarize the key BotE equations used in the preceding sections. Numerical results are given for important solenoid-kicker parameters, in particular the calculated magnitudes of ball and plunger velocity, maximum available work required, and the number of solenoid ampere-turns for our modeled kicker. For a nominal resistance, based on 713 turns of 17 gauge wire, we also calculate the required current and applied voltage.

Table 2.1 presents the key BotE results for the dynamics of a rolling soccer ball struck by a thin plunger. The required solenoid piston velocity is calculated for the minimum required input energy solution needed to produce a given natural roll velocity to readily score a goal.

Table 2.2 presents the BotE results for a basic solenoid actuator design that achieves a plunger velocity given by the results shown in Table 2.1. The primary calculated solenoid parameters are: (a) the required solenoid-stored magnetic energy (joules); (b) the required solenoid ampere-turns, NI; (c) the estimated solenoid current I and required voltage for a given solenoid wire size or gauge; and (d) the estimated kicker repetition time, based on estimated plunger travel times and the estimated R-L circuit power-up time scale.

Figure 2.11 (thanks to J. Jacobs, personal communication, October, 2010) is a candidate circuit schematic that shows how the 17 ampere current required to drive our pulsed solenoid can be turned "on" and "off" using a moderately priced commercial electronic solenoid driver. As per Table 2.2 the voltage applied to the driver is supplied by three 12 volt batteries. The solenoid driver is cued by a low (0 to 5 volt) digital control voltage signal supplied directly from the robot itself. The initiation signal is radioed to the robot from the student team that remotely operates the robot.

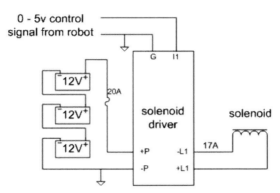

Figure 2.11. The schematic shown is based on the Model Si5SD1-50V-20A control board. This single, 50 V, 20 A solenoid driver (based on a 60 A MOSFET power semiconductor) is sold with an integrated heat sink by Signal Consulting on their website http://signalllc.com/products/Si5SD1-50V-20A.html.

2.3 APPENDIX: MODELING OF THE TEMPERATURE RISE PRODUCED BY OHMIC HEATING FROM SINGLE OR MULTIPLE SOLENOID-ACTUATOR KICKS

To illustrate how it is possible to obtain a BotE estimate quickly using a minimum amount of technical information we take as a sample assignment the problem of estimating the temperature rise produced by the current applied during one or more solenoid-actuator kicks.

2.3.1 Quick-Fire problem approach

As in Chapter 1, in the Quick-Fire approach to a BotE problem we develop our estimates by systematically following, in sequence, the following five steps:

(a) Define and/or conceptualize the problem using a sketch or brief mathematical description.
(b) Select a model or approach, either mathematical or empirical that describes the basic physics of the problem.
(c) Determine the input data parameters and their magnitudes as required to solve the problem, either from data sources or by scaling values by analogy, or simply by using an educated guess.

Table 2.1. BotE input parameters and summary of the calculations of the dynamics of a rolling soccer ball struck by a thin plunger (see Figure 2.2) to establish the requirements for our linear-actuator solenoid kicker.

	Required soccer ball and plunger velocity estimates	
	Given	*Data reference*
Soccer ball mass = m_b (kg)	0.45	Section 3.1
Ball diameter = $2R = d$ (m)	0.111	Section 3.1
Goal net width = w (m)	2.0	Section 3.1.1
Distance to goal-net edge X_{gn} (m)	3.16	Section 3.1.1
Lateral speed of goalie V_{goalie} (m/s)	2.0	Section 3.1.1
		Calculation
Minimum required natural roll ball velocity to score a goal = $V_{required} = V_{NR}$ (m/s)		$V_{NR} = \left(\dfrac{X_{gn}}{w/2}\right) \cdot V_{goalie} = \left(\dfrac{3.16}{1.0}\right) \cdot 2.0 = 6.32$ m/s
Initial ball spin after plunger hit at height h, ω_b (radian/s); initial ball velocity V_b is unknown		$\omega_b = \dfrac{3}{2}\left(\dfrac{h}{R}\right)\left(\dfrac{V_b}{R}\right)$ [Equation 2.6 based on piston impact on ball, Figure 2.2]
Velocity of the ball after impact, V_b, for a given piston velocity before impact, v_0 (m/s)		$V_b = \dfrac{2v_0}{\left[1 + \dfrac{m_b}{m_p} + \dfrac{3}{2}\left(\dfrac{h}{R}\right)^2\right]}$ [Equation 2.9; from conservation of momentum and energy, both linear and rotational]

Appendix: Modeling of the temperature rise produced by ohmic heating

Natural roll velocity in terms of V_b, for a given piston impact height, h	$V_{NR} = \frac{3}{5} V_b \left(1 + \frac{h}{R}\right)$ (Equation 2.15 using $\omega = V_{NR}/R$)
Piston velocity v_0 for a given natural roll velocity	$v_0 = \frac{5}{6} \underbrace{\left\{ \frac{\left(1 + \frac{m_b}{m_p} + \frac{3}{2}\left(\frac{h}{R}\right)^2\right)}{\left(1 + \frac{h}{R}\right)} \right\}}_{} V_{NR}$ [from the inverse of Equation 2.16; as in Figure 2.4]
Calculated piston velocity given $m_p = 4/3 m_b \approx 0.6$ kg for $h/R \approx 0.50$ necessary to obtain a minimum v_0; per Figure 2.4. For $h/R = 0.5$ the impact height $h = 0.5(0.111 \text{ m}/2) = 2.78$ cm above the ball's centerline	From above equation: $v_0 = \frac{5}{6} \underbrace{\left\{ \frac{\left(1 + \frac{3}{4} + \frac{3}{2}(0.50)^2\right)}{(1 + 0.50)} \right\}}_{} (6.32 \text{ m/s}) = 7.46 \text{ m/s}$ Rounding to nearest 0.1 m/s **7.5 m/s** is the estimated maximum velocity required of the push-type plunger in our linear-actuator solenoid (see Table 2.2).

Table 2.2. BotE results for a basic solenoid actuator design that achieves a plunger velocity of 7.5 m/s given by Table 2.1. The calculated solenoid parameters are: (a) the required solenoid-stored magnetic energy (joules); (b) the required ampere-turns, NI; (c) the estimated solenoid current I and required voltage for a given solenoid wire gauge; and (d) the estimated kicker repetition time based on estimated plunger travel times and R-L circuit power-up time scales.

	Required solenoid energy, ampere-turns, current etc. to meet given plunger velocity requirement	
	Given	Data source: [10, 11]. Solenoid geometrical parameters are typical of a large Push Type Tubular Solenoid (3.0 inch diameter by 4.13 inch in outer cylindrical shell length) produced by Magnetic Sensor Systems [www.solenoidcity.com].
Plunger mass, m_p	0.6 kg	Section 2.2.3.3.7
Plunger diameter = 1.68 inch	0.0426 m	Section 2.2.3.3.7
Plunger Area, A_p	1.425×10^{-3} m^2	Section 2.2.3.3.7
Maximum plunger stroke, x_0 = 60 mm	0.06 m	Section 2.2.3.3.7
Estimated empirical length scale "a" for the baseline large push-type tubular solenoid; $a \approx 7$ mm [10, 11]	0.007 m	Section 2.2.3.3.7 The empirical length scale "a" is defined as additional gap scale length scaled to the force vs gap (or stroke) data for commercial solenoid actuators. One typical contributor to "a" in commercial actuators is the annular air gap between the outer steel wall of the solenoid and the steel plunger itself.
Resistance for AWG 17 wire with 115 m length	1.9 ohm or resistance per length = 0.166 ohms/m	

Appendix: Modeling of the temperature rise produced by ohmic heating

$\mu_0 = 4\pi \times 10^{-7}$ newton/(ampere)^2	1.257×10^{-6}	Permeability of free space
		Calculation
Maximum energy, U_B, stored in a solenoid actuator with a given initial gap length, x_0		$U_{B_{max}} = \dfrac{Cx_0}{a(x_0 + a)}$ (Eq. 2.26), Section 3.2.2.2 where $C = \dfrac{\mu_0(NI)^2}{2}(A_p)$ and $NI =$ the number of ampere-turns. The quantitative value of NI for our model design is determined below to meet the plunger velocity requirement
Theoretical Maximum plunger velocity		$V_{P_{max}} = \left[\dfrac{2C}{m_p a}\right]^{1/2} \left[\dfrac{x_0}{(x_0 + a)}\right]^{1/2}$ (Eq. 2.34); Section 3.2.2.4
Theoretical value of NI needed to produce a given plunger velocity		$NI = V_{P_{max}} \left[\dfrac{m_p}{\mu_0 A_p}\right]^{1/2} \left[\dfrac{a(x_0 + a)}{x_0}\right]^{1/2}$ (Solve Equation 2.34 for C and then take the square-root)
Numerical value of NI needed to produce a given plunger velocity		$NI = 7.5 \left[\dfrac{0.6}{1.257 \times 10^{-6}(1.43 \times 10^{-3})}\right]^{1/2} \left[\dfrac{0.007(.067)}{.06}\right]^{1/2}$ $NI = 12{,}150$ ampere-turns See max velocity curve in Figure 2.10.
Maximum energy, U_B, stored in a solenoid actuator with $NI =$ value needed to produce a given plunger velocity (see calculation above)		$U_{B_{max}} = \dfrac{\mu_0(NI)^2(A_p)x_0}{2a(x_0 + a)}$ $= \dfrac{1.257 \times 10^{-6}(12{,}150)^2(1.43 \times 10^{-3}).06}{2(.007)(.067)}$ $= 17.0$ J

(continued)

Table 2.2 (cont.)

Description	Calculation
Required current, I and voltage, E assuming that the number of turns $N = 713$ and the resistance $= 1.9$ ohms for a wire size set at 17 gauge (AWG) as per the large push-type tubular solenoid with a 10% duty cycle	$I = \dfrac{NI}{N} = \dfrac{12{,}150}{713} = 17.1$ A
Estimated required supply voltage, E	$E = I \cdot R = 17.1$ A $\cdot (1.9 \text{ ohm}) = 32.5$ volts. Ohm's law. Use 3 twelve volt batteries.
Dissipated power due to Ohmic heating, P (watts)	$P = I^2 R = E \cdot I$ $P = (32.5)(17.1) \approx 555$ W the power dissipated by more than five 100 W bulbs (Ohmic heating law)
Total plunger travel time (sec)	$t_{\max} \approx \dfrac{\pi}{2}\left(\dfrac{m_p x_0^3}{2C}\right)^{1/2}$ where $C = \dfrac{\mu_0(NI)^2}{2}(A_p) = 0.132$ newton-m^2 $t_{\max} \approx \dfrac{\pi}{2}\left(\dfrac{m_p x_0^3}{2C}\right)^{1/2} = 1.57\left(\dfrac{0.6(.06)^3}{2(.132)}\right)^{1/2} = 34.8$ millisec (Equation 2.37) Section 3.2.2.5
Power up time scale, t_{RL}, for an R-L circuit; about $5L/R$ sec. Solenoid length is estimated to be 4 inch $= 0.1$ m for 4.13 inch push-type	$L \approx \dfrac{\mu_0 N^2 A_p}{\ell_{\text{solenoid}}} = \dfrac{1.257 \times 10^{-6}(713^2)1.43 \times 10^{-3}}{0.1}$ Inductance $L = 0.914 \times 10^{-3}$ henry $t_{RL} = 5L/R = 5(9.14 \times 10^{-3}/1.9) = 24$ millisec
Total time required for solenoid power-up plus the plunger travel time	$T_{\text{total}} = t_{\max} + t_{RL}$ $= 35 + 24 = 59$ millisec ≈ 0.06 s
Estimated kicks rate	• About 17 kicks per second for solenoid kicker if cooling not a factor (100% duty cycle) • Less than 2 kicks per second for a 10% duty cycle

Sec. 2.3]   **Appendix: Modeling of the temperature rise produced by ohmic heating**   69

(d) Substitute the input data into the model and compute a value or range of values for the estimate.
(e) Present the results in a simple form and provide a brief interpretation of their meaning.

We demonstrate here the use of the Quick-Fire approach in establishing the key thermal measures for our solenoid-actuator problem.

2.3.2 Problem definition and sketch

We pose the following set of questions for our solenoid thermal problem:

- Determine a simple equation for the temperature increase in a solenoid coil contained within a steel cylindrical shell as a function of total current activation time. Consider the small time period before reaching a steady state temperature. The solenoid is assumed to be convectively cooled by an air stream perpendicular to the long axis of our solenoid-activator.
- Find the temperature increase per "kick" and estimate the peak coil temperature for a typical robot soccer game.
- Determine the maximum number of powered solenoid cycles (i.e. the number of kicks) that the robot can initiate without the coil temperature reaching some specified thermal limit based primarily on the failure point of the coil's electrical insulation.

2.3.3 The baseline mathematical model

We first write down a thermal energy balance equation based on Figure 2.12. In order to reduce the complexity of the problem, we use the method of "lumped capacitance" to determine the spatially averaged temperature, $T(t)$, within the solenoid-actuator cylinder boundaries. This temperature is a function of time (but not space) because we do not account for heat conduction gradients within the solenoid components; in other words, we do not account for differences in the internal temperatures in the copper wire, steel, and air regions.

With this model, the internal temperature, $T(t)$, begins to increase with time as the solenoid current, i, starts to flow in the coil with resistance, R. The initial time rate of change of temperature is initially determined by a simple balance between the heat storage per unit time in the copper coil

$$\frac{d}{dt}[mC_{\mathrm{p}}(T - T_{\infty})]$$

and the rate of Ohmic or i^2R heating (measured in watts). Note that based on our assumptions, m and C_{p} are respectively the mass and specific heat of the copper wire only. We also assume that the solenoid is cooled slowly by air convection where the heat out of the cylinder is proportional to the temperature difference (i.e. $T(t) - T_{\infty}$)

70 Design of a high school science-fair electro-mechanical robot [Ch. 2

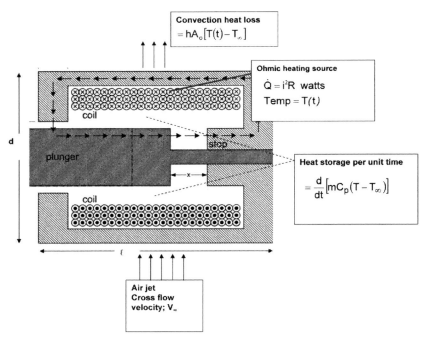

Figure 2.12. Sketch of cylindrical solenoid. Thermal balance "lumped capacitance" components: (a) Ohmic heating by the applied current flowing through the coil with a prescribed resistance per unit length, (b) heat storage in the coil mass, i.e. the thermal capacity, and (c) convective air jet cooling of the external cylinder. [Illustration courtesy of J. Jacobs]

multiplied by the product of the convective heat coefficient "h" in units of $\left[\dfrac{W}{m^2 \, °C}\right]$ and the outer surface area for convection A_0. We presume T_∞ is a constant ambient reference temperature for the stream of air flowing normal to the axis of the cylinder. We determine h empirically using well-known curve fits to laboratory experiments for flow over heated cylinders. We balance these three energy components, according to the first law of thermodynamics, to produce the following energy balance differential equation describing the time varying internal temperature $T(t)$,

$$\underbrace{\frac{d}{dt}[mC_p(T - T_\infty)]}_{\text{heat storage change}} = \underbrace{i^2 R}_{\text{ohmic heating}} - \underbrace{hA_0(T - T_\infty)}_{\text{convective heat loss}} \qquad (A.1)$$

We assume that the solenoid's internal temperature T is equal to T_∞ (i.e. the ambient air temperature) at $t = 0$ when the current "i" immediately jumps from a value of zero to its steady state dc value of 17 amperes chosen for our design solution.

The outer surface area A_0 for radial heat transport from within the cylinder enclosing the solenoid is calculated using the cylinder diameter d and cylinder length ℓ by the expression

$$A_0 = \pi d \ell$$

By defining $\theta \equiv (T - T_\infty)$ and $\dot{Q} \equiv i^2 R$, we can rewrite this energy balance differential equation in the following compact form

$$\left. \begin{array}{c} (mC_p)\dfrac{d\theta}{dt} = \dot{Q} - (hA_0)\theta \\ \text{initial condition } \theta(t=0) = 0 \end{array} \right\} \quad (A.2)$$

2.3.4 Physical parameters and data

Estimated copper wire mass [14]:

$$m = (\text{wire density } \rho, \text{mass per unit length}) \times (\text{wire length}) = \rho L$$

For 17 gauge wire, with $L = 150$ m, wire mass

$$m = (0.92 \times 10^{-2} \text{ kg/m})(150 \text{ m}) = 1.38 \text{ kg}$$

Specific heat of copper: $C_p = 390 \dfrac{\text{J}}{\text{kg} \, ^\circ\text{C}}$

Solenoid current:

$$i = 17 \text{ A [see Section 2.2.3.3.8, Figure 2.10]}$$

Resistance R (ohms) for about 150 m length of 17 gauge solenoid wire [14]:

$$R = \frac{(\text{copper resistivity})(L)}{(\text{wire} \times \text{sectional area})} = \frac{(1.678 \times 10^{-8} \text{ ohm} \cdot \text{m})(150 \text{ m})}{(1.1 \times 10^{-6} \text{ m}^2)} = 2.3 \text{ ohm} \approx 2 \, \Omega$$

Heating rate at full current levels:

$$\dot{Q} = i^2 R \approx 578 \text{ W}$$

Heat transfer coefficient, h, for cross flow of air over a cylinder of diameter d:

The value of $h \left[\dfrac{\text{W}}{\text{m}^2 \, ^\circ\text{C}}\right]$ is found from experimental curve fits of the related non-dimensional heat transfer parameter, called the Nusselt number, $Nu_d = hd/k$ plotted as a function of the flow Reynolds number. These fits of the data, Nu_d vs Re_d, are based on laboratory experiments that measure the heat loss from a heated cylinder in cross flow. In this expression, k is the thermal conductivity of the convective fluid (air in our case) measured far from the cylinder. "k" has units of $\left[\dfrac{\text{W}}{\text{m} \, ^\circ\text{C}}\right]$. Reynolds number, the ratio of kinematic forces to viscous forces, $Re_d = \dfrac{V_\infty \cdot d}{v_\infty}$ is directly proportional to flow velocity V_∞ and inversely proportional to the kinematic viscosity of air; $v_\infty \approx 10^{-6} \text{ m}^2/\text{s}$. In general the empirical curve fits are of the form $Nu_d = (C)(Re_d)^m$, where d is the diameter of the cylinder. The constant C is estimated (Table 9-1 of [15]) to be $C = 0.024$ and the exponent $m = 0.805$ for high Reynolds numbers air flows in the following range of Reynolds numbers: $Re_d = 4 \times 10^4$ to 4×10^5. Let's pick a low air-cooling velocity that might be produced by a fan operating in that Reynolds number range

$$V_\infty = 4 \text{ m/s (or 8.8 miles/hr)}$$

The corresponding Reynolds number for a 3 inch diameter cylinder is

$$Re_d = \frac{4 \cdot (0.076)}{10^{-6}} = 3.04 \times 10^5$$

Using the curve-fit coefficients, the Nusselt number is

$$Nu_d = 0.024(3.04 \times 10^5)^{0.805} = 622$$

and so the corresponding numerical value of the heat transfer coefficient h is

$$h = \frac{(Nu_d)(k_{\text{air}\cdot\infty})}{d} = \frac{(622) \cdot (0.026)}{(0.076)} = 213 \left[\frac{\text{W}}{\text{m}^2 \, ^\circ\text{C}}\right]$$

The area A_0 that scales the convective heat transfer from the outer surface of the cylinder encasing our solenoid coil (diameter $d = 0.076$ m and length $\ell = 0.1$ m) is

$$A_0 = \pi d\ell = \pi(0.076)(0.1) = 2.39 \times 10^{-2} \text{ m}^2$$

We now have calculated, or selected from our references, all the constants we require to estimate the temperature from our "lumped capacitance" solenoid model, Equation A.1. The values for $T(t)$ can be calculated directly from the approximate or exact solutions to the model energy balance differential equation Equation A.2.

2.3.5 Numerical results

We first set forth a simple approximate solution to our model differential equation, valid for small heating times, $t \ll \tau$.

The time scale τ is a characteristic time measure where the heat storage term in Equation A.2 is comparable in magnitude to the convective cooling term, and can readily be calculated using solenoid system and heat transfer parameters from the solution to Equation A.2. The following is the exact solution of this differential equation. It should be well known to calculus and electrical engineering students, and can be checked by direct substitution in Equation A.2.

The solution for θ, the temperature difference relative to the ambient, is

$$\theta = \left(\frac{\dot{Q}}{hA_0}\right)[1 - \exp(-t/\tau)] \quad \text{(A.3)}$$

where $\tau = mC_p/hA_0$.

Based on the parameters presented in Section 3.4, we find that τ is of order of 100 seconds for air cooling at 4 m/s (It is a longer time scale if we reduce the amount of air cooling). The exact value is given by

$$\tau = mC_p/hA_0 = (1.38)(390)/[(213)(0.0239)] = 106 \text{ s}$$

For heating times much greater than 100 s (i.e. large values of t/τ), Equation A.3 approaches the following steady state temperature determined by setting the outgoing cooling rate equal to the Ohmic heating rate; $\theta_{\text{steady}} \rightarrow \left(\dfrac{\dot{Q}}{hA_0}\right)$. For our solenoid pulse-heating problem we will be concerned primarily with much shorter times (indeed times less than 1 s) where the effects of convective cooling on the temperature change are quite small.

Short time solution:

When $t \ll \tau$, the heat storage term is much larger than the convective cooling term. In this case, Equation A.2 reduces to the simple expression

$$(mC_p)\frac{d\theta}{dt} = \dot{Q} \tag{A.4}$$

Integrating Equation A.4 over time t and applying the initial condition $\theta(t = 0) = 0$ gives us the following simple result

$$\theta = \left(\frac{\dot{Q}}{mC_p}\right)t \qquad t \ll \tau \tag{A.5}$$

Equation A.5 establishes that the predicted temperature change is directly proportional to t for short "constant current" heating times.

Note that this result can also be obtained from the Equation A.3 by taking its limit as $t/\tau \rightarrow 0$; simply expand the exponential term for small values of t/τ.

Using the constants $\dot{Q}, m, C_p$ determined in section 3.4, we calculate the proportionality constant in Equation A.5 and find that the short time temperature increase is given by the linear equation

$$T - T_\infty = \left(\frac{578}{(1.38)(390)}\right)t = (1.07°\text{C/s})t \qquad t \ll \tau \tag{A.6}$$

Let's assume that we want to calculate the temperature change for a single "firing" of the solenoid actuator. As previously shown, the duration of this single firing is approximately 0.06 s in total. That is, we assumed for this Quick-Fire calculation that the current is approximately constant for all 60 millisec, which isn't quite correct because the current rises rapidly and then levels off for large times during the estimated 24 millisec power-up R-L time period for the solenoid circuit.

For a "pulse" lasting $t = 0.06$ s (that is for 60 milliseconds of Ohmic heating)

$$T - T_\infty \equiv \Delta T_1 = 0.064°\text{C} \quad \text{single pulse} \tag{A.7}$$

2.3.6 Interpretation of results

In a given robot soccer game of 30 minutes duration, one can crudely estimate that, at most our solenoid actuator would kick the ball 3 to 4 times per minute, or about 100 kicks in all. We can estimate the *highest* solenoid temperature *during a game* by considering a worst case scenario; we will take as a given that there is absolutely *no cooling* between pulses. This is the same as a 100% duty cycle with no convective cooling at all, i.e. zero air speed. The estimated temperature increase for the game, ΔT_{game}, is then 100 times the value of the single pulse temperature increase ΔT_1

$$\Delta T_{\text{game}} \approx 100 \cdot (0.064°\text{C/pulse}) = 6.4°\text{C} \qquad (A.8)$$

If the temperature of the solenoid actuator was initially at a room temperature of 20°C, the final peak solenoid temperature would be 80°F

$$\underbrace{T_{\text{game}_{\text{peak}}}}_{\text{no cooling}} \approx 26.4°\text{C or about } 80°\text{F} \qquad (A.9)$$

which is reasonably low.

Estimate of total allowable kicks per game (no cooling)

For our simple thermal solution in the absence of cooling, the coil temperature increases by 64 millidegrees every time the ball is kicked. We must therefore ask how many kicks it would require to cause some degree of failure of the electrical system due to the higher temperatures?

Ledex corporation, a leading manufacturer of solenoid actuators states that wire temperatures should be no higher than 130°C to prevent failure of class B electrical insulation and no higher than 180°C for Class H insulation [16]. If we set 130°C, for the poorest of these two insulation classes, as an upper limit for our solenoid kicker, then $\Delta T_{\text{game}} = 110°\text{C}$. The total allowable kicks per game N is

$$N_{\text{upper limit}} = 110°\text{C}/0.064°\text{C per kick} = 1{,}719 \text{ kicks} \qquad (A.10)$$

The total time required by the solenoid actuator to perform 1,719 pulses without cooling between pulses is about 103 s, or a little less than 2 minutes of *continuous* operation. As 103 seconds is about equal to the time scale τ, the cooling effects should also begin to be taken into consideration as our kicker exceeds the 1,000 kick mark. We conclude that it would be unwise to run our solenoid to make over 1,700 kicks because this may cause serious damage to our electrical system.

From this worst case BotE *Quick-Fire* analysis it is clear that our robot kicker will be able to easily handle the temperature rise due to Ohmic heating if we keep the number of game kicks at or below several hundred kicks. Our analysis and calculations have provided the estimates required to address the Quick-Fire heating problem that we initially set for our robot kicking device.

But what if we had a more stressful solenoid system designed for a different use; a solenoid not limited to just a few hundred pulses, but one that might run for a very long time? Various industrial applications, like fuel injection systems or industrial hydraulic actuators operate continuously for significant periods of time. They therefore require active cooling, large heat sinks, or the use of *low* duty cycles to keep them from reaching potential high damaging temperatures. We don't have time to analyze these more stressing industrial applications, but the Back-of-the-Envelope approach could certainly be used to predict the long term temperatures for these industrial type solenoid-actuators.

2.4 REFERENCES

[1] Middle Size Robot League Rules and Regulations for 2010. Version 14.1. See FIFA Law 1, Field of Play (RC-1.1; official field size for 2010) May 12, 2010. web site: http://wiki.msl.robocup-federation.org/
[2] Massachusetts Institute of Technology, Dept. of Physics, Physics course 8.01T, Problem set 11: Angular momentum, Rotation, and Translation solutions. Fall term 2005. See also [3] Problem 3.4.
[3] Shepard, Ron. "Amateur Physics for the Amateur Pool Player," Problem 3.10, Third Edition, 1997.
[4] Shepard, Ron. "Amateur Physics for the Amateur Pool Player," Problem 2.2, Third Edition, 1997.
[5] Zandsteeg, C.J. "Design of a RoboCup Shooting Mechanism," Report 2005.147 Dynamics and Control Technology Group, Dept. of Mechanical Engineering, Eindhoven University of Technology, The Netherlands, 2005.
[6] Society of Robots website, "Actuators-Solenoids", 2010. http://www.societyofrobots.com/actuators_solenoids.shtml
[7] G.W. Lisk Co., Inc. website, "Design Guide, Linear Solenoids Technical Information". http://gwlisk.com/design-guide.aspx
[8] Magnetic Sensor Systems website, "Solenoid Construction". http://solenoidcity.com/solenoid/manual/construction/construction.htm
[9] Halliday, D. and R. Resnick, *Fundamentals of Physics*, 2nd Edition, Section 33-5 Energy Density and the Magnetic Field, John Wiley and Sons, 1981.
[10] Magnetic Sensor Systems website, "Specifications for Series S-41-300-H Push Type Tubular Solenoid, 3 inch diameter $\times$ 4.13 inch length". http://solenoidcity.com/solenoid/tubular/tubularcatalog.htm
[11] Magnetic Sensor Systems website, "Specifications for Series S-41-300-H Push Type Tubular Solenoid, 3 inch diameter $\times$ 4.13 inch length", Force vs stroke data. http://solenoidcity.com/solenoid/tubular/s-41-300hp2.htm
[12] Gogue, G. and J. Stupak, Jr., "Theory and Practice of Electromagnetic Design of DC Motors and Actuators", G2 Consulting, Beaverton, Oregon, 1990. http://consult-g2.com/course/chapter11/chapter.html
[13] *Top Chef*, Television show, Bravo Network, Season 1, March 2006.
[14] National Bureau of Standards Handbook 100, "Copper Wire Tables", U.S. Government Printing Office, February, 1966.
[15] Eckert, E. and R. Drake, *Heat and Mass Transfer*, 2nd edition, Table 9-1, McGraw-Hill Book Company, 1959.
[16] Ledex corporation website discussing solenoid design basics. http://ledex.com/basics/basics/html

3

Estimating Shuttle launch, orbit, and payload magnitudes

3.1 INTRODUCTION

The design, construction, and operation of the Space Shuttle ranks as one of the most significant engineering achievements of the twentieth century. Officially the Space Transportation System, referred to as the Shuttle, is a complex multi-purpose reusable spacecraft system capable of transporting humans into low earth orbit and returning them safely to the earth. The return is characterized by reentry into the earth's atmosphere followed by an airplane-like return of the Orbiter to the ground. In addition, the Shuttle system is designed to place heavy cargo into earth orbit for a variety of multi-national projects ranging from the deployment and maintenance of unmanned satellites, such as the Hubble Space Telescope, to the construction and servicing of the International Space Station.

The Shuttle is the ultimate Swiss Army knife of the aerospace world. It's a launch vehicle for the first $8\frac{1}{2}$ minutes (see Figure 3.1), a reconfigurable science platform for up to two weeks, a hypersonic reentry vehicle for an hour and then a piloted subsonic glider for about 5 minutes [1].

Along with many successes there have been two major mission failures: the *Challenger* launch disaster in 1986 due to a breach of one of its solid rocket motors, and the disintegration of *Columbia* during reentry in 2003 as a result of damage to its thermal protection system caused by a foam debris strike of the Orbiter wing during ascent.

In Chapter 4, we employ Back-of-the-Envelope techniques to estimate the foam strike velocity, the resultant force imposed by the impact on the leading edge of the wing, and the maximum stress set up in the wing's thin outer surface. Under certain conditions we show that this peak stress exceeds the maximum strength of the leading-edge material, which can result in structural failure of the wing.

I. E. Alber, *Aerospace Engineering on the Back of an Envelope*.
© Springer-Verlag Berlin Heidelberg 2012.

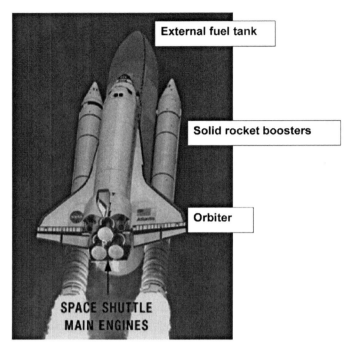

Figure 3.1. Shuttle Orbiter on launch showing the main engines, solid rocket boosters, and external tank.

3.1.1 Early Space Shuttle goals and the design phase

Early conceptual designs for the Space Shuttle were initiated by NASA in the late 1960s to provide a post-Apollo launch vehicle and spacecraft for the new space frontier challenges of the day. One of NASA's objectives was to construct a large manned space station in the mid-to-late 1970s. To keep the overall mission costs low "NASA intended to develop a fully reusable vehicle" [2].

In the early conceptual design phase of the project, "there was great debate about the optimal Shuttle design that best balanced capability, development cost, and operating cost". NASA's Associate Administrator for Manned Space Flight, George Mueller presented the cost and technology challenges for this period in the following introductory remarks to a Space Shuttle symposium in October 1969 [3].

> "The goal we have set for ourselves is the reduction of the present costs of operating in space from the current figure of $1,000 per pound for a payload delivered in orbit by the Saturn V, down to a level of somewhere between $20 and $50 per pound. By so doing we can open up a whole new era of space exploration."

But the number of launches per year and the cost per pound of payload that were achieved, fell considerably short of these figures. Since the launch of the first Orbiter, *Columbia*, on April 12, 1981, the Shuttle was much more costly to operate than projected. In 2002, the cost per pound of Shuttle payload into low earth orbit was roughly estimated to be around $1,000 per pound, measured in 1972 dollars, which is more than an order of magnitude higher than anticipated [4].

In addition, the Shuttle flew considerably fewer missions per year than originally anticipated, partly as a result of the two Shuttle accidents. The original goal was 12 flights per year, but actual launch rates reached only a maximum of 9 flights per year in 1985. Flight rates fell considerably below that level in the years that followed. The last mission, the 135th, was flown in the summer of 2011, some thirty years after the system was introduced.

3.1.2 The Shuttle testing philosophy and the need for modeling

Before the first flight, NASA conducted a variety of tests on the Space Shuttle that differed significantly in concept from tests performed in previous complex aircraft and spacecraft programs. According to the 2003 Columbia Accident Investigation Report "The Evolution of the Space Shuttle Program" [5], the strategy was to "ground-test key hardware elements such as the main engines, Orbiter, External Tank, and Solid Rocket Boosters separately, and then to use *analytical models*, not true flight testing, to certify the integrated Shuttle system". As a result of this potentially risky and cost-constrained philosophy, "the Space Shuttle was *not* flown on an unmanned orbital test flight prior to its first mission" [5]. Fortunately, the first manned mission was successful. The analytical models were implemented within large-scale-computer-algorithm design codes developed by experienced engineers and software programmers employed by NASA and its contractors. These codes were designed to emulate (at various levels of complexity) the Shuttle launch, flight trajectory, orbital maneuvers, insertion of the mission payload into a suitable final orbit, and the reentry and landing dynamics.

We are led to ask if it is possible to construct a *simplified analytical model* for the Shuttle that might allow any aerospace engineer ("neophyte" or experienced professional) to estimate the magnitude and system dependencies of key Shuttle launch, orbit, and payload metrics *without employing NASA's detailed system design codes*. In principle, such an analytical model could then be used for preliminary system design and initial test-planning purposes on a variety of launch system projects. In addition, this simplified model might be used to simulate *future* space transportation systems.

We will show, in the following sections, that a simple Back-of-the-Envelope analytical model can be constructed in a straightforward manner, without the use of large-scale computer algorithm-based design codes. This model will make it possible to rapidly estimate the magnitudes of basic launch, orbit, and payload parameters, and also to deduce operational trends for a wide range of system parameters.

3.1.3 Back-of-the-Envelope analysis of Shuttle launch, orbit, and payload magnitudes

We pose the following three part question: Can a physics-based "Back-of-the-Envelope" model provide a simple credible estimation of: (a) the orbital velocities required for the mission; (b) the complete spacecraft velocity (or Δv) budget requirements for multi-stage rocket accelerations of the Shuttle system; (c) the burn times for the Shuttle's first and second stages; and (d) a prediction of the payload or cargo mass delivered to orbit as a function of mission orbital altitude? As we will show, the answer to each of these questions is: yes!

In order to calculate such detailed mission performance measures, we must input into our baseline model a number of quantitative measures that uniquely characterize the multi-stage mass and rocket performance characteristics of the Shuttle. Specifically, the analysis that follows shows the need for the following inputs: (1) a quantitative measure of the "dry mass" of each rocket stage; (2) the masses and specific impulses of the rocket propellants used in a particular stage; and (3) the desired mission orbital altitudes. And, of course, we must have a basic description of the Shuttle launch sequence.

For preliminary design studies of *future* launch systems using Back-of-the-Envelope techniques, the masses, fuel, and rocket engine types for each stage would *not* be known *a-priori*, but would be determined as part of an overall solution for a system designed to satisfy one or more specific goals or constraints. One such goal might be to maximize the payload delivered into a range of orbits with the proviso that the system not exceed a specific total cost.

Returning to our focus on the baseline Space Shuttle design, we present in the following sections the information required in order to describe the specific operations, dimensions, and rocket propellant properties. This information is readily available from a variety of books, open-source NASA and contractor documents, and web-site postings.

3.2 SHUTTLE LAUNCH, ORBIT, AND REENTRY BASICS

As shown in Figure 3.1, the Space Shuttle has three major components: (1) the Orbiter, which houses the crew and is *partly* propelled by its three main liquid rocket engines; (2) the external fuel tank which supplies the liquid hydrogen and liquid oxygen to the Orbiter's main engines; and (3) the twin solid rocket boosters (SRBs) which provide most of the thrust during the first two minutes of flight (the "first stage"). All components are reusable with the exception of the external fuel tank, which is jettisoned after all its fuel is expended (at the end of the "second stage") and burns up as it reenters the atmosphere over the ocean [6].[1]

3.2.1 The liftoff to orbit sequence

The launch sequence: liftoff, jettison of the SRBs, external tank separation, orbit insertion, deorbit, reentry, and landing are shown schematically in Figure 3.2.

---

[1] This chapter was written during the time when the Space Shuttle was still in operation. The STS program ended in the summer of 2011 with flight STS-135. The remainder of this chapter is written as if the Space Shuttle is still in service.

Sec. 3.2]	Shuttle launch, orbit, and reentry basics 81

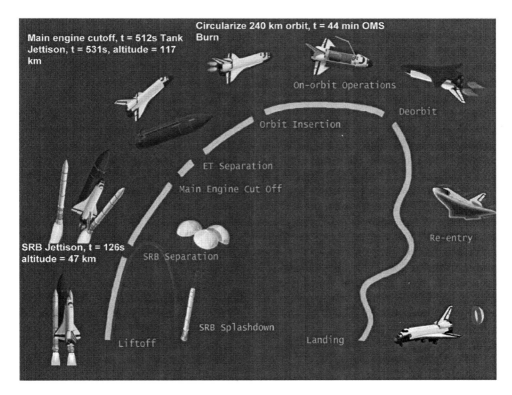

Figure 3.2. Shuttle launch, orbit, and reentry mission elements. Typical altitude and time scales are shown. After the solid rocket fuel is exhausted, the SRB casings are jettisoned to parachute into the Atlantic Ocean, from where they are recovered, the casings refilled with fuel, and used on a later Shuttle mission (drawing from [7]).

At liftoff, the mass of the entire Shuttle system, including its propellant and payload, is approximately 2.06 million kg (or 4.54 million pounds mass). Initially, the solid rocket boosters and the main liquid propellant engines of the Orbiter work *in parallel* to provide the thrust required to lift this large vehicle off the launch pad. The total thrust at launch is 7.77 million pounds-force (lbf), or 34.6×10^6 Newtons. The two SRBs jointly supply 6.60×10^6 lbf, some 85% of the total takeoff thrust, and the three main engines of the Orbiter supply an additional 1.17×10^6 lbf, the remaining 15% of the total thrust.

One minute after liftoff, the Shuttle is traveling at more than 1,000 miles per hour and has consumed more than 1.5×10^6 pounds-mass (lbm) of propellant, or about 1/3rd of its total takeoff mass. After about 2 minutes, by now at an altitude of 45 kilometers (148,000 feet), the solid propellant in the two SRBs is exhausted and the casings are jettisoned, terminating the first stage of the flight. During the second stage, the Shuttle increases its acceleration. Towards the end of the ascent the Orbiter engines are throttled to somewhat lower thrust levels in order not to exceed the $3g$ physical limit set for the vehicle.

About 8.5 minutes after liftoff, the main engines of the Orbiter are shut down (referred to as main-engine-cutoff, or MECO) as the available external fuel tank propellants are depleted, and about 30 seconds later the tank is jettisoned. The Orbiter's structural mass sent into orbit is about 100,000 kg (not accounting for the delivered payload). The payload mass (often called the cargo mass) carried in the Orbiter's cargo bay, is typically of order 25,000 kg for very low earth orbit missions. This payload or cargo mass is therefore about 1% of the total Shuttle takeoff mass. At MECO, the Orbiter's orbit is elliptical with its high point (apogee) significantly farther from the earth than its low point (perigee). For STS-7, which was a typical early mission, the initial orbit had an apogee of 160 nautical miles (296 km) and a perigee of 52 nautical miles (96 km).

The elliptical orbit is circularized by firing the two low thrust orbital maneuvering systems (OMS) engines on the Orbiter for several minutes. In the case of STS-7, the orbit was circularized at an altitude of 160 nautical miles. The OMS engines are set on the left and right sides of the Orbiter's vertical tail. The engines burn monomethyl hydrazine fuel with a nitrogen tetroxide oxidizer, and they deliver a thrust of about 6,000 lbf (26,700 N) per engine. The propellants are carried in tanks within each OMS pod. The total OMS propellant mass is approximately 5,400 kg per engine, or 11,900 lbm. Consequently, the maximum propellant mass for the two pods is approximately 10,800 kg. As the initial orbit has a period of about 90 minutes, the OMS firing to raise its perigee is typically made about 44 minutes after launch.

If a higher orbit is required, the OMS engines can also be used to raise the orbit (typically to an altitude of 200 nautical miles or 400 km). This OMS burn is conducted prior to the circularization maneuver. The orbital phase of a mission can last up to 16 days, depending on its complexity.

3.2.2 Reentry

After completing its space operations, the Orbiter begins the rapid return to earth. At the start of the deorbit sequence, the vehicle is rotated *tail-first* in the direction of its orbital velocity vector, using its low-thrust reaction control system engines. After this rotation, the OMS engines are fired to decrease its velocity below the mission's orbital speed. It then rotates *nose-first, with its nose pitched up*. As the Orbiter starts its initial reentry into the upper atmosphere, the air density begins to increase, aerodynamic pressures on the vehicle build up, and the spacecraft's flight control surfaces (wings, tail rudder, etc) begin to influence its motion. A high "angle of attack" on the fuselage allows aerodynamic drag to significantly reduce its high initial velocity (about 8,000 m/s) to speeds low enough for a subsequent runway landing.

During this high-drag reentry period, the thermal protection system (or TPS) provides the heat transfer protection required to enable the structure to survive the resultant heating from the extremely high air temperatures surrounding the hypersonic Orbiter. The thermal protection system is made up primarily of the tiles on the wing and fuselage. It also includes surfaces constructed from very important "heat-resistant reinforced carbon-carbon materials" that comprise the nose-tip and wing leading edges. These critical surfaces are heated to extremely high temperatures midway through the atmospheric reentry period.

In the lower part of the atmosphere, after much of the initial velocity has been dissipated, the unpowered Orbiter glides to Earth and lands on a runway like an ordinary airplane, although it makes its approach at a much steeper angle and a faster speed leading to a nominal touchdown speed of about 190 knots or about 100 m/s.

Thus, during reentry, the Orbiter's speed decreases from 8,000 m/s to 100 m/s, a factor of 80, or a factor of 6,400 in kinetic energy. Basically, the dissipated kinetic energy heats the air around the vehicle, which in-turn leads to convective heating of its thermally protected surfaces (We present a Back-of-the-Envelope analysis of this reentry heating problem in Chapter 5).

3.3 INVENTORY OF THE SHUTTLE'S MASS AND THRUST AS INPUT TO THE CALCULATION OF BURNOUT VELOCITY

The above overall description of the Shuttle launch and orbit scenarios shown in Figure 3.2, provides us the primary *conceptual* information needed to predict the velocity that can be achieved by the Orbiter spacecraft at the end of the powered launch phase. One of the primary goals of our Back-of-the-Envelope analysis is to demonstrate that our estimated Orbiter burnout velocity at MECO is sufficient to place a prescribed payload mass into a high-speed orbit around the earth.

3.3.1 Burnout velocity

To first order, the achievable rocket burnout velocity must be equal to the velocity required by the laws of orbital mechanics for a given satellite mission. An Orbiter can make small corrections using the small engines of either its OMS or reaction control system.

The estimated burnout velocity for each stage of the Shuttle can be readily calculated from a simplified form of Newton's second law of motion applied to the rocket flight problem, as we will show below. To calculate this, we need to know, or at least be able to estimate, the following information: the structural, payload, and propellant masses for each Shuttle system stage, along with the *effective* exit (or exhaust) velocity with respect to the rocket, C_i, for each rocket engine stage. Thrust T_i, for each of the "i" stages, is approximately related to the corresponding mass flow rate $\dot{m}_i$ and the effective exit velocity C_i by the simple equation: $T_i = \dot{m}C_i$. This equation is readily derived by applying the principle of the conservation of momentum for a point mass rocket which is losing propellant mass at the rate $\dot{m}$. This is discussed in more detail in Section 3.6. This expression is exact when the rocket engine exhaust exit pressure is equal to the local ambient air pressure. The final achievable burnout velocity, v_{burnout}, at the end of the Shuttle's powered launch phase is, as we will show, simply equal to the *sum* of the "ideal" burnout velocities for each stage corrected by several smaller velocity "loss" terms, such as (a) gravity loss and (b) drag loss.

3.3.2 The velocity budget

An *inventory* of each of these contributing velocity increments, followed by their overall summation, determines the final predicted burnout velocity. This *inventory* is often referred to as the Δv *budget*. We will estimate the magnitude of each of the major Δv budget elements in the launch analysis that follows this introduction, including the so-called Δv losses due to gravity and drag. Our BotE estimated Δv budget is tabulated later in Section 3.11.

3.3.3 Mass inventory

Figure 3.3 is a schematic of the Shuttle that specifies the key mass components for the first and second stages of the Shuttle's flight. We also summarize these mass elements in Table 3.1. The table includes the corresponding mass ratios, or MR_i, for each stage as $MR_i \equiv$ (final mass at the end of stage i)/(initial mass for that stage). In the companion Table 3.2, we provide (for each stage) the following quantities: the solid and liquid rocket thrusts, the individual engine mass flow rates, and the average exit velocities, C_i.

In Table 3.1 as well as in Figure 3.3, a nominal payload mass of 25,000 kg (or cargo mass delivered to orbit in the Orbiter cargo bay) is used to calculate the Shuttle mass summaries. The achievable burnout velocity as a function of actual payload or cargo mass will be assessed later in our analysis (see Section 3.14.1).

3.3.4 Thrust and specific impulse inventory

Rocket thrust, specific impulse, and mass flow rate parameters are presented in Table 3.2. Specific impulse, Isp, is defined as the rocket impulse "I" per unit weight of rocket fuel; where, $I = \int_0^{\Delta t}$ Thrust $dt \approx \bar{T} \Delta t$, generated over a short period of time, Δt. This definition can also be written either in terms of thrust and mass flow rate, or as the ratio of effective exit velocity to the acceleration of gravity

$$\text{Isp} = \frac{\text{Thrust}}{(\text{fuel weight}/\Delta t)} = \frac{\text{Thrust}}{\text{fuel weight flow rate}} = \frac{\bar{T}}{\dot{m}g} = \frac{\text{effective exit velocity } C}{g}$$

where g is the acceleration of gravity $= 9.81 \text{ m/s}^2$ and "effective" exit velocity C is defined more fully in Section 3.6 as Equations 3.1a and 3.1b.

Notice in Table 3.2 that the specific impulse for the *liquid* hydrogen and liquid oxygen propellants (feeding the main engines, or SSMEs, of the Orbiter) is more than 50% greater than the specific impulse for the higher molecular weight *solid* propellant used by the SRBs. This is because both the properties of low molecular weight and high heat of combustion produce *high* chemical energy combustion levels in LOX/LH2 systems, typically at combustion temperatures above 3,000°K. Such

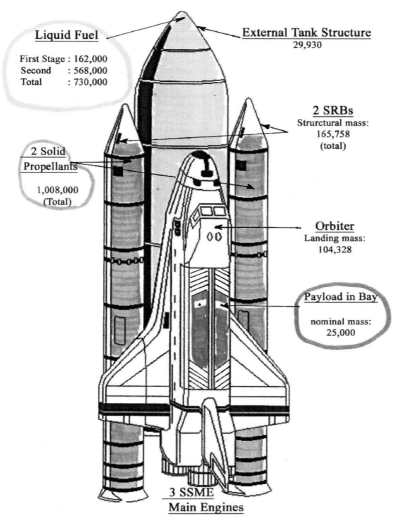

Figure 3.3. Schematic of the Shuttle with mass components given in kg, as annotated. Overall Shuttle height ≈ 56 m (see [8] for data on masses of individual Shuttle components).

high chemical energy levels are converted, by expansion in the attached rocket nozzle, into high levels of kinetic energy and therefore high exit velocities C.

NASA's historical records state that *solid propellant* engines were selected by the government as the major contributor to the Space Transportation System's takeoff thrust *primarily* because of their lower cost in comparison to that of liquid propellant engines. This cost-based decision was made despite the fact that the specific impulse levels for solid propellants are considerably lower than those for LOX/LH2 liquid propellant systems.

Table 3.1. Breakdown of Shuttle mass into propellant, structure and payload components for the Shuttle first and second stages. Mass in kg. Mass ratios presented for the components of each stage and for the Shuttle treated as a single composite entity [8].

	First stage	Second stage	Total: Shuttle treated as a composite entity
Propellant m_p	1,170,000 (Solid + first stage Liquid)	568,000 (Second stage Liquid)	1,738,000 $m_p/m_{total} = 0.843$
Structure m_s	165,758 (SRB casing)	134,258 (External tank + Orbiter)	300,016 $m_s/m_{total} = 0.145$
Payload $m_{payload}$	727,258 Second stage fuel + Tank + Orbiter + m (p/l in bay)	25,000 nominal = m (p/l in bay)	25,000 payload in cargo bay $m_{payload}/m_{total} = 0.012$
Initial mass, $m_i = m_p + m_s + m_{payload}$	2,063,016	727,258	2,063,016 Total mass
Final mass, $m_f = m_i - m_p$	893,016	159,258 (includes external tank)	129,326 Orbiter + payload delivered to orbit
Mass ratio, $MR = m_f/m_i$	0.433 MR1	0.219 MR2	0.063 $(m_{p/l+Orbiter})/m_{total}$ $m_{Orbiter} = 104,326$

Table 3.2. Thrust (T), specific impulse (Isp), average exhaust velocities (C), and mass flow rates (dm_e/dt) for Shuttle first stage and second stage engines [8] (see also: NASA Space Shuttle & Facilities information summary [9]).

	First stage	Second stage
Solid Rocket Booster thrust (2 SRBs at sea level)	5.6×10^6 lbf 24.91×10^6 N	SRB thrust = 0
Isp—solid engine (ammonium perchlorate + aluminum + iron oxide) Exit velocity, C	268 s (vac) 242 s (sea level) [8] 2,374 m/s (sea level)	
Liquid Fuel Main Engines thrust (three SSMEs) Isp—H2/O2 engine Exit velocity, C	1.18×10^6 lbf 5.25×10^6 N [7] 363 s (sea level) 3,561 m/s	1.47×10^6 lbf 6.55×10^6 N 452.5 s (vac) 4,439 m/s
Total thrust (combined thrust for all stage operating engines)	6.80×10^6 lbf 30.16×10^6 N (takeoff total)	1.41×10^6 lbf 6.27×10^6 N (47 km altitude)
Thrust to Weight ratio (T/W) at stage initiation	1.49	0.89
Average Specific impulse, Isp$_{avg}$ combined engines Avg exit velocity, C_{avg} combined engines	277 s 2,717 m/s	452.5 s 4,439 m/s
Total average mass flow rate	10.64×10^3 kg/s for 110 s [see calculations Section 3.7.1]	1.474×10^3 kg/s for 385 s [see calculations Section 3.7.1]

3.4 MASS FRACTION RULES OF THUMB

Sometimes a quick BotE analysis for a rocket concept is needed even if detailed design information for a given system is not readily available. In particular, for the launch problem being considered here, the detailed rocket mass components, presented in Table 3.1, are not always readily available. (While the internet can provide information about a well-known launch system such as the Space Shuttle, component mass information is quite difficult to find for many of the lesser-known launch systems). Rather than conclude that an estimate of an important Shuttle system parameter, for example a propellant mass or final burnout velocity, cannot be made without detailed input data, a resourceful engineer can generate a crude measure of the data for such an estimate based on *rules of thumb* that have been established by the "Space Missions" community. In Table 3.3 we present rule of thumb values for propellant, structure, and payload mass fractions, for typical launch systems (see column 2). These mass fractions are cited in Wertz's 1999

Table 3.3. Mass fractions for typical spacecraft compared to Shuttle fractions. Rules of thumb from column 2 are used to estimate the masses for the Ariane 1 rocket system in column 4. Published data for Ariane 1 are given in parentheses [11, p. 160, Table 7.2 for Ariane 1].

Mass Fraction Elements Masses are normalized by the total mass for the entire flight vehicle not distinguishing individual stages	Typical Mass fractions Wertz: Space Mission Analysis and Design (SMAD, 3rd ed., 1999) page 723.	Shuttle Overall Mass fractions (Shuttle treated as a composite entity)	Estimate of masses for three stage liquid Ariane 1 elements based on SMAD typical MFs (Ariane 1): total mass = 210×10^3 kg 10% of Shuttle total mass Estimate (Actual) Mass units, 10^3 kg
Propellant m_p/m_{total}	85%	84.2%	178.5 (188.7)
Structure m_s/m_{total}	14%	14.5%	29.4 (19.5)
Payload $m_{payload}/m_{total}$ based on payload mass in cargo bay	1%	1.2% Payload mass ratio = 6.3% if the nominal Shuttle Orbiter mass, placed into orbit, is included	2.1 (1.8 into GTO**)

**GTO: Geostationary transfer orbit. Apogee = 36,000 km, Perigee = 200 km.

book "Space Mission Analysis and Design" ([10]; also called SMAD). We observe, for example, that the biggest single contributor to the overall launch mass, by far, is the total propellant mass. According to Wertz, the average propellant mass for a typical launch system is 85% of the overall total system mass. In contrast, the useful payload mass delivered to orbit is quite small, typically only about 1% of the total. The remaining significant mass component is defined to be the structural mass (i.e. the mass of the propellant tanks, rocket engines, etc, without propellants), which is 14% of the total. In column 3, the "exact" mass fractions for the entire Space Transportation System are shown for comparison. For this column, the Shuttle is treated as a single composite entity with mass elements selected from Table 3.1. The Shuttle mass ratios are seen to be quite close to the overall "rule of thumb" averages given by Wertz in column 2.

In column 4 of Table 3.3, Wertz's rule of thumb mass ratios have been used to estimate the masses for a completely different system, the Ariane 1 three stage rocket. Here we use these rule of thumb mass ratios to predict the propellant mass, structural mass, and payload mass (in kg) for the Ariane 1 with the only information given *a-priori* being the total rocket system mass = 210,000 kg. Note that in making this estimate we ignore the fact that the geostationary transfer orbit for the Ariane 1 mission differs considerably in its Δv requirement from that required for lower-earth-orbit Shuttle missions. We see that our predicted rule of thumb estimate for *propellant* mass, 178.5×10^3 kg is in good agreement with the actual Ariane 1 propellant mass of 188.7×10^3 kg (stated in parentheses in column 4). While our estimated rule of thumb propellant mass differs by only about 5% from the actual data, our predicted *structural* mass estimate for Ariane 1 exceeds the published structural mass by a considerable amount, specifically 29.4×10^3 kg versus 19.5×10^3 kg. This is one example of the limitations inherent in rule of thumb estimates. But in the absence of real data, sometimes an engineer must use the *crude* tools and information sources that are available, particularly if the time for analysis is severely limited.

3.5 QUICK-FIRE MODELING OF THE TAKEOFF MASS COMPONENTS AND TAKEOFF THRUST USING SMAD RULES OF THUMB

To illustrate how it is possible to obtain a BotE answer quickly using a minimum amount of technical information, coupled with the SMAD *rules of thumb* of Table 3.3, we will take as a sample assignment the problem of estimating the mass and takeoff-thrust parameters for the Space Shuttle. We will "blindfold" ourselves and assume that the NASA data for these quantities are not available from Tables 3.1 and 3.2. We call this a Quick-Fire BotE problem, and in Sections 3.5.1 to 3.5.4 we will apply this technique to determine the key physical inputs to our Shuttle launch problem.

3.5.1 Quick-Fire problem approach

As previously presented in Chapter 1, in the Quick-Fire approach to a BotE problem we develop our estimates by systematically following, in sequence, the following five steps:

(a) Define and/or conceptualize the problem using a sketch or brief mathematical description.
(b) Select a model or approach, either mathematical or empirical that describes the basic physics of the problem.
(c) Determine the input data parameters and their magnitudes as required to solve the problem, either from data sources or by scaling values by analogy, or simply by using an educated guess.
(d) Substitute the input data into the model and compute a value or range of values for the estimate.
(e) Present the results in a simple form and provide a brief interpretation of their meaning.

Let's get started with our first order BotE Quick-Fire "Shuttle launch" problem.

90   Estimating Shuttle launch, orbit, and payload magnitudes   [Ch. 3

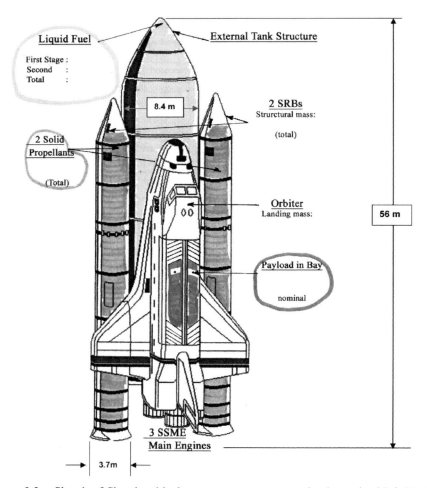

Figure 3.3a. Sketch of Shuttle with the component masses to be determined left blank.

3.5.2 Problem definition and sketch

- Estimate the Space Shuttle's total takeoff mass and total takeoff thrust.
- Estimate the mass of each major component including payload mass.

We will generate these estimates based on the *rules of thumb* in Table 3.3 for the three major rocket mass components: structure, propellant, and payload. We start off by drawing a sketch of the problem that specifies the known dimensions and lists the unknown variables whose magnitudes are to be determined.

3.5.3 Mathematical/"Rule of Thumb" empirical models

Before writing down the rule of thumb empirical models, we must define all of the relevant mass components and their interdependent relationships.

3.5.3.1 The initial total rocket mass, m_i, is the sum of the principal component masses

$$m_i = m_{propellant} + m_{structure} + m_{orbiter} + m_{cargo}$$

where

$$m_{propellant} = m_{p/solid} + m_{p/liquid}$$

and

$$m_{structure} \approx m_{s/srb} + m_{s/external\ tank} + m_{orbiter}$$

The *rule of thumb* rocket mass in Table 3.3 for structural mass as a fraction of initial mass is

$$m_{structure} = 0.14(m_i)$$

We can use this *rule of thumb* equation to calculate m_i, if we can develop estimates for the mass components contributing to the structural mass for the complete Shuttle system, $m_{structure}$.

The following are the structural mass components requiring crude estimates:

(a) the structural mass of the SRB casings; we can estimate the SRB casing mass using a similar rule of thumb; namely that $m_{s/srb} \approx 0.14 \cdot$ (solid propellant mass $m_{s/srb}$) based on an estimate of solid propellant mass
(b) the structural mass of the external fuel tank; we estimate $m_{s/external\ tank}$ based on a thin-wall cylindrical model
(c) the Orbiter mass; we can estimate $m_{orbiter}$ by a crude scaling from an aircraft of similar size.

The calculated value for total initial mass, $m_i = m_{structure}/0.14$, can then be substituted into a similar "rule of thumb" expression for the total liquid propellant mass, $m_{propellant\ total} = 0.85 m_i$, in order to obtain an estimate for the mass of liquid propellant stored at takeoff in the external fuel tank, $m_{p/liquid}$, as follows:

$$m_{p/liquid} = 0.85(m_i) - m_{p/solid}$$

As is evident from this rule of thumb expression, liquid propellant mass can only be evaluated from the above equation if we are able to reasonably estimate the solid propellant mass term, $m_{p/solid}$, used in our above estimate for item (a).

Fortunately, in the following section of our Quick-Fire problem, in which key physical parameters and data are defined, we directly calculate $m_{p/solid}$. This requires us first to estimate the volume V_{srb} of two solid rocket booster "cylinders" from the height and diameter dimensions given in Figure 3.3a, and then multiply this volume by an *assumed* propellant density, ρ_{srb}, to obtain the solid propellant mass.

Finally, our rule of thumb for payload or cargo mass states that cargo mass is approximately 1% of m_i. Using the above estimate for initial total rocket mass, the cargo mass is then simply given by

$$m_{cargo} = 0.01(m_i)$$

3.5.3.2 The initial total SRB engine thrust, T_{srb}, is calculated from the total solid propellant mass for two SRBs, $m_{p/solid}$, using the approximate 2 minute burnout time for the SRBs (t_b) given in Figure 3.2 and an estimate of the specific impulse for the solid propellant ($\mathrm{Isp}_{s/srb}$). Utilizing the rocket thrust equation derived from Newton's second law of motion (this will be derived later in this chapter), T_{srb}, is equal to the product of the rocket exhaust mass flow rate $\dot{m}$ and the exit velocity C ($= \mathrm{Isp} \cdot g$).

$$T_{srb} = \dot{m}C = \left(\frac{m_{p/solid}}{t_b}\right)(\mathrm{Isp}_{s/srb})g$$

As is evident in the above expression, we estimate that $\dot{m}$ (the solid rocket mass flow rate at takeoff) is the average rate of consumption of propellant during the 2 minute SRB burn period; i.e. $\dot{m}_{avg} =$ total solid propellant mass/(total burn time of 120 s).

3.5.3.3 The total combined takeoff thrust for both liquid and solid engines, less the weight of the rocket ($m_i g$) is modeled from Newton's second law of motion as the product of the estimated initial vehicle acceleration, a_i, and the initial mass m_i, as

$$T_{total} - m_i g = m_i a_i$$

We model the initial acceleration, a_i, based on observations of the Shuttle's altitude a few seconds after launch using video images of the Shuttle's take-off.

3.5.3.4 The takeoff thrust associated with the Orbiter's three liquid propellant SSME engines is then simply $T_{ssme} = T_{total} - T_{srb}$.

3.5.4 Physical parameters and data

Before calculating the structural mass of the two solid propellant rocket boosters we must first calculate the solid propellant mass of a single SRB.

3.5.4.1 Solid propellant mass term, $m_{p/solid}$. As outlined in Section 3.2, the mass of the propellant encased in a single SRB, can be calculated utilizing an estimate of the volume of the SRB casing, V_{srb}, and the propellant's average density, ρ_{srb}.

From the idealized SRB in Figure 3.3a, we estimate the height L and diameter of the cylinder as follows:

Height of SRB, $L \approx 0.8$(Height of Shuttle) $= 0.8(56) = 44.8$ m

Diameter of SRB, $D \approx 3.7$ m

$$V_{srb} = \left(\frac{\pi D^2}{4}\right)L = 482 \text{ m}^3$$

The average density of the propellant over the whole SRB case volume is not easy to guess. The inside of the SRB has a hollow center with a diameter on the order of 50% of the SRB outer diameter. The propellant is a complex mixture of materials; ammonium perchlorate is the oxidizer, it has some kind of binder, and a rubber insulation layer on the outer wall to keep it from burning through the thin steel walls. It has been said that the cured propellant looks and feels like a *hard rubber typewriter eraser*. A very crude guess is that a portion of the SRB rocket has a density above that of water while some parts have a density less than that of water, in particular the hollow star-shaped core and the rubber-insulation. We estimate that the average density of the SRB propellant over the whole case is probably on the order of that of water, namely 1,000 kg/m^3. So let's crudely set the density equal to that value.

$$\rho_{srb} \approx 1,000 \text{ kg/m}^3$$

The propellant mass of one SRB is then

$$m_{p/1srb} \approx \rho_{srb} \cdot V_{1srb} = (1,000)482 = 482,000 \text{ kg}$$

The propellant mass of a pair of SRBs is twice that value, or

$$\boxed{m_{p/2srb} \approx 964,000 \text{ kg}}$$

For reference (Figure 3.3), NASA quotes the total propellant mass for a pair of SRBs as $m_{p/2srb} \approx 1.008$ million kg.

Note that $m_{p/2srb}$ is equivalent to $m_{p/solid}$ as defined in Section 3.5.3.1.

3.5.4.2 Structural mass of the solid rocket boosters; $m_{s/srb}$. Let's assume that an individual SRB can be considered a "complete rocket system" in its own right, as if it were operating independent of any other Shuttle component. With this assumption we then can use the *rule of thumb* for the structural mass of a rocket system as a fraction of the total mass, m_{total}, to estimate $m_{s/srb}$; namely $m_{structure} = 0.14(m_{total})$.

Since m_{total} for our hypothetical independent solid rocket system is the sum of the propellant mass and structural mass, the resulting equation for $m_{s/2srb}$ is

$$m_{s/2srb} = 0.14(m_{p/2srb} + m_{s/2srb})$$

Using our estimate for the propellant mass of two SRBs, we calculate the *structural* mass for the Shuttle's two solid rocket boosters as

$$\boxed{m_{s/2srb} = \frac{0.14}{0.86}(m_{p/2srb}) = 0.163(964,000) = 157,000 \text{ kg}}$$

For comparison (Table 3.1) the *NASA-quoted* mass is: $m_{s/2srb} \approx 166,000$ kg.

3.5.4.3 External tank structural mass, $m_{s/external\ tank}$. For the external tank we assume that it has a cylindrical shape. Let's evaluate the mass of the external fuel tank in a "dry" configuration. From Figure 3.3a we see that it has a length about 80% of the Shuttle's overall height. Assume it is a thin-walled aluminum cylinder of diameter of 8.4 m, and has a wall thickness of the order of 1 cm. (We will seek to justify this assumption below by scaling up a soda can based on a physical analogy argument.) Thus we use the following dimensions for our calculation:

Height $L \approx 0.8(\text{Height of Shuttle}) = 0.8(56) = 44.8$ m

Diameter $D = 8.4$ m

Thickness $t = 1$ cm $= 0.01$ m

Cylindrical Volume $\approx$ (circumference of circular ring) · (length) · (thickness)

$$V_{\text{s/external tank}} = \pi D L t = 11.8 \text{ m}^3$$

For basic aluminum walls, such as used in traditional aluminum structures, a "material properties" handbook gives a typical density of $\rho_{\text{aluminum}} = 2{,}700$ kg/m^3, which makes our estimate of the structural mass of the external tank (without its propellant)

$$m_{\text{s/external tank}} \approx \rho_{\text{aluminum}} \cdot V_{\text{s/external tank}} = (2{,}700)(11.8) = 31{,}900 \text{ kg}$$

A soda can analogy estimate of wall thickness

Let's see if we can justify a 1 cm wall thickness, t, for the external tank. We will base our external tank wall thickness estimate on the thickness of a "scaled" aluminum soda can as suggested in [13]. A typical soda can has a diameter of 2.6 inches (66 mm) and a wall thickness of only 0.005 inches (0.127 mm). To "scale up" the soda can to the size of our external tank (to 8.4 m or 8,400 mm), requires a diameter scale increase of $8{,}400/66 = 127$. If we similarly scale up the soda can wall thickness "t" by the same factor then we obtain a wall thickness for an external tank of $t = (127)(0.127 \text{ mm}) = 16 \text{ mm} = 1.6$ cm. This scaling logic is based on the equation for the circumferential hoop stress in a cylindrical vessel under pressure, σ_{hoop}, being proportional to the pressure inside the vessel, p, multiplied by the ratio of the diameter/thickness, d/t, i.e. $\sigma_{\text{hoop}} = pd/2t$. Thus, to maintain the hoop stress (relative to pressure p) at the same level in differently sized cylindrical cans one should increase the thickness t in proportion to the diameter d, which is equivalent to holding the ratio t/d fixed with an increase in scale.

One other factor comes into play in assessing the appropriate wall thickness, namely the yield strength or the modulus of elasticity of the aluminum material. The later generation of "super-lightweight" Shuttle external tanks exploited a high strength aluminum-lithium alloy. The modulus of elasticity of this alloy is shown by material handbooks to be about 20% higher than ordinary aluminum. So the wall thickness for our stronger "scaled soda-can" should be able to be reduced by a factor of approximately 1.2 (for a given allowable hoop strain) in comparison to the required thickness for an ordinary aluminum vessel. This additional material strength factor reduces our estimate of external tank wall thickness from 1.6 cm to 1.3 cm. [13] states that the external tank skin thickness is less than 0.5 inches or <1.27 cm, while [14] estimates the super-lightweight tank walls as less than 0.25 inches or <0.64 cm.

Thus $t = 1.0$ cm appears to be the right order of magnitude for the Shuttle's external tank wall thickness in the above estimate of $m_{\text{s/external tank}}$.

3.5.4.4 Orbiter mass. In the 1970s, the crucial design factors that helped determine the size, shape, and mass of the Shuttle Orbiter were the requirements for the Orbiter to be able to deliver the largest Air Force satellites into low earth orbit, including into polar orbit, and that it be capable of performing an emergency "once-around-the-earth abort" for the crew after a possibly failed Shuttle launch into polar orbit. The first requirement led to the incorporation of high-thrust/high-specific-impulse engines into the Orbiter. The latter requirement necessitated wings with a large surface area to enable the Orbiter to aerodynamically glide over an extended distance, prior to a safe abort landing [2]. The need for powerful engines, large wing areas, and a large cargo bay for the delivered payloads, significantly increased the mass of the Orbiter beyond that initially considered during the early Shuttle design period.

To obtain an order of magnitude estimate of the Orbiter's mass, we note that the spacecraft is roughly comparable in length to the old McDonnell Douglas DC-9 commercial jetliner (38 m), which had a takeoff mass, loaded with fuel, of about 55,000 kg, and a wing area comparable to that of the Orbiter [www.airliners.net]. The DC-9's jet engines were considerably lower in mass than their SSME rocket counterparts. The exterior fuselage width of the Orbiter's payload bay is about 6 meters. In comparison, the fuselage of the cylindrical DC-9 passenger-carrying compartment was a little over 3 meters. Thus in the absence of direct information, one might assume that due to its greater width the Orbiter has a mass roughly twice that of the DC-9. In fact, the Orbiter's mass is about twice that of the DC-9 ($2 \times 55,000 = 110,000$ kg). NASA lists the maximum takeoff mass for the Orbiter as 109,000 kg and the landing mass as 104,000 kg [7].

For the crude calculations we are performing here in our Quick-Fire problem, we simply set the Orbiter mass to be $\approx 100,000$ kg.

3.5.5 Numerical calculation of total takeoff mass, cargo bay mass, and total takeoff thrust

3.5.5.1 The total initial Shuttle mass, m_i, is calculated using our *rule of thumb* for structural mass with the simple model presented in Sections 3.5.3 and 3.5.4 of our Quick-Fire problem

$$m_i = m_{\text{structure}}/0.14 = \frac{\{m_{\text{s/2srb}} + m_{\text{s/external tank}} + m_{\text{orbiter}}\}}{0.14}$$

$$\boxed{m_i = \frac{\{157,000 + 31,900 + 100,000\}}{0.14} = 2,063,570 \text{ kg}}$$

This estimated total mass is quite close to the initial mass in Table 3.1 gained from a number of sources. Our estimated structural mass for the pair of SRBs (157,000 kg) is somewhat lower than that given by NASA sources in Figure 3.3 (165,758 kg). However, our estimated external fuel tank mass (31,900 kg) is fairly close to the NASA supplied external fuel tank mass (29,930 kg) listed in Figure 3.3.

3.5.5.2 Estimated cargo bay mass (from the 1% rule of thumb) is given by

$$m_{\text{cargo}} = 0.01(m_i) \text{ and thus found to be}$$

$$\boxed{m_{\text{cargo}} = 0.01(2{,}063{,}570) = 20{,}636 \text{ kg}}$$

The nominal cargo or payload mass that the Shuttle can deliver to a low earth orbit at an altitude of about 100 km to 200 km is of the order of 25,000 kg (about 17% higher than this estimate). It should be noted that delivered payload masses are sensitive (as we will see later) to the exact orbital altitude set for the mission and to whether or not additional OMS engine burns are required to complete that mission. So a range of payload solutions, of this order of magnitude, are possible depending upon specific mission requirements.

3.5.5.3 The estimated mass of liquid propellant stored at takeoff in the external fuel tank, $m_{\text{p/liquid}}$, is given by

$$m_{\text{p/liquid}} = 0.85(m_i) - m_{\text{p/solid}}$$

$$\boxed{m_{\text{p/liquid}} = 0.85(2{,}063{,}570) - 964{,}000 = 790{,}035 \text{ kg}}$$

This is from the SMAD 85% *rule of thumb*, less the propellant mass of the two SRBs. Our first-order estimate for liquid propellant mass is about 8% greater than NASA's quoted value in Table 3.2 of $m_{\text{p/liquid}} = 730{,}000$ kg.

3.5.5.4 The initial total SRB engine thrust, T_{srb}, is calculated from the total solid propellant mass for two SRBs, $m_{\text{p/solid}}$, using the approximate 2 minute burnout time for the SRBs (t_b) shown in Figure 3.2 and an estimate of the specific impulse for the solid propellant ($\text{Isp}_{\text{s/srb}} \approx 260$ s) using

$$T_{\text{srb}} = \dot{m}C = \left(\frac{m_{\text{p/solid}}}{t_b}\right)(\text{Isp}_{\text{s/srb}})g$$

$$\boxed{T_{\text{srb}} = \left(\frac{m_{\text{p/solid}}}{t_b}\right)(\text{Isp}_{\text{s/srb}})g = \left(\frac{964{,}000 \text{ kg}}{120 \text{ s}}\right)(260 \text{ s})(9.81 \text{ m/s}^2) = 20.5 \times 10^6 \text{ N}}$$

This first order estimate for total 2SRB thrust is about 18% less than NASA's quoted value $T_{\text{srb}} = 24.91 \times 10^6$ N (or 5.6×10^6 lbf) given in Table 3.2.

3.5.5.5 The combined takeoff thrust for both liquid and solid engines was modeled in Section 3.5.3.3 of our Quick-Fire problem as the product of an estimate for the initial vehicle acceleration, a_i, and the initial mass m_i plus the initial takeoff weight ($m_i g$), i.e. $T_{\text{total}} = m_i(a_i + g)$.

Videos of the Shuttle's liftoff indicate that it initially accelerates vertically, for about 7 seconds, until the base of the vehicle has cleared the top of the adjacent launch tower, a height approximately twice that of the Shuttle stack, or $2 \times 56 \text{ m} = 112 \text{ m}$.

Sec. 3.5]   **Quick-Fire modeling of the takeoff mass components**   97

For a point mass under constant acceleration "a_i", the distance traveled in time t is $\frac{1}{2}a_i t^2$. Solving for "a_i" we obtain the following estimate for the Shuttle's takeoff acceleration

$$a_i = \frac{2x}{t^2} = \frac{2(112 \text{ m})}{(7 \text{ s})^2} = 4.57 \text{ m/s}^2$$

From Newton's second law of motion, and this acceleration, we obtain the corresponding estimated total takeoff thrust

$$T_{\text{total}} = m_i(a_i + g)$$

$$\boxed{T_{\text{total}} = (2{,}063{,}570 \text{ kg})(4.57 + 9.81) = 29.7 \times 10^6 \text{ N} \quad \text{or} \quad 6.68 \times 10^6 \text{ lbf}}$$

3.5.5.6 The thrust associated with the Orbiter's engines at takeoff is simply the difference between T_{total} and T_{srb}

$$T_{\text{ssme}} = T_{\text{total}} - T_{\text{srb}}$$

$$\boxed{T_{\text{ssme}} = 29.7 \times 10^6 \text{ N} - 20.5 \times 10^6 \text{ N} = 9.2 \times 10^6 \text{ N}}$$

This first order estimate for the total SSME thrust is about 75% *higher* than NASA's value of $T_{\text{3ssme}} = 5.25 \times 10^6$ N (or 1.18×10^6 lbf) cited in Table 3.2. The large error is due to the fact that our estimate of SRB thrust is low by 18% or 5.4 million Newtons. As we are estimating liquid fuel thrust as the difference between total and SRB thrust, and since SSME thrust is a small fraction of the total thrust, it is clear that a small error in SRB thrust can be significantly magnified in this final liquid fuel thrust calculation. This example highlights one of the difficulties with BotE estimates when an estimated value is to be based on the *difference* of two large numbers.

3.5.6 Interpretation of the Quick-Fire results

Overall our Quick-Fire estimates for the masses of the Shuttle's major mass components (structure, propellant, and payload) are within 10 to 20% of the mass values listed by NASA. Interestingly enough, our total mass estimate is quite good, within 1% of the actual NASA value. Our results for individual solid and liquid thrust levels are not as good, but our total thrust estimate (based on our video observations of the takeoff acceleration) is within 2% of the total NASA takeoff thrust.

In summary, a Quick-Fire estimate helps an engineer to start working on a possibly difficult problem when little "true" data is available. As more accurate data is uncovered (through *a-priori* knowledge or by researching available databases), the accuracy of these estimates can typically improve quite significantly.

3.5.7 From Quick-Fire estimates to Shuttle solutions using more accurate inputs

In the next section we set aside the Quick-Fire *rule of thumb* estimates that were developed using crude parameter guesses for key physical quantities such as material density or tank dimensions. In the more complete analysis that follows, we replace these Quick-Fire estimates with available and more accurate NASA data specified in Tables 3.1 and 3.2 for structural mass, propellant mass, solid and liquid propellant thrust and specific impulses as input to an "expanded set" of launch model calculations. We will input these more accurate data into our subsequent BotE model equations in order to calculate more accurate estimates of Shuttle burnout velocity, altitude as a function of time, and cargo payload delivered to orbit. At the conclusion of this more complete treatment, we will compare our calculated flight velocities, altitudes, and delivered cargo payloads to specific flight measurements culled from a variety of NASA databases.

We start this more complete analysis of the multi-stage launch problem with a presentation of the fundamental equation for the velocity change, Δv, for each of the stages of an ideal rocket system.

We will summarize all our detailed BotE equations and numerical results in tabular format at the end of this chapter.

3.6 IDEAL VELOCITY CHANGE Δv FOR EACH STAGE OF AN IDEAL ROCKET SYSTEM

Applying Newton's second law of motion to the problem of a point-mass rocket with a time varying mass $m(t)$, constant thrust T, exit velocity V_e, and exit mass flow rate, $\dot{m}_e = dm_e/dt$ (and for the time being, ignoring gravity and aerodynamic forces), leads us to the following *ideal rocket* differential equation, i.e. the rate of change of rocket velocity with time [15]

$$m\frac{dv}{dt} = T = V_e\left(\frac{dm_e}{dt}\right) + (P_e - P_\infty)A_e \qquad (3.1)$$

In Equation 3.1, $(P_e - P_\infty)A_e$ is the net pressure force on the rocket. The pressure force is equal to the exit area of the rocket nozzle, A_e, multiplied by the difference between the rocket nozzle exit pressure P_e and the ambient or free stream pressure P_∞. The pressure force is typically much smaller than the rocket momentum term in Equation 3.1 and in fact vanishes when $P_e = P_\infty$. While small, the pressure force term can be important under some conditions.

To simplify Equation 3.1 we rewrite the right-hand side as the product of an "effective" exit velocity, C, and the mass flow rate $\dot{m}_e$

$$m\frac{dv}{dt} = T = C\dot{m}_e \qquad (3.1a)$$

where, following convention, we define "effective" exit velocity C as

$$C \equiv V_e + \frac{(P_e - P_\infty)A_e}{\dot{m}_e} \tag{3.1b}$$

From Equation 3.1b, we observe that when $P_e = P_\infty$ the effective exit velocity C is identically equal to the actual gas exit velocity V_e. In general C is not constant throughout the launch period. This is because the pressure force term varies with altitude, or the corresponding P_∞, for a given atmosphere.

We will later reconsider the impact of atmospheric pressure on rocket engine performance, particularly for the liquid fuel SSME engines when operating at or near sea level. Notice in Table 3.2, that the sea-level thrust for the Shuttle's main liquid propellant engines is about 80% of the SSME vacuum thrust. Hence, at low altitudes the pressure force term in Equation 3.1a is negative, since for large-area-ratio engines like the SSME, $P_e \ll P_\infty$ and $P_\infty A_e$ is not an insignificant force in Equation 3.1.

Applying the principle of *conservation of mass* for the entire rocket system, it is apparent that the engine exit mass flow rate, dm_e/dt, is equal to the negative rate of change of mass for the overall rocket ($-dm/dt$). So for every 1 kg of mass out of the rocket exhaust in a given time interval, there is a corresponding 1 kg *reduction* in propellant mass for the rocket system.

Substituting $\dot{m}_e = -dm/dt$ into Equation 3.1a produces the following idealized form of Newton's "rocket" equation of motion

$$m\left(\frac{dv}{dt}\right) = -C\left(\frac{dm}{dt}\right) \tag{3.2}$$

Multiplying both sides of Equation 3.2 by dt, and integrating the resulting equation from an initial rocket mass m_{initial} to a "final" rocket mass, m_{final}, assuming (to first order) that C is a constant, yields the single stage rocket *performance* equation for the resulting *ideal* velocity change

$$\Delta v_{\text{ideal}} = -C \ln\left(\frac{m_{\text{final}}}{m_{\text{initial}}}\right) = -C \ln\left(\frac{m_{\text{initial}} - m_{\text{propellant}}}{m_{\text{initial}}}\right) \tag{3.3}$$

This equation is also referred to as the "Tsiolkovsky rocket equation", named after the pioneering scientist/mathematician who initially derived it in 1903. We can also express Δv_{ideal} as a simple function of flight time, $\Delta v_{\text{ideal}}(t)$ based on the following analysis.

3.6.1 Propellant mass versus time

The mass of the propellant burned between the time of launch at $t = 0$ and any later time, t, is the integral of the mass flow rate over time. For a *constant mass flow rate* defined by $\dot{m}_e = \dot{M}_e$, the amount of propellant consumed up to time t is equal to the product of the constant exit mass flow rate, $\dot{M}_e$, and the net flight or burn time, t, according to

$$m_{\text{propellant burned}} = \int \dot{m}_e(t)\, dt = \dot{M}_e t \qquad (3.4)$$

Since mass flow rate, $\dot{M}_e$, is equal to the initial thrust T_i divided by the effective exit velocity C [see Equation 3.1a], we can also rewrite Equation 3.4 as follows

$$m_{\text{propellant burned}} = (m_i)\left[\frac{(T/W)_i}{\text{Isp}}\right] t \qquad (3.5)$$

where $(T/W)_i$ = initial *thrust* to *weight* ratio of the Shuttle, $\text{Isp} = C/g$ = propellant specific impulse, and m_i = initial rocket mass. Again we note that mass flow rate is assumed constant in this approximation.

Here, we also have assumed that, $\dot{m}_e$, has the same mass flow rate as at takeoff. In general, this is not strictly true, particularly for the SRB engines where the solid fuel geometry and attendant combustion processes cause the thrust to vary significantly with time over the full burn period, as noted experimentally in ground test firings. In fact, SRB thrust and $\dot{m}_e$ magnitudes can vary by as much as 50% in the first two minutes of first stage flight, with the highest thrust levels achieved within the first 10 seconds after ignition and the lowest thrust levels reached near the two minute mark. We will consider the impact of this variable flow-rate later when we estimate the "effective *average* flow-rate" for the SRBs.

3.6.2 Time varying velocity change

Substituting Equation 3.4 into Equation 3.3, we obtain the following simple logarithmic dependence of Δv_{ideal} on time t, assuming a constant mass flow rate constant $\dot{M}_e$

$$\Delta v_{\text{ideal}}(t) = -C \ln\left[1 - \left(\frac{\dot{M}_e}{m_i}\right) t\right] \quad \text{for } 0 \le t \le t_{\text{burnout}} \qquad (3.6)$$

Another expression for Δv_{ideal}, using the "propellant burned" relationship of Equation 3.5 is given by

$$\Delta v_{\text{ideal}}(t) = -C \ln\left[1 - \left(\frac{(T/W)_i}{\text{Isp}}\right) t\right] \quad \text{for } 0 \le t \le t_{\text{burnout}} \qquad (3.6a)$$

where

$$t_{\text{burnout}} = \frac{\left(\dfrac{m_{\text{total propellant consumed}}}{m_i}\right)(\text{Isp})}{(T/W)_i} \qquad (3.7)$$

3.6.3 Effective burnout time and average flow rate

For the general situation in which the thrust and flow rate is *not* constant we can calculate an "effective burnout time" from the following definition

$$t_{b,\text{eff}} \equiv \frac{\int \dot{m}(t)\, dt}{\dot{m}_{\text{avg}}} = \frac{m_{\text{propellant burned}}}{\dot{m}_{\text{avg}}} \qquad (3.7a)$$

provided we can estimate the average mass flow rate (say from experimental data).

This equation can also be inverted to solve for the average flow rate, $\dot{m}_{avg}$, if we have a reasonable estimate of the effective burnout time and if we also know the total amount of propellant burned

$$\dot{m}_{avg} = \frac{m_{\text{propellant burned}}}{t_{b,\text{eff}}} \qquad (3.7b)$$

Figure 3.2 indicates that the SRB solid propellant is fully expended close to the 2 minute mark. We set $t_{b,\text{eff}} \approx 110$ seconds for the effective solid booster burn time based on test-firing data made available from ATK Thiokol, manufacturer of the SRBs, that shows that SRB thrust drops rapidly after 110 s and declines even further in the 16 seconds preceding the jettison of the SRBs at 126 s [16, "Solid Propellant Rocket Motor Fundamentals"].

If we are given the total amount of *solid propellant* initially in the two SRBs, $m_{\text{propellant}}$, we can then readily solve for an *average flow rate* from Equation 3.7b. Using the solid propellant mass given in Figure 3.3 for two SRBs, the average mass flow rate, $\dot{m}_{avg,2SRBs}$, is calculated to be

$$\dot{m}_{avg,2SRBs} = \frac{1.008 \times 10^6 \text{ kg}}{110 \text{ s}} = 9164 \text{ kg/s}$$

SRB ground test firing data [17] also shows that the measured *maximum* mass flow rate for two SRBs is approximately 19% higher than the average burn rate, and that the maximum occurs about 20 seconds after ignition. Thiokol data from 2005 [17] shows that

$$\dot{m}_{max,2SRBs} = 10{,}900 \text{ kg/s}$$

The *average* solid fuel flow rate for *two* SRBs and an estimate of the average liquid propellant flow rate for the three SSME engines [see section 3.7.1] can be used to estimate the velocity change in Equation 3.6. The combined total average mass flow rate $\dot{M}_e$ is then

$$\dot{M}_e = \dot{m}_{avg,2SRBs} + \dot{m}_{avg,3SSMEs}$$

3.6.4 Ideal altitude or height for each rocket stage

Let's take our simple analysis one step further to obtain a closed-form expression for the *height* of an *ideal* rocket at time t after launch.

If our *ideal rocket* is accelerated *vertically* upward on a straight-line trajectory, then, ignoring for the moment the effects of gravity, the time-varying height of this rocket, $h_{\text{ideal}}(t)$, is the integral over time of the ideal vertical velocity in Equation 3.6

$$h(t) = \int_0^t \Delta v_{\text{ideal}}(t)\, dt = -C \int_0^t \ln\left[1 - \left(\frac{(T/W)_i}{\text{Isp}}\right)t\right] dt \qquad (3.8)$$

Notice that instead of using the constant $\left(\frac{(T/W)_i}{\text{Isp}}\right)$ that appears in the ideal velocity expression, we could also replace it with the equivalent constant $\left(\frac{\dot{M}_e}{m_i}\right)$, the average mass flow rate to initial mass ratio employed in Equation 3.6.

A closed form integral expression for $h(t)$ can readily be obtained using the well known integral

$$\int \ln x \, dx = x \ln x - x$$

along with the relationship $C = (\text{Isp}) \cdot g$ defined in the previous section. Evaluating the integral in Equation 3.8 (which is suggested as an exercise for the reader) provides the following expression for $h(t)$

$$h(t) = \frac{g(\text{Isp})^2}{(T/W)_i}\left[1 - \left(\frac{(T/W)_i}{\text{Isp}}\right)t\right] \cdot \ln\left[1 - \left(\frac{(T/W)_i}{\text{Isp}}\right)t\right] + (\text{Isp} \cdot g)t \quad (3.9)$$

For a short time t after launch, when $t \ll \text{Isp}$, we find, utilizing a Taylor's series expansion of the logarithm function $[\ln(1-x) = -(x + x^2/2 + \cdots), x \leq 1]$, valid to second order, that Equation 3.9 yields the following familiar form

$$h(t) \approx \tfrac{1}{2}(T/W)_i \cdot gt^2 = \tfrac{1}{2}(T/m)_i \cdot t^2 = \tfrac{1}{2}at^2 \quad (3.10)$$

This equation is, of course, the well-known expression for the displacement of a point mass m subject to a constant acceleration "a", where "a" is the ratio of the rocket's initial thrust T to its initial mass m, i.e. $a = (T/m)_i$. If we also include the acceleration due to gravity in our short time analysis, then the net acceleration at takeoff will be $a_{\text{net}} = [(T/W)_i - 1]g$.

A more detailed analysis of the influence of gravity on the vehicle's motion is presented in Section 8 after we perform a numerical evaluation of Δv_{ideal}.

3.7 Δv_{ideal} ESTIMATE FOR SHUTTLE FIRST STAGE, WITHOUT GRAVITY LOSS

For the initial entry in our Shuttle's Δv *budget*, we now calculate the ideal velocity achieved by the Shuttle at the end of the first stage of its flight ($\Delta v_{1,\text{ideal}}$). This is the velocity achieved just at the time of SRB burnout, i.e. when the solid fuel is fully consumed. Since the liquid-propellant-driven SSME engines become fully operational at takeoff and burn in tandem with the SRBs, we find that about 1/4 of the fuel in the external tank is also consumed in the 2 minute period from launch to SRB burnout. We calculate $\Delta v_{1,\text{ideal}}$ from Equation 3.3 using the values listed in Table 3.1 for the initial (or total) mass at liftoff, the mass of propellant consumed during the first stage burn period, and the effective average exit velocity C_{avg} characterizing the thrust performance for the combined solid and liquid engine burns.

Sec. 3.7] Δv_{ideal} **estimate for Shuttle first stage, without gravity loss**   103

With this definition for C_{avg}, the calculated ideal velocity increment for the first stage is given by

$$\Delta v_{1,\text{ideal}} = -C_{\text{avg}} \ln\left(\frac{m_{\text{final},1}}{m_{\text{initial},1}}\right) = -C_{\text{avg}} \ln\left(\frac{m_{\text{initial},1} - m_{\text{propellant},1}}{m_{\text{initial},1}}\right) \quad (3.3\text{a})$$

The following mass parameters are used in the numerical evaluation of $\Delta v_{1,\text{ideal}}$.

$$m_{\text{initial},1} = m_{\text{total@takeoff}} = 2{,}063{,}016 \text{ kg}$$
$$m_{\text{final},1} = m_{\text{initial},1} - m_{\text{propellant},1} = 2{,}063{,}016 - 1{,}170{,}000 = 893{,}016 \text{ kg} \quad (3.3\text{b})$$

The total propellant mass for stage 1 used here (1,170,000 kg) is equal to the sum of the masses of both the solid rocket motor propellants (1,008,000 kg) and the liquid rocket propellants burned during the first flight period. In the following section we estimate the first stage liquid propellant mass burned as 162,000 kg. The propellant masses burned by the liquid SSME engines for the Shuttle's first and second stages of operation are summarized in Figure 3.3, based on the calculations presented below.

3.7.1 Estimate of SSME propellant mass burned during first stage

Employing Equation 3.7a we calculate the SSME liquid propellant mass burned from liftoff to the time of SRB separation as the product of the "average" first stage SSME flow rate ($\dot{m}_{\text{avg}}$) and the effective burn time $t_{b,\text{eff}} = 110 \text{ s}$. Our initial estimate for $\dot{m}_{\text{avg}}$ is based on the following thrust-based calculation for the total mass flow rate for the 3 SSME engines at liftoff, $\dot{m}_i$, using the tabulated values for thrust and sea level specific impulse given in Table 3.2

$$\dot{m}_i = \frac{T_i}{C} = \frac{T_i}{(\text{Isp}_i)g} = \frac{5.25 \times 10^6 \text{ N}}{(363 \text{ s})\left(9.81 \frac{\text{m}}{\text{s}^2}\right)} = 1{,}474 \text{ kg/s}$$

Dividing $\dot{m}_i$ by 3 gives the mass flow rate *per* SSME engine as 491.3 kg/s. If we assume that the SSME flow rate is nearly constant during the first stage burn period, the total SSME liquid propellant mass burned (or consumed) from liftoff to the time of SRB separation is

$$m_{\text{SSME propellant burned,1st stage}} = (\dot{m}_{\text{avg}})t_{b,\text{eff}} = (1{,}474 \text{ kg/s})(110 \text{ s}) \approx 162{,}000 \text{ kg}$$

Since the mass of liquid propellant in the external tank at takeoff is 730,000 kg, the liquid propellant *not* consumed in the first stage is available for the second stage burn, i.e.

$$m_{\text{SSME propellant available,2nd stage}} \approx 730{,}000 - 162{,}000 = 568{,}000 \text{ kg}$$

These values are summarized in Figure 3.3 and in Table 3.1 for both the Shuttle's first and second stages of operation.

The combined total mass flow rate for both the solid and liquid propellant engines burning at the same time, $\dot{M}_{e,1st}$, is

$$\dot{M}_{e,1st} = \dot{m}_{avg,2SRBs} + \dot{m}_{avg,3SSMEs} = 9{,}164 \text{ kg/s} + 1{,}474 \text{ kg/s} = 10{,}638 \text{ kg/s}$$

For the first stage burn period, the total propellant expended in 110 s is more than 1 million kg, i.e.

$$m_{\text{propellant},1st} = (\dot{m}_{avg,2SRBs} + \dot{m}_{avg,3SSMEs})(t_{b,\text{eff}}) = (10{,}638 \text{ kg/s})(110 \text{ s}) = 1{,}170{,}000 \text{ kg}$$

This is the propellant mass we used in Equation 3.3b for our computation of the first stage mass ratio.

3.7.2 First stage mass ratio and average effective exhaust velocity

Using the initial and final masses given in Equation 3.3b, the overall mass ratio, MR_1, for the first stage (required in Equation 3.3a) is

$$MR_1 = m_{\text{final},1}/m_{\text{initial},1} = (893{,}016)/(2{,}063{,}016) = 0.433$$

The *average* exit velocity, C_{avg}, to be used in Equation 3.3a is, by definition of the specific impulse for a given rocket propellant (see Table 3.2 notes) equal to the *average* specific impulse Isp_{avg} listed in Table 3.2 multiplied by the acceleration of gravity

$$C_{avg} = (\text{Isp}_{avg})g \qquad (3.11)$$

We calculate Isp_{avg} and C_{avg} in the following section.

3.7.3 Average specific impulse for the "parallel" (solid + liquid) first stage burn

A basic expression for the *average* specific impulse for the first stage, Isp_{avg}, can readily be derived from Equation 3.1. It is written here for a dual propulsion system of solid and liquid rockets working in parallel to provide the total first stage thrust. We follow the approach of Ryan and Townsend [18] using the expression

$$m\frac{dv}{dt} = T_{\text{total}} = C_{\text{solid}}\left(\frac{dm_e}{dt}\right)_{\text{Solid}} + C_{\text{Liquid}}\left(\frac{dm_e}{dt}\right)_{\text{Liquid}}$$

$$= C_S \dot{m}_S + C_L \dot{m}_L = \left[C_S \frac{\dot{m}_S}{\dot{m}_{\text{total},1}} + C_L \frac{\dot{m}_L}{\dot{m}_{\text{total},1}}\right]\dot{m}_{\text{total},1} \qquad (3.12)$$

To evaluate C_{avg} we use the *average* flow rate values, calculated previously for the solid and liquid propellant flow rates:

$$\dot{m}_{avg,2SRBs} = 9{,}164 \text{ kg/s}; \quad \dot{m}_{avg,3SSMEs} = 1{,}474 \text{ kg/s}$$

We now define C_{avg} to be a single *constant* velocity equal to the bracketed expression in Equation 3.12. The mass-rate averaged value of the exit velocities

Sec. 3.7] Δv_{ideal} **estimate for Shuttle first stage, without gravity loss**   105

for the two propellants, C_{avg}, can then be written as follows,

$$\left.\begin{aligned} C_{\text{avg}} &= C_S \frac{\dot{m}_S}{(\dot{m}_S + \dot{m}_L)} + C_L \frac{\dot{m}_L}{(\dot{m}_S + \dot{m}_L)} \\ C_{\text{avg}} &= C_S \alpha + C_L(1-\alpha) \\ \text{where} \quad \alpha &\equiv \frac{\dot{m}_S}{(\dot{m}_S + \dot{m}_L)} \end{aligned}\right\} \quad (3.13)$$

From Equation 3.11, the corresponding expression for the average specific impulse, Isp_{avg}, is

$$\text{Isp}_{\text{avg}} = \text{Isp}_S \alpha + \text{Isp}_L(1-\alpha) \quad (3.14)$$

Assuming that this "constant" C_{avg} applies (in an average sense) to the *entire* first stage flight period, we can insert C_{avg} into Equation 3.3a in order to obtain the equivalent evaluation for $\Delta v_{1,\text{ideal}}$.

The magnitude of the average exit velocity C_{avg} for the parallel-burn stage 1 propulsion system is evaluated using the data from Tables 3.1 and 3.2. First, we calculate the solid/liquid mass ratio constant α defined in Equation 3.13

$$\alpha = \frac{\dot{m}_S}{(\dot{m}_S + \dot{m}_L)} = \frac{9{,}164 \text{ kg/s}}{(9{,}164 \text{ kg/s} + 1{,}474 \text{ kg/s})} = 0.86 \quad (3.15)$$

The combined average specific impulse for the first stage is then given by the following expression,

$$\begin{aligned} \text{Isp}_{\text{avg}} &= \text{Isp}_S \alpha + \text{Isp}_L(1-\alpha) \\ &= (\text{Isp}_{S,\text{avg}})(0.86) + (\text{Isp}_{L,\text{avg}})(0.14) \end{aligned} \quad (3.16)$$

The thrust and specific impulses for the SRB and SSME engines vary with altitude due to the pressure varying force term, $(P_e - P_\infty)A_e$, given in Equation 3.1. As a result, engine thrust T increases with altitude for a rocket nozzle which is under-expanded at sea level. For an under-expanded nozzle, the exit pressure, P_e, is less than the altitude varying ambient pressure P_∞. For a fixed flow rate, specific impulse is proportional to T. Hence, Isp will also increase with altitude. From Table 3.2 we observe that the sea-level Isp magnitudes for both solid and liquid engines are less than their corresponding vacuum levels, Isp_{vac} by about 10% for the SRBs and 20% for the SSMEs.

To numerically evaluate Equation 3.16, we need to specify average Isp values for both the solid and liquid engines that are representative of the full range of specific impulses for the first stage flight period. We very crudely approximate the flight-average specific impulse for each type of engine as the average of the values given for $\text{Isp}_{\text{sea level}}$ and $\text{Isp}_{\text{vacuum}}$ according to

$$\text{Isp}_{\text{avg}} \approx \frac{(\text{Isp}_{\text{sea level}} + \text{Isp}_{\text{vacuum}})}{2} \quad (3.16a)$$

Using the sea-level and vacuum specific impulses in Table 3.2, the numerical values for the average specific impulses are then

$$\text{Isp}_{S,avg} \approx \frac{(242\,\text{s} + 268\,\text{s})}{2} = 255\,\text{s}$$

$$\text{Isp}_{L,avg} \approx \frac{(363\,\text{s} + 453\,\text{s})}{2} = 408\,\text{s}$$

Inserting these values into Equation 3.16 yields the combined average specific impulse for both engines running as a composite rocket system

$$\text{Isp}_{avg,\text{both engines}} = (\text{Isp}_{S,avg})(0.861) + (\text{Isp}_{L,avg})(0.139)$$
$$= (255)(0.86) + (408)(0.14) \approx 277\,\text{s}$$

Although this calculated specific impulse Isp_{avg} of 277 seconds is somewhat greater than the Isp for a first stage solid propellant, operating alone near sea level, it is considerably less than the Isp for an all-liquid propellant system.

Using $\text{Isp}_{avg} = 277\,\text{s}$, the corresponding effective-average exit velocity for the first stage is

$$C_{avg} = (\text{Isp}_{avg})g = (277\,\text{s})\left(9.81\,\frac{\text{m}}{\text{s}^2}\right) = 2{,}717\,\text{m/s} = 2.72\,\text{km/s} \quad (3.16\text{b})$$

3.7.4 Δv_{ideal} estimate for Shuttle first stage

Using Equation 3.3a, the calculated ideal velocity ($\Delta v_{1,\text{ideal}}$) for the Shuttle at the end of the first stage of its flight is

$$\Delta v_{1,\text{ideal}} = -C_{avg}\ln(MR_1) = -2.72\ln(0.433) = 2.28\,\text{km/s} \quad (3.17)$$

3.7.5 Δv_{ideal} and altitude as functions of time, for the Shuttle first stage

Figure 3.4a is a plot of the "ideal" first stage velocity as a function of time, for all times $t \leq t_{\text{burnout}}$, based on Equation 3.6; with the ratio of mass flow rate to initial takeoff mass given by

$$(\dot{M}_{e,\text{1st}}/m_i) = (10{,}638\,\text{kg/s})/(2{,}063{,}016\,\text{kg}) = 5.156 \times 10^{-3}\,\text{s}^{-1}$$

Note from this figure that the maximum first stage velocity of 2,275 m/s is reached at the burnout time, $t = 110\,\text{s}$, in accordance with Equation 3.17.

Figure 3.4b is the corresponding plot of "ideal" first stage *altitude* (or height) as a function of time, for all times $t \leq t_{\text{burnout}}$, based on Equation 3.9. Note from this figure that the maximum first stage altitude at burnout is *very* large; $h_{\max} = 108\,\text{km}$ or 354,000 feet. However, the effect of gravity has not been included in this calculation. This will be considered in the next section.

Sec. 3.7]   Δv_ideal **estimate for Shuttle first stage, without gravity loss**   107

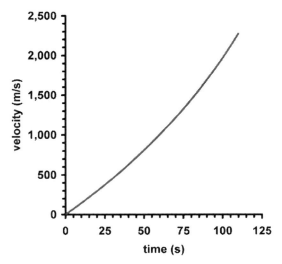

Figure 3.4a. "Ideal" Shuttle velocity as a function of flight time based on Equation 3.6, employing the parameters used in Equations 3.16b and 3.17; e.g. $C_{avg} = 2.72$ km/s. Total mass flow rate divided by initial takeoff mass is $\dot{M}_{e,1st}/m_i) = (10{,}638 \text{ kg/s})/(2{,}063{,}016 \text{ kg}) = 5.156 \times 10^{-3} \text{ s}^{-1}$. The effect of gravity on Shuttle velocity is not included in this plot.

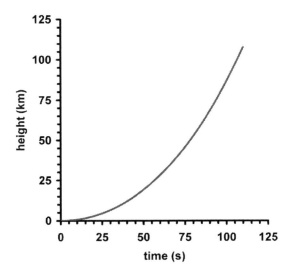

Figure 3.4b. "Ideal" Shuttle altitude as a function of time using Equation 3.9 based on the parameters adopted in Equations 3.16b and 3.17; e.g. $C_{avg} = 2.72$ km/s. The effect of gravity and pitch angle variation is not included in this calculation. This "ideal" rocket trajectory is vertical over the entire flight period.

3.8 THE EFFECT OF GRAVITY ON VELOCITY DURING FIRST STAGE FLIGHT

$\Delta v_{1,\text{ideal}}$ at burnout, as defined above, is not the true value of velocity at the end of the first stage burn. It is the *ideal* velocity that can be achieved by an accelerating rocket system maintaining a constant thrust, *in the absence of real-world gravity and drag forces*. In particular, this ideal velocity does not take into account the significant effect of gravity on the rocket's velocity during the Shuttle's first stage of acceleration. At takeoff, the rocket initially moves straight upwards, but shortly thereafter it begins to rotate away from the vertical; rotating towards zero pitch angle as it heads towards a circular orbit around the earth, *paralleling* the local horizontal.

In this section, we estimate the effect of gravity on the Shuttle's final speed at the end of the first stage of flight. We also determine the reduced burnout altitude that results from the downward force on the rocket due to gravity.

The scenario used to set up our *approximate* BotE model is presented in the following paragraph. The resulting mathematical equation is then solved to obtain expressions for velocity and altitude as a function of time for a rocket with a time-*varying* pitch angle in the presence of gravity.

Let's first consider that the Shuttle lifts off in a vertical position, which has an initial pitch angle β of 90° from the horizontal (see Figure 3.5). Shortly thereafter, the Shuttle begins to "nose over" by enacting pre-programmed commands that mechanically alter the rocket engine thrust vector, as a result of which the pitch angle begins to decrease over time. By design, the Shuttle's pitch angle is a pre-programmed function of flight time, $\beta(t)$, where β is reduced from 90° to between 30°–45° in the first 2 minutes of flight due to the application of thrust vector or "gimbal" control. The pitch or flight path angle time history is programmed so as to minimize both aerodynamic loads and main engine propellant usage en-route to orbit.

3.8.1 Modeling the effects of gravity for a curved flight trajectory

To account for the gravitational force on the rocket's motion, we need to include in our formulation of Newton's law (Equation 3.1) an added force, *along* the flight path, $= -mg \sin \beta(t)$ as indicated by the force balance depicted in Figure 3.5 for a curved trajectory. For the moment, we will ignore the aerodynamic effects of drag and lift on the Shuttle during its initial pitch and roll sequence, and simply assume that pitching during the flight trajectory is confined to a prescribed plane above a flat earth as shown in the figure. This sketch illustrates the thrust and gravity forces on our simplified rocket model in which the thrust vector $\vec{T}$ is aligned with the velocity vector $\vec{v}(t)$ along the flight trajectory.

The revised equation of motion, expressed in terms of the forces *along* the curved flight path, is

$$m\left(\frac{dv}{dt}\right) = -C\left(\frac{dm}{dt}\right) - mg \sin \beta \qquad (3.18)$$

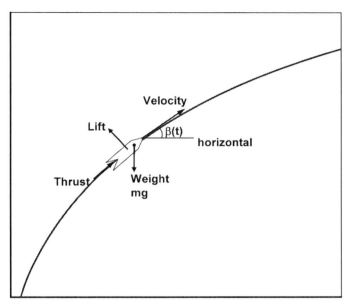

Figure 3.5. Force balance on a rocket with mass m moving on a curved trajectory. At time t, the tangent to the trajectory is defined equal to the pitch angle $\beta(t)$ where β is measured relative to the local horizontal. The gravity vector always points to the center of the earth. The thrust direction is taken to be the same as that of the velocity vector.

Integrating Equation 3.18 yields the following expression for the velocity change Δv at time t at the end of the first stage flight period

$$\Delta v = -C \ln\left(\frac{m_{\text{final}}}{m_{\text{initial}}}\right) - \int_0^t g \sin \beta(\tau)\, d\tau \qquad (3.19)$$

The integral on the right hand side of Equation 3.19 is often called the gravity-loss term, and we can crudely estimate its magnitude by assuming a particular functional form for $\beta(t)$. When calculating $\Delta v_{\text{1st stage with gravity loss}}$ at the end of the first stage of flight, the upper limit of the integral is set to the first stage burnout time $t_{\max}$.

An exact evaluation of rocket velocity as a function of time, $\Delta v(t)$, properly requires a simultaneous solution of the following two coupled-ordinary-differential equations: (a) the equation of motion based on the forces along the flight path (i.e. Equation 3.18), and (b) the following equation of motion describing the acceleration and forces that are *perpendicular* to the curved flight path

$$mv\left(\frac{d\beta(t)}{dt}\right) = -m\left(g - \frac{v^2}{r}\right)\cos \beta(t) + L \qquad (3.20)$$

Notice that Equation 3.20 directly relates the rate of change of pitch angle β to the gravitational constant g, the opposing *effective* centrifugal force mv^2/r, where r is the distance to the center of the earth, and to the aerodynamic lift force L ([18, p. 180] and [19, Equation 3-59]). Except for very high-speed near-orbital and reentry conditions, the centrifugal force term (mv^2/r) is small relative to the other terms and is usually ignored in an analysis of the lower velocity portions of the launch trajectory. In general, there is no closed-form analytical solution to the coupled problem (Equations 3.18 and 3.20) even with the assumption of zero lift, $L = 0$, and assuming a negligible geo-centrifugal force.

In the spirit of the Back-of-the-Envelope estimation approach, we will simplify our first stage gravity-loss estimate by completely "bypassing" Equation 3.20. This approximation eliminates the need to calculate a full numerical solution of the two simultaneous differential equations for $v(t)$ and $\beta(t)$. We accomplish this "bypass" approximation by assuming that $\beta(t)$ is *not* an independent variable but rather is a *prescribed* function of time created by the Shuttle's thrust-vector control system. Such a *prescribed* function can be inserted directly into Equation 3.19 for a BotE estimate of the gravity loss.

3.8.2 Time-varying pitch angle model

Let's set up a simple temporal model for the pitch angle, in which β decreases linearly in time from a pitch angle of 90° or $\pi/2$ radians down to $\beta_{\text{final}} = 30°$ or $\pi/6$ radians over a time period up to $\tau = t_{\text{max}}$. Using our model parameters for the Shuttle, the first stage burnout time is approximately $t_{\text{max}} \approx 110$ seconds. Our model equation for $\beta(t)$ is then

$$\beta(\tau) = \frac{\pi}{2} - \left(\frac{\pi}{2} - \beta_{\text{final}}\right)\left(\frac{\tau}{t_{\text{max}}}\right) \qquad (3.21)$$

By substituting this expression for β into the gravity-loss term in Equation 3.19 (let's call it Δv_{loss}) and then integrating the sine function from 0 to t_{max} we obtain the following closed-form estimate for Δv_{loss}

$$\Delta v_{\text{gravity-loss}} = \left[\frac{\sin(\pi/2 - \beta_{\text{final}})}{(\pi/2 - \beta_{\text{final}})}\right] g t_{\text{max}} = \left[\frac{\sin(\pi/3)}{\pi/3}\right] g t_{\text{max}} \qquad (3.22)$$

where we assume that the final pitch angle is a known constant.

In evaluating Equation 3.22 we set the final pitch angle at time t_{max} equal to $\pi/6$; i.e. $\beta_{\text{final}} = \pi/6$ radians (or 30°).

Evaluating Δv_{loss} with $t_{\text{max}} = 110\,\text{s}$ yields a gravity loss estimate for first stage flight of just under $1\,\text{km/s}$

$$\Delta v_{\text{gravity-loss}} = \left[\frac{3\sqrt{3}}{2\pi}\right] g t_{\text{max}} = (0.827)(9.81)(110.0) = 892\,\text{m/s} = 0.89\,\text{km/s} \qquad (3.23)$$

Sec. 3.8] **The effect of gravity on velocity during first stage flight** 111

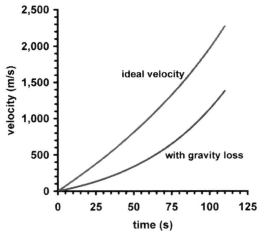

Figure 3.6. Shuttle velocity as a function of flight time: $t_{max} = 110$ s; $\beta_{final} = 30° = \pi/6$; $Isp_{avg} = 277$ s. The green curve (Equation 3.24) includes gravity loss. For comparison, the ideal velocity solution (Equation 3.6) is shown as the higher velocity red curve. Total mass flow rate divided by initial takeoff mass is $\dot{M}_{e,1st}/m_i) = (10{,}638 \text{ kg/s})/(2{,}063{,}016 \text{ kg}) = 5.156 \times 10^{-3} \text{ s}^{-1}$.

3.8.3 Effect of gravity on rocket velocity during first stage flight

For times $t \leq t_{max}$, the complete gravity-corrected first stage solution for Shuttle velocity as a function of time t is (from Equations 3.6, 3.19, 3.21, and 3.22)

$$v(t) = -C \ln\left[1 - \left(\frac{(T/W)_i}{Isp}\right)t\right] - \left[\frac{\sin[(\pi/2 - \beta_{final})(t/t_{max})]}{(\pi/2 - \beta_{final})}\right]gt_{max} \quad (3.24)$$

This first stage gravity-corrected velocity is plotted in Figure 3.6 as a function of time. Notice that the gravity-corrected "green" curve falls significantly below the "red" ideal velocity solution, yielding a velocity at burnout ($t = 110$ s) about 1/3 smaller than the ideal. Also, the effect of gravity produces reduced net vehicle accelerations throughout the first stage. Hence the velocity difference between the two curves grows with increasing time.

3.8.4 Effect of gravity on rocket height during first stage flight

The corresponding time varying height, $h(t)$, of our gravity-corrected rocket, is equal to the time integral of the *vertical* velocity component, $v(t) \sin \beta(t)$, as expressed by

$$\left.\begin{aligned} h(t) &= \int_0^t v(t) \sin \beta(t) \, dt \\ &= -C \int_0^t \ln\left[1 - \left(\frac{(T/W)_i}{Isp}\right)t\right] \cos\left(a\frac{t}{t_{max}}\right) dt \\ &\quad - \frac{gt_{max}}{a} \int_0^t \sin\left(a\frac{t}{t_{max}}\right) \cos\left(a\frac{t}{t_{max}}\right) dt \end{aligned}\right\} \quad (3.25)$$

where $a \equiv \left(\frac{\pi}{2} - \beta_{final}\right)$

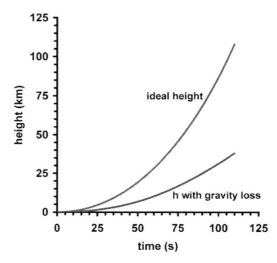

Figure 3.7. Shuttle altitude as a function of time evaluated from the integrals in Equation 3.25 using the parameters adopted in Equations 3.16 and 3.17. The lower green height curve includes gravity loss. For comparison, the ideal height solution for a vertical trajectory without the effects of gravity (Equation 3.9) is shown as the red curve.

This first stage gravity-corrected altitude is plotted in Figure 3.7 as a function of time. It can be evaluated either by numerical quadrature from the integrals in Equation 3.25 or from a corresponding closed-form solution expressed in terms of the natural logarithm function and the "SinIntegral" and "CosIntegral" functions, $Si(z)$ and $Ci(z)$ (see [20, Wolfram.com] for a definition of these trigonometric integral functions).

Note in Figure 3.7 that the gravity-loss-corrected altitude, with the pitch dropping to $30°$ in 110 s, lies considerably below the ideal altitude solution for a *vertical* trajectory. Our calculated gravity-corrected height at first stage burnout, $h_{burnout} \approx 38.0$ km, is about 20% less than the nominal SRB "jettison" altitude of 47 km, as annotated in Figure 3.2.

3.8.5 Comparing model velocity and altitude with Shuttle data

In Figure 3.8, model height as a function of time is compared with Shuttle time-dependent altitudes derived from NASA ambient pressure data obtained during the first Shuttle flight (see [21] data for the STS-1 mission flown April 12, 1981). These measured static pressures were converted into corresponding altitudes using U.S. Standard Atmosphere tables of static pressure as a function of altitude.

We observe that our predicted altitude curve follows quite closely the NASA data over the reported time interval from 30 s to 100 s.

Sec. 3.8] The effect of gravity on velocity during first stage flight 113

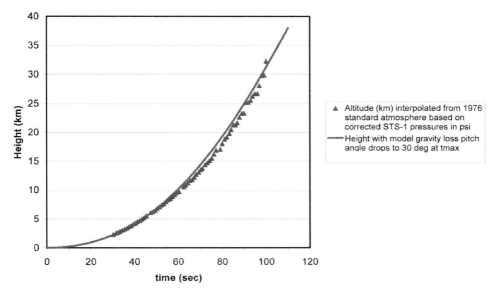

Figure 3.8. Shuttle altitude as a function of time evaluated from the gravity loss model integrals in Equation 3.25 compared to measured altitude data from STS-1. The model curve is calculated using the parameters adopted in Equations 3.16 and 3.17. The altitude data points are obtained from U.S. Standard Atmospheric scaling of the NASA ambient static pressure measurements. These pressure data were obtained from the pressure sensors located on a boom extending forward of the external tank (see the measured ambient pressure vs. time plot on p. 116 of [21]).

3.8.6 Gravity loss magnitudes for previously flown launch systems

Estimates of the gravity losses for several launch vehicles have been calculated using recorded flight data; e.g. according to Fortescue and Stark [11, p. 149] the gravity loss for the Ariane 2 rocket at the end of its first stage burn is

$$\Delta v_{\text{gravity loss, Ariane 1st stage}} = 1.08 \text{ km/s}$$

Ariane is a three stage rocket with a takeoff mass about 1/10th that of the Shuttle system, and the primary function of its third stage is to place a satellite payload directly into a highly elliptical transfer orbit with a low perigee and an apogee at the geostationary altitude of 35,786 km, with the orbit later being circularized at that high altitude [11, p. 88].

Note that our crude BotE estimate for first stage gravity loss (0.89 km/s) is within 18% of Ariane 2's first stage gravity loss.

We recognize, however, that the magnitude of any gravity loss is actually dependent on a rocket's initial thrust-to-weight ratio $(T/W)_i$. A *low* T/W launch vehicle (such as for a massive system like the Shuttle, where initial $T/W \approx 1.5$) generates a larger gravity loss than does a launch vehicle with a *high* T/W (like an unmanned missile where $T/W > 3$) because a vehicle which has a low T/W

spends more time in ascent [a larger t_{max} in Equation 3.22] than does one with a high T/W, before reaching orbit.

For typical medium to large launch vehicles, with nominal flight trajectories, the overall gravity losses (all stages included in the loss) range from 0.75 to 1.5 km/s [10, Wertz, "Space Mission Analysis and Design", 3rd ed. 1999]. Our estimate of 0.84 km/s for the Shuttle first stage in Equation 3.22 falls within this nominal range.

This gravity loss is also seen in the difference between our gravity-corrected velocity vs time curve and the corresponding "ideal" curve, when evaluated at burnout. Observe that the velocity difference between the two curves at $t = 110$ seconds in Figure 3.6 is exactly equal to 0.89 km/s.

3.8.7 Model velocity, with gravity loss, compared with flight data

Another test of our ability to predict gravity velocity losses is seen in Figure 3.9, where our model velocity function (i.e. Equation 3.24) is compared with the time-dependent Shuttle velocity derived from the flight Mach number data obtained by NASA during flight STS-1. The Mach number "time-histories" deduced from pressure data were converted into corresponding flight velocities using U.S. Standard Atmosphere tables for the speed of sound as a function of altitude [22].

Our predicted velocity solution (blue line) is closely aligned with the NASA data over the time period from 30 to about 45 s. Beyond this time, however, our model

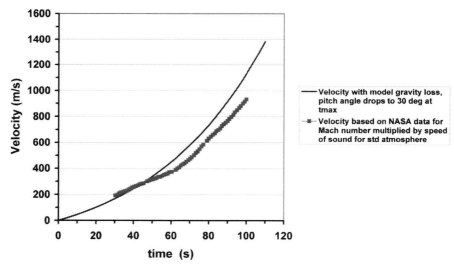

Figure 3.9. Shuttle first stage velocity as a function of time evaluated from the gravity loss model (Equation 3.24) compared to NASA velocity data deduced from STS-1 Mach number measurements (30 s < t < 100 s; the available data do not extend to the effective burnout time of 110 s). The velocity data points were obtained from a U.S. standard atmospheric scaling of NASA measured Mach numbers, M, [21] using the "1976 U.S. standard atmosphere" table of speed of sound as a function of altitude, $a(h)$, given in [22]. Note, that velocity = $M \cdot a$.

Sec. 3.8]   **The effect of gravity on velocity during first stage flight**   115

predictions for velocity exceed the data due to the fact that the measured acceleration begins to decline for times between 45 and 70 s, in contrast to our expectations. For $t > 70$ s, measured acceleration increases, partly because of the continual reduction in rocket mass, and approaches the acceleration (or the velocity derivative) of our model solution. There is a "permanent velocity deficit" caused by the noticeably slower acceleration observed in the NASA data during the period from 45 s to 70 s.

Perhaps the most important cause for the reduced velocity of the Shuttle (for $45 < t < 70$ s) is the increase in *aerodynamic drag* experienced by the vehicle in this high dynamic pressure (or high q) regime. In Section 3.9 it will be shown that drag losses can account for more than a 100 m/s decrease in Shuttle velocity in the region of high q. A small but significant part of the velocity difference can be attributed to a *throttling back* of the SSME engines during this interval (this throttling interval is sometimes called the "thrust bucket"). This programmed reduction in thrust is initiated at $t = 45$ s in order to reduce the aerodynamic loads on the vehicle as q reaches its highest level (see Section 3.9.2). For $t > 70$ s, the SSME flow-rate and thrust levels are returned to their full launch settings as q begins to decrease.

3.8.8 Calculation of gravity-loss corrected velocity at first stage burnout

Using our estimate for "first stage gravity loss" (Equation 3.23) to correct the ideal Shuttle velocity at SRB burnout (Equation 3.17) we find that the magnitude of Δv_1 at $t = 110$ seconds is

$$\Delta v_{\text{1st stage with gravity loss}} = 2.28 - 0.89 = 1.39 \text{ km/s} \qquad (3.26)$$

This gravity-corrected velocity estimate, or burnout speed, is about 64% of the *ideal* first stage burnout velocity in the absence of gravity.

$\Delta v_{\text{1st stage with gravity loss}}$ is, of course, exactly equal to our *time-dependent* velocity at the time of burnout, i.e. the value of $\Delta v(t)$ calculated from Equation 3.24 at $t = 110$ s (see Figures 3.6 and 3.9).

NASA tabulated data for the Shuttle velocity at the time of SRB separation lists values for first stage burnout velocity in the range of 1.24 to 1.62 km/s for a range of missions in the range STS-28 to STS-65. [23, "Shuttle Ascent and Orbit Insertion Data"]. Our estimate, Equation 3.26, clearly falls within this range.

Fleming and Kemp's experience in analyzing a number of ascent trajectories ([24]; cited in Griffin and French's Space Vehicle Design text [25]) shows that typical gravity and drag losses lead to first stage burnout velocities on the order of 70% of $\Delta v_{1,\text{ideal}}$. Hence our estimate of 64% for the burnout velocity as a percent of $\Delta v_{1,\text{ideal}}$ compares well with this rule of thumb.

As we will see upon calculating the velocity at the end of the *second* stage of flight, when all of the liquid fuel is consumed and the pitch angle approximates a value of 0° as orbit is reached, the gravity losses in the second stage are a smaller, but still significant fraction, of the ideal burnout velocity of the second stage.

3.9 THE EFFECT OF DRAG ON SHUTTLE VELOCITY AT END OF FIRST STAGE FLIGHT

Another velocity correction of importance to the overall Δv velocity budget is the aerodynamic drag during the Shuttle's atmospheric ascent where q is high. The drag-loss correction to our ideal solution is smaller in magnitude than the gravity-loss correction but (as demonstrated in Figure 3.9) it is not entirely negligible.

As an example of the magnitude of this particular loss, consider the estimated drag loss published for the Ariane 2 rocket. For the first stage, the drag loss is approximately, $\Delta v_{\text{drag loss}} = 0.22$ km/s, based on values cited by Fortescue and Stark [11, p. 149]. For the Ariane 44L variant $\Delta v_{\text{drag loss}} \approx 0.135$ km/s; as given in the velocity budget of Humble, *et al.* [26, p. 66].

3.9.1 Modeling the effects of drag in the equation of motion

When we include the aerodynamic drag force D in a revised equation of motion, expressed in terms of forces on the rocket along the curved flight path, Equation 3.18 becomes

$$m\left(\frac{dv}{dt}\right) = -C\left(\frac{dm}{dt}\right) - mg\sin\beta - D \qquad (3.27)$$

The additional velocity loss at the end of the first stage, due to the drag term alone, is given approximately by the integral expression

$$\Delta v_{\text{drag loss}} = -\int_0^t \frac{D(\tau)}{m(\tau)} d\tau \qquad (3.28)$$

However, this is only an approximate estimate of the drag contribution to the solution of Equation 3.27. This is primarily due to the fact that $D(t)$ depends on the square of the time-varying rocket velocity $v(t)$, making Equation 3.27 a non-linear equation for velocity. The exact solution of such a non-linear equation cannot formally be written as a simple sum of individual thrust, gravity, and drag terms. Nevertheless, Equation 3.28 helps the engineer obtain a rough estimate of the velocity loss due to drag.

To estimate $\Delta v_{\text{drag loss}}$, let us assume that the drag D can be calculated by the usual fluid dynamics drag force expression

$$D = (\tfrac{1}{2}\rho v^2)C_{\text{d}}A \qquad (3.29)$$

where q = dynamic pressure = $\tfrac{1}{2}\rho v^2$ (a function of flight time t).

Note in Equation 3.29, that the term "A" represents the vehicle drag *reference* area, also referred to as the *frontal* area exposed to the oncoming flow. C_{d} is the drag coefficient, typically of order 1.0 for a blunt body, particularly for flight speeds at near-sonic and low supersonic Mach numbers.

A *simple* first-order estimate for drag near the time of maximum dynamic pressure can readily be obtained by setting the rocket velocity v in Equation 3.29 equal to the average Shuttle velocity that is representative of that portion of the ascent flight where drag is most significant (say at time $t = t_{\text{max drag}}$). We then also need an estimate of the air density ρ at that time. We take ρ to be the average "free-stream" air density determined solely by the Shuttle's gravity-corrected altitude at $t = t_{\text{max-drag}}$.

3.9.2 Estimating first stage drag loss

In this section we obtain a crude numerical estimate of the first stage drag loss. We calculate this estimate using a Back-of-the-Envelope approximation to the integral of Equation 3.28. To perform this calculation, we must first determine the variation of dynamic pressure with time, $q(t)$. We estimate the flight period when there is a substantial rise, and subsequent fall, of $q(t)$, in order to determine the time interval during the first stage of flight when aerodynamic drag becomes significant. Using Equation 3.24 for $v(t)$, which is an expression that includes the effects of gravity loss, but not drag, and an exponential atmospheric model for the dependence of density ρ on altitude h

$$\rho = \rho_0 \exp(-h/H) \tag{3.30}$$

allows us to generate a plot of dynamic pressure ($q = \frac{1}{2}\rho v^2$) as a function of flight time in Figure 3.10.

In Equation 3.30, the exponential scale height constants $H \approx 22{,}000$ ft $= 6.73$ km and $\rho_0 = 1.23$ kg/m^3 are based on curve fits to atmospheric data by Allen and Eggers ([27], dating back to 1958. An equivalent rule of thumb, corresponding to Equation 3.30, is that the density of the Earth's atmosphere drops off by a factor of 10 for every 10 miles (16.1 km) of altitude [28], which can be written as $\rho = \rho_0 10^{-h/(16.1 \text{ km})}$ or in exponential form as $\rho = \rho_0 e^{-h/(7.0 \text{ km})}$.

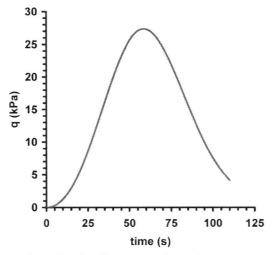

Figure 3.10. Estimate of the Shuttle's free stream dynamic pressure, q, as a function of flight time after launch, t.

The gravity-loss-corrected altitude plot of Figure 3.7 gives us the approximate Shuttle height, h as a function of time. This is the altitude that is to be used in the expression for density in Equation 3.30.

From Figure 3.10, we observe that dynamic pressure q reaches a maximum of 27.3 kPa at a time $t = 59$ s, i.e. about 1 minute after launch.

A similar dynamic pressure vs time plot for the Shuttle, given in a paper by Edinger [29] indicates that $q_{max} \approx 31.1$ kPa at $t \approx 50$ seconds after launch. Also NASA's data for q_{max} span the range from 30.7 to 34.9 kPa ([23], citing data from missions STS-28 to STS-65). Our estimate for q_{max}, at the peak of the curve in Figure 3.10, lies about 12% below NASA's reported dynamic pressure range. At $t = 59$ s, the time of maximum drag (or max q in Figure 3.10), the Shuttle velocity (as shown in Figure 3.6) is $v \approx 0.44$ km/s. At that time the Shuttle is traveling at a Mach number of about 1.3 since the speed of sound for an air temperature of 0°C is approximately 0.34 km/s, or 760 mph. Our estimated Mach number is quite close to those values reported by NASA for the Shuttle at "q max" [23]. Our predicted altitude at the time of q_{max} is $h = 9.8$ km or 32,000 feet (as shown in Figures 3.7 and 3.8).

The air density value at max q is calculated using the simple exponential atmospheric model for the dependence of ρ on altitude h. For $h = 9.8$ km at max q, Equation 3.30 gives an air density at max q of

$$\rho \approx 0.29 \text{ kg/m}^3$$

Using Equation 3.29, the peak drag force on the Shuttle (at max q) is estimated to be

$$\left. \begin{array}{l} D = (\tfrac{1}{2}\rho v^2)C_d A = (0.5)(0.29)(437.0)^2(1.0)(126.7) \\ = 3.50 \times 10^6 \text{ N} \end{array} \right\} \quad (3.31)$$

To get a feel for the magnitude of this drag force we note that 3.50×10^6 N (or 787,000 lbf) is about 7% of the Shuttle's total takeoff thrust.

Using a nose-on drawing of the STS stack (Figure 3.11, showing the external fuel tank diameter = 8.4 m) we are able to estimate the equivalent "frontal" drag area for the Shuttle that was inserted into Equation 3.31 as

$$A = 126.7 \text{ m}^2 \quad (3.32)$$

This drag area is based on the sum of the approximate cross-sectional areas of: (1) the external tank, 8.4 m in diameter; (2) the circular SRB casings, each of which is 3.7 m in diameter; and (3) the frontal area of the Orbiter, estimated from our drawing at about 7×7 m. The total area of all of these drag surfaces from the nose-on perspective is $A = 126.7 \text{ m}^2$, yielding the approximate overall maximum drag given in Equation 3.31, $D = 3.50 \times 10^6$ N.

To complete our drag loss calculation in Equation 3.28, we must estimate the Shuttle's mass at the time of max q. Shuttle mass as a function of time, derived from the propellant mass vs time relationship of Equation 3.6, is calculated from the

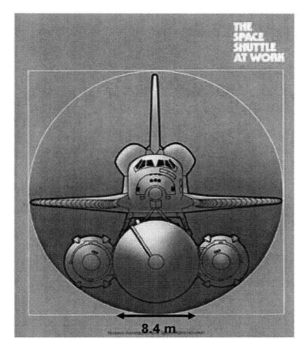

Figure 3.11. Nose on view of Shuttle. External tank diameter = 8.4 m. SRB casing diameter = 3.7 m (drawing from [30]).

expression

$$m(t) = m_i \left[1 - \left(\frac{\dot{M}_{e,1st}}{m_i} \right) t \right] \quad (3.33)$$

where for the first stage of flight the average mass flow rate is

$$(\dot{M}_{e,1st}/m_i) = 5.156 \times 10^{-3} \text{ s}^{-1}$$

Substituting the above first stage mass flow-rate ratio into Equation 3.33, we find that the Shuttle's mass at max q ($t = 59$ s) is

$$m(59 \text{ sec}) = 1.43 \times 10^6 \text{ kg} \quad (3.34)$$

Note that the Shuttle's mass at q_{max} has dropped 31% from its takeoff mass, $m_i = 2.06$ million kg. This drop in mass is, of course, due to the consumption of a significant portion of the SRB and SSME propellants during the first 59 seconds of flight.

Inserting the above estimated values for drag D and mass m into the following approximation to the full drag loss integral given in Equation 3.28

$$\text{Drag loss} \approx D \cdot [\Delta t_{(\text{significant drag duration})}]/m \quad (3.35)$$

and using an estimate of the time interval during which there is significant drag,

120   Estimating Shuttle launch, orbit, and payload magnitudes   [Ch. 3

$\Delta t_{\text{(significant drag duration)}} \approx 50\,\text{s}$ based on Figure 3.10, we estimate the first stage drag loss as

$$\text{Drag loss} \approx D(\Delta t)/m$$
$$= (3.50 \times 10^6)(50)/(1.43 \times 10^6) = 122\,\text{m/s} = 0.12\,\text{km/s} \quad (3.36)$$

This drag loss of 0.12 km/s accounts for a good fraction (about 3/4) of the difference in velocity between our gravity-loss velocity model and NASA's post-flight velocity data (Figure 3.9) for first stage flight times greater than 1 minute. While our estimated *drag loss* of 0.12 km/s is considerably smaller in magnitude than the *gravity loss* of 0.89 km/s estimate in Equation 3.23, it cannot be neglected in an overall estimate of the Shuttle's second stage burnout velocity. As we will subsequently show, a velocity loss of even this low magnitude translates into a considerable reduction in the possible payload that can be delivered to orbit by the Shuttle.

3.9.3   Final drag and gravity-corrected velocity at first stage burnout; key elements of the overall "velocity budget" for the first stage

If we use our first stage estimate for "gravity loss" (Equation 3.23) and first stage estimate for "drag loss" (Equation 3.36) in the calculation of the corrected velocity at SRB burnout, we find that Δv_1 (at $t = 110\,\text{s}$) is approximately

$$\boxed{\Delta v_{\text{1st stage with gravity and drag losses}} = 2.28 - 0.89 - 0.12 = 1.27\,\text{km/s} \quad (3.37)}$$

This *final* gravity-corrected and drag-corrected velocity estimate at burnout is approximately 56% of the *ideal* first stage velocity (Equation 3.17). If we perform a straight-line extrapolation of the NASA velocity data in Figure 3.9 at $t = 100\,\text{s}$ out to $t = 110\,\text{s}$, the *measured* Shuttle velocity at our effective burnout time would be approximately 1.10 km/s. Our prediction of 1.27 km/s exceeds that extrapolated experimental value of 1.10 km/s by 0.17 km/s. Thus our estimation error for first stage burnout velocity is about 14%. The errors associated with estimating the key parameters (e.g. average burn-rate or approximate drag) that contribute to the final "velocity budget" are typically of the same order.

The next section will calculate the second stage velocity-time history, taking into account second stage gravity losses. Due to the very high altitudes during this phase of the launch trajectory, $h > 37\,\text{km}$, aerodynamic drag will be a negligible contributor to velocity losses.

3.10   CALCULATION OF SECOND STAGE VELOCITIES AND GRAVITY LOSSES

In this section we continue our modeling of Shuttle dynamics for the time period beyond SRB burnout, which we will define to be the second stage of the flight trajectory.

First, we determine the equations that model the "ideal" accelerations and velocities of the second stage of the Shuttle. We evaluate the changes in flight dynamics due to: (a) the reduced structural mass of the Shuttle after the SRB casings are jettisoned; (b) the loss of the SRB thrust component; and (c) the continuous consumption of liquid-propellant mass by the Orbiter's main engines, for all times prior to main-engine cutoff (MECO).

To calculate "ideal" second stage velocities, Newton's second law of motion for a point mass rocket with time-varying propellant mass is again employed (i.e. Equations 3.1 and 3.6). We utilize as input to these calculations our estimate of the "initial" mass of the second stage of the vehicle, and an estimate of the flow-rate magnitudes and specific impulse levels characterizing the three SSME engines burning LOX/LH2.

Second, we determine the time-varying "gravity-corrected" velocities for the second stage, utilizing the same simplified model for the time-dependent pitch angle that was employed in our analysis of the first stage. We then determine a numerical estimate for the second stage "gravity loss", the magnitude of which is important in assessing the Shuttle's overall velocity budget and its final burnout velocity prior to entering low-earth orbit.

Finally, in order to assess the accuracy of our predictions, we compare our simplified velocity and altitude histories, for both stages, to NASA's detailed computer simulations for the entire launch trajectory.

3.10.1 Pitch and gravity loss modeling for the second stage flight period

We previously constructed a simple time-dependent analytical model for Shuttle pitch angle, $\beta(t)$, for the first stage of its powered flight. With $\beta = 90°$ at launch, we assumed that pitch drops linearly over the time scale interval, $\Delta t_1 = 110$ s. We set Shuttle pitch to be $30°$ at the end of this flight period. Using this model, we calculated the gravity-loss-correction integral shown in Equation 3.19, and used this to correct the "ideal" rocket velocity solution (Equation 3.6) for the effects of gravity. The resulting time-dependent gravity-corrected velocity for the first stage of the Shuttle is given by Equation 3.24. The closed-form expression for the gravity-loss reduction *at the end of the first stage* is given by Equation 3.22. We showed that the magnitude of that final first stage gravity loss, $\Delta v_{\text{gravity loss}, 1}$, is approximately 0.89 km/s (Equation 3.23).

In the second stage of flight, $t > 110$ s, the pitch angle continues to decrease towards $\beta = 0$, as the Shuttle heads towards a near-circular orbit parallel to the surface of the earth. During the second stage, the absolute value of the negative pitch rate, $|-d\beta/dt|$, is less than that in the first stage. Figure 3.2 implies that the "bending" of the flight path actually progresses at a *slower rate* after SRB burnout than before burnout. The basic time scale for changes in the pitch angle during the second stage appears to be longer than for pitch changes in the first stage, by about a factor of two. After take-off, the pitch angle is reduced as quickly as possible in order to minimize the gravity losses at the lower flight speeds for the first stage. Pitch and roll maneuvers are designed to minimize the aerodynamic loads imposed upon the

122 Estimating Shuttle launch, orbit, and payload magnitudes [Ch. 3]

vehicle's structure. These loads significantly increase during the high dynamic pressure (high q) portion of the flight, with an intensity that peaks about one minute after launch.

For the second stage, we now estimate the magnitude of the *time scale* that characterizes the time interval during which there is an overall drop in pitch angle β from 30° down to 0°, where our modeled Shuttle is at its maximum altitude. This time scale is an input parameter to our second stage gravity-loss model.

We estimate this second-stage time scale to be on the order of $\frac{1}{2}$ of the 2nd stage SSME burn period (The full 2nd stage SSME burn period extends from 110 s to about 500 s at SSME main-engine cutoff as we will show in Section 10.3).

Hence, for our second stage pitch model, initiated at $t = 110$ s, we crudely set $\beta = 0°$ at about $t = 300$ s. The estimated time scale for pitch reduction to 0° in this initial portion of second stage flight (let's call this time period *region 2a*), is then given by

$$\Delta t_{2a} = 300\,\text{s} - 110\,\text{s} = 190\,\text{s} \quad \text{[region 2a, pitch time scale]}$$

For flight times $t > 300$ s (call this time period *region 2b*) we assume that the Shuttle's pitch angle remains constant at 0°, even though the vehicle continues to accelerate. We will maintain this zero degree pitch until second stage burnout.

We summarize here the three regions chosen for our pitch model analysis:

region 1: $0\,\text{s} \leq t \leq 110\,\text{s}$; $30° \leq \beta \leq 90°$ first stage

region 2a: $110\,\text{s} \leq t \leq 300\,\text{s}$; $0° \leq \beta \leq 30°$ second stage

region 2b: $300\,\text{s} \leq t \leq 500\,\text{s}$; $\beta = 0°$ second stage

3.10.2 Time-varying gravity loss solution, region 2a

Following the approach employed in region 1 (the first stage of flight), we choose the following region 2a model equation for a *linearly decreasing* pitch angle $\beta(t)$

$$\left.\begin{array}{c} \beta(\tau) = \left(\dfrac{\pi}{6}\right)\left[1 - \left(\dfrac{(\tau - 110)}{\Delta t_2}\right)\right] \\ 110 \leq \tau \leq (110 + \Delta t_2); \quad \text{for region 2a, } \Delta t_2 = 190\,\text{s} \end{array}\right\} \quad (3.38)$$

Inserting this expression into Equation 3.19 and integrating from $t = 110$ to $110 + \Delta t_2$ yields the following closed-form estimate for the time varying gravity loss,

$$\Delta v_{\text{loss},2a}(t) = -\left(\dfrac{1}{\pi/6}\right)\left\{\cos\left[\dfrac{\pi}{6}\left(1 - \dfrac{(t-110)}{\Delta t_2}\right)\right] - \cos\left[\dfrac{\pi}{6}\right]\right\}(g\,\Delta t_2) \quad (3.39)$$

Recall that for our simple model, the region 2a pitch angle is set to $\pi/6$ (i.e. 30°), and the final pitch angle is required to be 0° at $t = 300$ s.

Sec. 3.10]   Calculation of second stage velocities and gravity losses   123

The *complete* gravity-corrected second region solution for Shuttle *velocity* as a function of time in region 2a, based on Equations 3.6 and 3.39, is then given by

$$v_{\text{region 2a}}(t) = -C_2 \ln\left[1 - \left(\frac{\dot{M}_{e_2}}{m_{i_{2a}}}\right)(t - 110)\right] + \Delta v_{\text{loss, 2a}}(t) + (\Delta v_{\text{region 1 with gravity loss}})$$

(3.40)

Note that we have added to our incremental region 2a solution the "ideal" velocity magnitude achieved by the Shuttle at the end of its first stage of flight $\Delta v_{\text{region 1 with gravity loss}} = 1{,}382 \text{ m/s} \approx 1.39 \text{ km/s}$, as per Equation 3.26. This will serve as a "constant of integration" added to the region 2a *incremental* time-varying velocity function (with gravity correction). This constant region 2a "initial velocity" is needed to complete the full region 2a velocity solution specified by Equation 3.40.

To numerically evaluate Equation 3.40, we must specify the magnitude of the average rocket exit velocity, C_2, for region 2a. This exit velocity is calculated directly from the vacuum specific impulse values presented in Table 3.2 for the LOX/LH2 propellants.

$$C_2 = (\text{Isp}_{\text{SSME:vacuum}})g = (452.5 \text{ s})(9.81 \text{ m/s}^2) = 4{,}439 \text{ m/s} \quad (3.41\text{a})$$

The corresponding second stage SSME mass flow rate is based on the flow-rate analysis in Section 3.7.1. We simply assume that, for these engines operating at a nominal 100% thrust rating, the mass flow rates are constant throughout the first and second stages of flight. Hence

$$\dot{M}_{e_2} = 1{,}474 \text{ kg/s} \quad (3.41\text{b})$$

The initial region 2a mass, $m_{i_{2a}}$, is estimated from Equation 3.3b and Table 3.1, and is equal to the Shuttle's mass after SRB burnout (i.e. the final mass of the first stage) *less the structural mass of the jettisoned first stage SRB casings*,

$$m_{i_{2a}} = 893{,}016 \text{ kg} - 165{,}758 \text{ kg} = 727{,}258 \text{ kg} \quad (3.41\text{c})$$

We substitute the parameters from Equations 3.41a to 3.41c into Equation 3.40 to calculate and plot the resulting region 2a gravity-corrected velocity, $v_{\text{region 2a}}(t)$. This velocity is displayed as the red curve in Figure 3.12. We append this region 2a curve to the first stage solution (the blue curve, from Equation 3.24) to create a continuous *piece-wise* velocity solution, allowing for a necessary abrupt change in slope (i.e. acceleration) at the terminus of the first stage and the beginning of the second stage.

For Region 1 in Figure 3.12, the "blue" velocity function does *not* include the time-varying effects of drag on velocity.

The overall impact of the cumulative first stage drag loss can be represented, at least in a gross sense, on the second stage burnout velocity magnitude simply by subtracting the estimated single drag velocity loss of 0.12 km/s (Equation 3.36) from our modeled gravity-corrected second stage burnout velocity at the time of MECO. This overall velocity drag loss is also included in the final velocity budget summary presented below in Section 3.11.

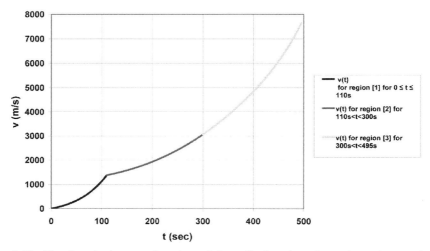

Figure 3.12. Shuttle velocity as a function of time, displayed as three piece-wise continuous segments: Region 1, blue curve (Equation 3.24, first stage), Region 2a, red curve (second stage, Equation 40), Region 2b, yellow curve (second stage, Equation 3.42). Gravity losses vanish in Region 2b since the Shuttle flight path angle (or pitch angle β) is set to zero. The effect of first stage drag loss is not included in these curves.

3.10.3 Time-varying velocity solution, region 2b

The expression for the time-varying velocity in region 2b (i.e. $t \geq 300$ s) is much simpler than that given for either regions 1 or 2a. Because the pitch angle β is assumed to vanish in region 2b ($\beta = 0°$), there is no gravity loss there. Also, the impact of drag is zero because the second stage of the Shuttle's flight trajectory occurs at such a high altitude that the air density is very, very low. As a result the dynamic pressure, q, is quite low, even though the Shuttle's velocity is very high, $v > 3,000$ m/s. Thus, the equation for the time-dependent velocity in region 2b is given by the "ideal" velocity expression (Equation 3.6) combined with the final Δv increments previously calculated for regions 1 and 2a (the added constants of integration), i.e.

$$v_{\text{region 2b}}(t) = -C_2 \ln\left[1 - \left(\frac{\dot{M}_{e_{2b}}}{m_{i_{2b}}}\right)(t - 300)\right] + (\Delta v_{\text{region 1}}) + (\Delta v_{\text{region 2a}}) \quad (3.42)$$

We use the following Region 2b parameters in evaluating Equation 3.42

- Flow rate $\dot{M}_{e_3} = \dot{M}_{e_2} = 1{,}474$ kg/s
- Initial mass $m_{i_{2b}} = m_{\text{final}_{2a}} = [m_{i_{2a}} - \dot{M}_{e_{2a}}(300 - 110)]$
  $= 727{,}258 - (1{,}474 \text{ kg/s})(190 \text{ s}) = 447{,}198$ kg
- $(\Delta v_{\text{region 2a}}) = $ ideal velocity increment for region 2a [Equation 3.40, $t = 300$ s]
  $+$ "gravity loss" for region 2a [Equation 3.39, $t = 300$ s]
  $= 2{,}159$ m/s $- 477$ m/s $= 1{,}682$ m/s

- $(\Delta v_{\text{region 1,Eq. 3.26}}) + (\Delta v_{\text{region 2a}}) = 1{,}382 \text{ m/s} + 1{,}682 \text{ m/s} = 3064 \text{ m/s}$
  - The velocity at the end of region 2a is the sum of the two Δv's for regions 1 and 2a. Both Δv's include the effects of gravity loss.
  - Note that this combined Δv of 3,064 m/s is plotted in Figure 3.12 as the final region 2a velocity; graphically it is the velocity point at the far right end of the red curve (at $t = 300$ s).

Equation 3.42, the expression for the "ideal velocity" for *region 2b*, is plotted as the yellow curve in Figure 3.12. The final velocity for region 2b is at MECO, $t \approx 495$ s.

Footnote   The time to MECO is calculated from the following expression

$$t_{\text{MECO}} = \text{second stage burnout time } interval + \text{first stage burnout time}$$

The full second stage burnout time *interval* (comprised of regions 2a + 2b) is readily estimated from Equation 3.7a using the values of liquid propellant mass at the start of the second stage (Table 3.1) and the total SSME flow-rate given by Equation 3.41b, i.e.

$$t_{\text{MECO}} = \frac{m_{\text{2nd stage propellant}}}{\dot{m}_{\text{3SSMEs}}} + t_{\text{1st stage burnout}} = \frac{568{,}000 \text{ kg}}{1474 \text{ kg/s}} + 110 \text{ s} = 495.3 \text{ s}$$

The final MECO time is thus approximately 495 s.

3.10.4 Combined velocity solution for regions 1, 2a, and 2b and v(MECO)

The combined piece-wise velocity function encompassing all three model regions is plotted in Figure 3.12. Note the change in Shuttle acceleration (i.e. a change in dv/dt) at the time of first stage (SRB) burnout, when the SRBs are jettisoned. The final velocity at $t = 495$ s is 7,645 m/s ≈ 7.65 km/s. This is the velocity at the end of the second stage burnout (or MECO). Accounting for a drag loss of 0.12 km/s in the first stage of flight, the net drag-corrected velocity at MECO is estimated to be

$$v(\text{MECO}) \approx 7.53 \text{ km/s}$$

As we will show in Section 3.13, where we review the equations of orbital mechanics, a velocity of 7.53 km/s is insufficient for the Shuttle to maintain a minimum 115 km altitude circular orbit; a velocity of 7.84 km/s is required.

However, if we include the earth's *rotational* velocity in the Shuttle's overall velocity budget, assuming an *eastward launch* at a latitude of 28.5° (Kennedy Space Center)

$$\Delta v_{\text{rotational}} = \frac{2\pi r_e}{(24 \text{ hr})(3{,}600 \text{ s/hr})} \cos(\text{latitude}) = (0.464 \text{ km/s}) \cos(28.5°) = 0.408 \text{ km/s}$$

126 Estimating Shuttle launch, orbit, and payload magnitudes   [Ch. 3

then our model Shuttle's final burnout velocity, in an inertial framework, will be increased to

v(final including: gravity, drag, and earth's rotation) $\approx 7.53 + 0.41 = 7.94$ km/s

which is more than adequate for a 115 km circular orbit.

A table summarizing the Shuttle's overall spacecraft velocity (or Δv) budget for the first and second stages, based on our Back-of-the-Envelope models, is presented in the next section.

3.11 SUMMARY OF PREDICTED Δv BUDGET FOR THE SHUTTLE

Table 3.4 summarizes the incremental velocity increases for each of the Shuttle's two stages and for the entire launch sequence. In this "Δv budget" we break out the losses due to gravity and drag and include the *gain* in inertial velocity due to the earth's rotation. We combine together all of these Δv elements to yield the total vehicle Δv for the Shuttle at second stage burnout. This total vehicle Δv is then compared to the velocity "requirements" for two different orbital missions.

To finely-tune this Table 3.4 Δv budget, we should also include other possible velocity losses (e.g. steering losses, where the Shuttle thrust vector is not aligned with the flight path) and consider any likely Δv propulsion *gains* by firing the low-thrust on-board maneuvering propulsion system prior to final orbit insertion (Note the comment and sketch related to the OMS burn, in Figure 3.2). Such additional Δv components are not included here in our first cut at the summary Δv budget for the Shuttle.

Up until this point in our estimation effort, all of our Δv calculations are based on an assumed Shuttle payload of 25,000 kg (Table 3.1). If we redo our Δv (total) calculations for a 20% higher Shuttle payload, say 30,000 kg, the total burnout velocity v(MECO) would drop by about 0.11 km/s to approximately 7.82 km/s. To more precisely estimate the payload that can be placed in a 115 km circular orbit, we must iterate on the magnitude of that assumed payload in our calculations for both Shuttle stages, until we arrive at a calculated final Shuttle inertial velocity that exactly matches the requirement for a 115 km circular orbit, namely that the final burnout velocity be equal to 7.84 km/s (i.e. the figure of 7,835 m/s given in Table 3.4).

Section 3.14 presents a more general payload vs orbital altitude analysis. Such an analysis makes it possible to determine the payload that the Shuttle can safely lift to the orbit for any prescribed Shuttle mission (e.g. to a 610 km orbit, with a 28.5° inclination, typical of a Hubble Telescope repair mission).

However, before initiating such a payload analysis we must first assess the accuracy of our model for the maximum Shuttle burnout speed by comparing its velocity predictions to the best available "data" published by NASA for the full 8 minute launch period.

Sec. 3.11] **Summary of predicted Δv budget for the Shuttle** 127

Table 3.4. Shuttle velocity budget, by stage and region, for a 25,000 kg payload. It contrasts two example orbital Δv requirements for orbits of 115 km and 610 km (see Section 3.13).

	First stage Region 1	Second stage Region 2a	Second stage Region 2b	Total, two stages and three regions
Ideal velocity Δv, m/s (25,000 kg payload)	2,274	2,159	4,581	9,014
Gravity Losses, m/s	−892	−477	0	−1,369
Drag Loss, m/s	−120	0	0	−120
Total Δv overall for non-rotating earth, m/s		6,740		7,525
Added Δv due to rotation of earth at 28.5° latitude, for an east-ward launch, m/s				408
TOTAL overall Δv at second stage burnout for a rotating earth, m/s				7,933
Requirements from orbital mechanics; 115 km circular orbit				7,835
Excess Δv of launch burnout velocity relative to 115 km circular orbital velocity requirement				+98 Can add extra payload to 25,000 kg
Requirements from orbital mechanics; 115 km circular orbit + Hohmann transfer to a 610 km circular orbit				7,978
Excess Δv from launch relative to 610 km orbital requirement				−45 Need to reduce payload from 25,000 kg

3.12 COMPARISON OF BACK-OF-THE-ENVELOPE MODELED SHUTTLE VELOCITY AND ALTITUDE AS A FUNCTION OF TIME TO NASA'S NUMERICAL PREDICTION FOR ALL STAGES

Previously we tested our ability to predict velocity as a function of time (including gravity losses) for the first stage of flight, as shown in Figure 3.9. In that figure our estimated first stage velocity vs time curve was compared with NASA's time-dependent Shuttle velocity data. These data were derived from Mach number measurements obtained during flight STS-1. We observed that estimated drag losses of about 120 m/s account for a good part of the difference between our model calculations and the NASA measurements in the region of high dynamic pressure.

In a search of the available Shuttle launch literature we were unable to find any *recorded* flight velocity vs time data, taken during the second stage flight period. However, we did find an Excel spreadsheet of *predicted* Shuttle flight velocities, and other flight variables such as altitude vs time t. These velocities and altitudes were calculated using NASA's large-scale flight-simulation codes and output at prescribed time intervals. Several of these numerical-output flight-prediction sets, made available to reporters and the public just prior to recent Shuttle launches, were compiled by William Harwood, a reporter affiliated with CBS News [31], and are available on the website spaceflightnow.com.

3.12.1 Comparison of model velocity with NASA's numerical prediction

A comparison of our modeled velocity flight history, neglecting drag (the pink curve), with the NASA numerically-predicted flight velocities for the STS-119 mission to the ISS (blue curve) is shown in Figure 3.13.

Overall, our simple analytic model agrees quite well with the NASA numerical predictions in both shape and magnitude for both the first and stages. When we look in more detail, we observe that our model over-predicts velocities by about 100 m/s near the time of first stage burnout. This is consistent with the neglect of the drag loss in our approximate time-dependent model. The acceleration seems to be accurately modeled by our analytic solution throughout most of the second stage of flight. Our predicted second stage burnout time of 495 s is sooner than NASA's calculated burnout time (514 s), and our final model estimated velocity is faster by about 54 m/s. When we subtract our drag loss estimate of 120 m/s from our final result, we find that our model's final velocity estimate (7,513 m/s) is 66 m/s less than the final burnout velocity predicted by NASA of 7,579 m/s, which is an error of less than 1%. Not bad!

NASA's predicted velocity curve and our three-region model for earth-relative velocities in Figure 3.13 do not incorporate the effect of the earth's rotation at the latitude of the Kennedy Space Center, which, as noted above, is 408 m/s at that latitude. For an eastward launch, this rotational velocity must be added to both curves in Figure 3.13 in order to transform the earth-relative velocity values (the ordinate values) into "inertial" velocities. (For a launch not oriented due east, as was

Sec. 3.12]   Comparison of Back-of-the-Envelope modeled Shuttle velocity   129

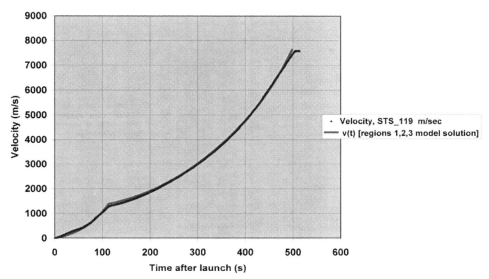

Figure 3.13. Comparison of our modeled three-region model velocity flight history, pink curve, with the NASA numerically-predicted flight velocities relative to earth, blue points, calculated by NASA for the conditions set up for flight STS-119. The predicted flight output velocities were compiled by CBS reporter William Harwood in a spreadsheet named: *SpaceCalc* [31].

the case for STS-119, the velocity correction diminishes approximately as the cosine of the launch azimuth relative to the eastward direction). The inertial velocity magnitudes and directions are needed to initialize any orbital mechanics calculation in order to be able to solve for the primary elliptical orbital properties like apogee and perigee and to assess whether a given burnout velocity can produce a viable elliptical or circular orbit.

3.12.2 Comparison of model altitude with NASA's numerical prediction

In addition to a comparison of velocities between our model and NASA's data, we can also compare a BotE calculated Shuttle *altitude* ascent profile with the altitudes output from NASA's predictive code.

Figure 3.14 shows our modeled height as a function of time, $h(t)$, calculated from our modeled velocities. Note that our maximum calculated altitude is ≈ 125 km.

Our solution for time varying height, $h(t)$, was calculated from the integral of the *vertical* velocity component, $v(t) \sin \beta(t)$ over time (Equation 3.25), i.e.

$$h(t) = \int_0^t v(\tau) \sin \beta(\tau) \, d\tau \qquad (3.43)$$

Due to the significant algebraic complexity of the functional form for $v(t)$ in our gravity-loss corrected velocities, we chose to numerically evaluate the integral in

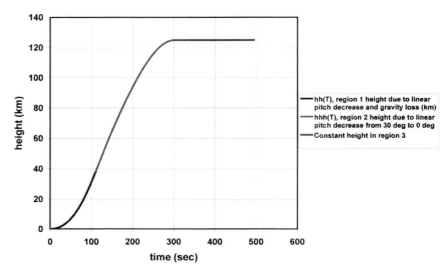

Figure 3.14. Shuttle altitude as a function of time evaluated from the gravity-loss model for region 1 (blue curve) and region 2a (red curve). Since pitch and gravity loss are zero in region 2b, h(t) is a constant in that region (green curve).

Equation 3.43 using a standard-numerical-quadrature algorithm, rather than seek an analytic form for the height integral.

Figure 3.15 compares our calculated model altitude history (as in Figure 3.14) to NASA's numerically-predicted time-dependent altitude output. NASA tailored its results to predict the ascent of STS-119. Our model-predicted altitude curve reaches a maximum of 125 km at exactly $t = 300$ s. This is a direct result of our assumption that pitch angle should reach zero at this time. And because we have set $\beta = 0$ throughout region 2b, our region 2b model solution yields altitudes that remain constant at the region 2a maximum. NASA's prediction indicates that the altitude drops to a local minimum of 103 km at 450 s after reaching a maximum of 109 km at $t = 340$ s. Thereafter, NASA's time-dependent altitude increases to 105 km just at second stage burnout. It is evident that our BotE model does not capture this near-orbital undulating altitude behavior. It is also clear, based on our simple pitch model that our calculated altitudes depend upon the particular time scale we chose for β to drop to zero.

3.12.3 Modeled altitude sensitivity to pitch time scale

We briefly assess here the impact on our altitude solution of recalculating $h(t)$ using a revised "time to zero pitch" time scale, Δt_{max}. For illustrative purposes we use a time scale reduced from its nominal value of 300 s down to 250 s. Our quadrature results for this time scale parameter are depicted in Figure 3.16 and compared both to our 300 s scale solution and to NASA's altitude predictions. When the shorter time scale

Sec. 3.12]   Comparison of Back-of-the-Envelope modeled Shuttle velocity   131

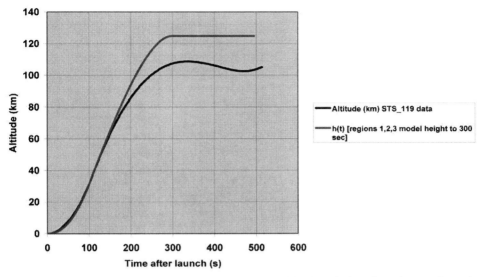

Figure 3.15. Comparison of our three-region model altitude solution (purple curve) to the NASA numerically predicted flight altitude curve (blue curve) for flight STS-119 [31].

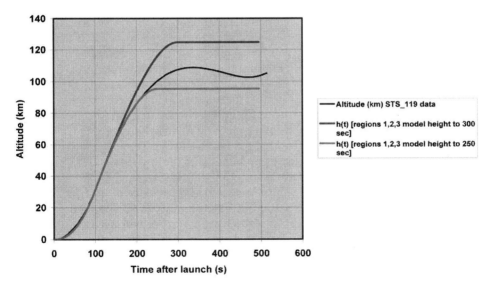

Figure 3.16. Comparison of two versions of our three-region model altitude solution in comparison to the NASA numerically predicted flight curve (blue curve) for flight STS-119 [31]. The purple curve is our model altitude solution when the "time for β to drop to zero", Δt_{max}, is set to 300 s. The green curve is the altitude solution when Δt_{max} is set to a smaller estimate of 250 s. This plot shows the sensitivity of our altitude solution to the selected input second stage pitch time-scale.

of 250 s is used, our altitude solution (green curve) achieves a smaller maximum, as expected; about 95 km instead of the longer time scale value of 125 km. So we see that our maximum *altitude* prediction is fairly sensitive to the selection of the Region 2a time scale. We also note that our two semi-analytical results presented here do reasonably "bracket" the NASA numerical predictions for altitude.

In contrast, our predicted model altitude at the time of *first-stage burnout* is, as expected, totally insensitive to the choice of a second stage zero-pitch time scale. Our first stage burnout altitude of about 38 km is in good agreement with the corresponding NASA altitude prediction at a burnout time $t \approx 110$ s.

3.13 ESTIMATING MISSION ORBITAL VELOCITY REQUIREMENTS FOR THE SHUTTLE

For a successful delivery of the Shuttle payload to orbit we need to demonstrate that the final Shuttle velocity, calculated at second stage burnout, is sufficient to place a given payload into a high speed orbit around the earth.

In general, the achievable burnout velocity for any rocket system must be at least equal to the velocity required by the laws of orbital mechanics for a given satellite mission.

In the two-part section to follow, we adopt a basic set of orbital mechanics equations in order to calculate the *orbital velocity requirements* for a variety of missions for which the orbital geometric parameters (e.g. apogee, perigee, and orbital inclination) are known input parameters. One can also solve the inverse problem and calculate the Shuttle's resulting *orbit and time-dependent position* from its altitude, velocity, and orientation at the time of burnout, but we will not present a detailed solution to this problem here.

In part two of this section, we estimate the additional impulsive velocity changes, Δv, required to move the Shuttle from an initial circular parking orbit, say at an altitude of 110 km, to another non-intersecting circular orbit at a greater altitude, such as at the altitude ≈ 610 km required for Hubble Space Telescope missions. This orbital transfer maneuver is initiated by an impulse of the low-thrust OMS engines in a manner calculated to put the Orbiter into an elliptical transfer path between the two required orbits. Another OMS burn, executed at the apogee of the transfer orbit, increases the Orbiter's speed to "settle-out" at the altitude of the higher circular orbit. To analyze these more complex orbital transfers we will need to review the classical mathematical equations and the basic solutions developed for *elliptical* orbital mechanics problems (see Section 3.13.2). But first let's consider simple circular orbits.

3.13.1 Part 1: circular orbital velocity

In the first part of this section, we consider the following problem: Calculate the velocity of a spacecraft traveling in a steady circular orbit about the earth, such as a Shuttle in its initial "parking" orbit.

Sec. 3.13] Estimating mission orbital velocity requirements for the Shuttle 133

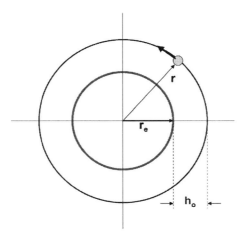

Figure 3.17. Point mass in circular orbit at altitude h_0 above the earth (green circle). The orbital radius $r = r_e + h_0$.

Consider a point mass orbiting, at altitude h_0, above a "green" earth (with radius $r_e = 6{,}378$ km) as depicted in Figure 3.17. Assume that the point mass (our model spacecraft) moves at constant velocity v along a circular path with radius $r = r_e + h_0$ measured from the center of the earth. We assume that this circular motion is solely driven by the gravitational force that the earth exerts on our point satellite having a mass m, where $m \ll m_{\text{earth}}$. The gravitational force between these two bodies, F_{gravity}, is directed *inward* towards the center of the earth. From the "law of universal gravitational attraction" [11, p. 61]

$$F_{\text{gravity}} = \frac{m(Gm_e)}{r^2} = \frac{m\mu_e}{r^2} \tag{3.44}$$

where $G \equiv$ universal constant of gravitation $= 6.670 \times 10^{-11} \frac{\text{m}^3}{(\text{kg})\text{s}^2}$.

Since the mass of the earth ($m_e \approx 6 \times 10^{24}$ kg), the "gravitational" parameter μ_e is $\mu_e \equiv Gm_e = 3.986 \times 10^5 \text{ km}^3/\text{s}^2$.

From elementary classical mechanics, a radially-inward *centripetal* acceleration, $a_{\text{centripetal}}$, arises from the rate of change of the position vector of a body's motion along a prescribed curved path. For circular motion, that vector's change is measured solely by the product of the velocity magnitude, v, and the change in the body's angular orientation, θ, over the incremental time dt, i.e. $v\, d\theta$. The differential vector component, with magnitude $v\, d\theta$, points towards the center of the circle. That change in velocity per unit time over the time interval dt is the radial inward *centripetal* acceleration. Hence $a_{\text{centripetal}}$ is

$$a_{\text{centripetal}}(\text{radially inward}) = v\frac{d\theta}{dt} = \frac{v^2}{r} \tag{3.45}$$

where v^2/r is the acceleration of a body moving at constant speed, v, along a circle with radius of curvature, r. The acceleration vector *tangential* to the circle, the tangential acceleration, vanishes for the case of constant circular velocity.

The inward-directed *force* on a body for that centripetal acceleration is, from Newton's second law of motion

$$F_{\text{centripetal}} = m\frac{v^2}{r} \qquad (3.46)$$

Equating this centripetal force to the gravitational force in Equation 3.44 allows us to solve for the orbital velocity v of a point mass as a function of radius r from the center of the earth

$$\text{orbital velocity} = v = \sqrt{\frac{\mu_e}{r}} = \sqrt{\frac{\mu_e}{r_e + h_0}} \qquad (3.47)$$

For a Shuttle circular "parking" orbit with altitude $h_0 = 115$ km, the magnitude of the orbital velocity (Equation 3.47) is

$$v = \sqrt{\frac{\mu_e}{r}} = 7.835 \text{ km/s} = 7{,}835 \text{ m/s} \qquad (3.48)$$

Note for a burnout height $h_0 = 115$ km, $r = h_0 + r_e = 6{,}493$ km.

3.13.1.1 *Orbital period*

The corresponding time to complete a single circular orbit, the orbital period τ_c, is simply the orbital circumference of a circle with radius r_c, divided by the velocity

$$\tau_c = \frac{2\pi r_c}{v} = 2\pi\sqrt{\frac{r_c^3}{\mu_e}} = 5{,}207 \text{ s} = 86.8 \text{ min} \qquad (3.49)$$

Typical low earth orbits have periods of the order of 90 minutes, consistent with the value calculated here.

Equation 3.49 is a special form of Kepler's third law, which formally states "The square of the period of any planet about the sun is proportional to the cube of the planet's mean distance from the sun". Of course it also applies to objects orbiting around the earth.

For elliptical orbits, the length of the major axis, designated as "2a", becomes the characteristic length scale. When the vertical and horizontal length scales of the ellipse are equal, the radius of the resulting circle is simply equal to "a", the length of the "semi-major axis" of an ellipse whose major and minor axes are the same length. It can be shown that the period for an *elliptical* orbital is identical in form to Equation 3.49 [11, p. 65]. To obtain the exact expression for the period of an elliptical orbit, one need merely replace r_c in Equation 3.49 with the semi-major axis length "*a*" for the corresponding elliptical orbit. For the problems we are studying here, the length "*a*" is equal to the mean distance of a satellite from the center of the earth.

The resulting expression for the period of an elliptical orbit is

$$\tau_{\text{ellipse}} = 2\pi\sqrt{\frac{a^3}{\mu_e}} \tag{3.49a}$$

3.13.2 Part 2: elliptical orbits and the Hohmann transfer Δv's

In the second part of our orbital analysis, we calculate the Orbiter's impulsive velocity change (using the OMS engines) that is required for a planar orbital transfer along an *elliptical* path from a 115 km parking orbit, corresponding to an initial perigee radius r_p, to a higher orbital altitude corresponding to an apogee radius, r_a, as shown in Figure 3.18. This orbital transfer is named after Walter Hohmann who first described the maneuver in 1925. Hohmann further showed that this two-maneuver transfer is the most energy efficient means of changing from one orbit to another.

This orbital transfer maneuver is initiated by a short burst of the low-thrust OMS engines that forces the Shuttle to follow an elliptical transfer path between the two orbits. To perform this calculation we need the equations of motion, for two *elliptical* Keplerian orbits in a common plane, which generalizes the circular motion analysis previously presented. The two major equations supporting our elliptical solution are: (1) the conservation of energy in a gravitational field, and (2) the conservation of angular momentum.

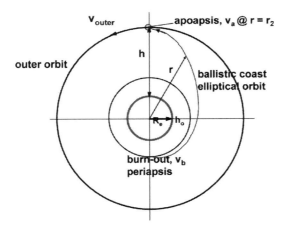

Figure 3.18. Hohmann transfer orbit; between inner orbit, $r = r_p = (r_e + h_0)$, and an outer orbit, $r = r_a = (r_e + h)$. Ballistic coast from perigee to apogee, after a first burn at point p of the inner parking orbit that raises the velocity by Δv_1. A second burn at the apogee of the transfer orbit, which increases the velocity by Δv_2 is used to circularize the spacecraft at radius r_2 (sketch based on [11, p. 149]).

3.13.2.1 Conservation of energy

For the motion of a point mass at speed v in a gravitational field, the equations of motion can be integrated to give the following "Conservation of Energy Equation" (as per Figure 3.18 and [11, p. 63])

$$E_{\text{total}} = \text{Kinetic energy}_{\text{per unit mass}} + \text{Potential energy}_{\text{per unit mass}}$$

$$= v^2/2 - \frac{\mu}{r} \equiv \text{constant} = -\frac{\mu}{(r_p + r_a)} \quad (3.50)$$

where

$$\boxed{\begin{array}{l} r_p = \text{inner orbit radius (perigee of ellipse)} = r_e + h_0 \\ r_a = \text{outer orbit radius (apogee of ellipse)} = r_e + h \end{array}}$$

3.13.2.2 Conservation of angular momentum, H [11, page 62]

$$H = r \cdot (v_\theta) = r^2 \cdot (d\theta/dt) = \text{constant} \quad (3.51)$$

where the angle and the transverse velocity are defined as

- θ is the polar angle of the ellipse; measured in the direction of motion from the perigee direction to the vector pointing toward the spacecraft
- v_θ is the component of the spacecraft velocity vector in the direction $\perp$ to the radial vector, i.e. in the "transverse direction"

At perigee and apogee the angular momentum H assumes a particularly simple form

- At perigee: $\theta = 0$; $\vec{v}_p$ is solely a transverse vector; and $H = r_p \cdot v_p$
- At apogee: $\theta = \pi$; $\vec{v}_a$ is solely a transverse vector; and $H = r_a \cdot v_a$

Since it can be shown that angular momentum is constant for this problem (i.e. $H = $ constant), then it follows that

$$r_a \cdot v_a = r_p \cdot v_p$$

or

$$\boxed{v_a/v_p = r_p/r_a} \quad (3.52)$$

We now determine an equation for the initial (post-burn) spacecraft velocity, v_p, required to establish a dynamically viable elliptical-transfer-orbit between the prescribed perigee and apogee positions. The conservation of energy equation (Equation 3.50) is used to calculate an expression for v_p that is consistent with the specified geometry parameters in Figure 3.18. In particular, we define the initial transfer-orbit position of the spacecraft to be at the perigee of the orbit where the radial distance (measured from the center of the earth) is r_p. At apogee the location of the spacecraft will be at the radial distance r_a.

3.13.2.3 Velocity at perigee

The velocity at perigee, v_p, is the solution of Equation 3.50 with $r = r_p$

$$v_p^2/2 - \frac{\mu}{r_p} = -\frac{\mu}{(r_p + r_a)} \tag{3.53}$$

Solving Equation 3.53 for v_p

$$v_p = \sqrt{2\mu\left(\frac{r_a}{r_p(r_p + r_a)}\right)} = \sqrt{\frac{\mu}{r_p}}\sqrt{\left(\frac{2}{(1 + r_p/r_a)}\right)} \tag{3.54}$$

In Equation 3.54, we note that the term $\sqrt{\frac{\mu}{r_p}}$ is the *circular* orbit velocity for the initial parking orbit. The second square root expression on the right of Equation 3.54 is a simple function of the ratio of the perigee and apogee distances, r_p/r_a. Since $r_p/r_a < 1$, we see that our initial transfer velocity v_p must be *greater* than $v_{\text{parking-circular}}$. The required increase in velocity from $v_{\text{parking-circular}}$ needed to send the Shuttle outward on its elliptical transfer orbit is then given by

$$\Delta v_1 = v_p - v_{\text{circular}; r_p} = \sqrt{\frac{\mu}{r_p}}\left(\sqrt{\left(\frac{2}{(1 + r_p/r_a)}\right)} - 1\right) \tag{3.55}$$

3.13.2.4 Velocity at apogee

The velocity at the apogee of the ellipse, v_a, shown in Figure 3.18, can also be obtained from the energy equation, but now with $r = r_a$

$$v_a^2/2 - \frac{\mu}{r_a} = -\frac{\mu}{(r_p + r_a)} \tag{3.56}$$

Thus

$$v_a = \sqrt{2\mu\left(\frac{r_p}{r_a(r_p + r_a)}\right)} = \sqrt{\frac{\mu}{r_a}}\sqrt{\left(\frac{2}{(1 + r_a/r_p)}\right)} \tag{3.57}$$

Note that Equation 3.57 is consistent with the simple conservation of angular momentum expression derived in Equation 3.52; namely, $v_a = v_p(r_p/r_a)$.

From Equation 3.57 we also observe that since $r_a/r_p > 1$, it is clear that the apogee velocity v_a must be less than the outer orbit circular velocity, $v_{\text{outer-circular}} = \sqrt{\frac{\mu}{r_a}}$. A second burn is required to increase the apogee velocity of the spacecraft to $v_{\text{outer-circular}}$. The equation for this increment, Δv_2, is

$$\Delta v_2 = v_{\text{circular; } r_a} - v_a = \sqrt{\frac{\mu}{r_a}}\left(1 - \sqrt{\left(\frac{2}{(1 + r_a/r_p)}\right)}\right) \quad (3.58)$$

3.13.3 Numerical values for transfer orbit Δv's

As a numerical example, let's calculate the magnitude of the two Δv's required to perform a minimum-energy Hohmann transfer of the Orbiter from a 115 km parking orbit to the 610 km orbit typical of Hubble Space Telescope missions.

From Equation 3.55 the magnitude of Δv_1 for the initial burn is

$$\Delta v_1 = \sqrt{\frac{\mu}{r_p}}\left(\sqrt{\left(\frac{2}{(1 + r_p/r_a)}\right)} - 1\right)$$

$$= \sqrt{\frac{3.986 \times 10^5 \text{ km}^3/\text{s}^2}{(6{,}378 + 115) \text{ km}}}\left(\sqrt{\left(\frac{2}{(1 + (6{,}493)/6{,}988)}\right)} - 1\right)$$

Hence

$$\Delta v_1 = (7.835 \text{ km/s})(0.0182) = 142.6 \text{ m/s} \quad (3.59)$$

and

$$v_p = 7{,}835 \text{ m/s} + 143 \text{ m/s} = 7{,}978 \text{ m/s} \quad (3.60)$$

From Equation 3.58 the magnitude of Δv_2 for the circularization burn is

$$\Delta v_2 = \sqrt{\frac{\mu}{r_a}}\left(1 - \sqrt{\left(\frac{2}{(1 + r_a/r_p)}\right)}\right)$$

$$= \sqrt{\frac{3.986 \times 10^5 \text{ km}^3/\text{s}^2}{(6{,}378 + 610) \text{ km}}}\left(1 - \sqrt{\left(\frac{2}{(1 + (6{,}988)/6493)}\right)}\right)$$

$$\Delta v_2 = (7.553 \text{ km/s})(0.0185) = 140.0 \text{ m/s} \quad (3.61)$$

$$v_a = 7533 \text{ m/s} - 140 \text{ m/s} = 7413 \text{ m/s} \quad (3.62)$$

To assess our elliptical orbit results, the conservation of angular momentum gives

$$v_a = v_p(r_p/r_a) = 7{,}978(6{,}493/6{,}988) = 7{,}413 \text{ m/s}$$

which, as expected, is identical to the value of the apogee velocity obtained from Equation 3.62.

Sec. 3.13]   Estimating mission orbital velocity requirements for the Shuttle   139

After the second burn that circularizes the orbit at 610 km, we find that the final circular velocity is $v_{\text{circular; } r_a} = \sqrt{\mu/r_a} = 7{,}553$ m/s. This, of course, is less than the orbital velocity at the lower parking orbit, in accord with the circular orbital velocity solution (Equation 3.47) where it has been shown that $v \sim r^{-1/2}$.

In carrying out the two Δv increases for the Hohmann transfer, it is clear that while we have reduced our spacecraft's kinetic energy we have also increased its potential energy. This transformation, in moving to an orbit farther from the earth, required the chemical energy supplied by the two Shuttle OMS burns in addition to the kinetic to potential energy exchange inherent in the transit from perigee to apogee.

The total Δv requirement for the Hohmann transfer is by definition the sum of the two individual Δv's

$$\Delta v_{\text{total}} = \Delta v_1 + \Delta v_2 = 142.6 + 140.0 = 282.6 \text{ m/s} \qquad (3.63)$$

3.13.4 Time of flight for a Hohmann transfer

Equation 3.49a is the equation used to calculate the full period of an elliptical orbit (i.e. a full circumnavigation of the earth). During a Hohmann transfer the satellite (or spacecraft) travels a distance that is exactly one half of the perimeter of an ellipse. Therefore the time of flight for the Hohmann transfer is, correspondingly, one half of that for a full period

$$\tau_{\text{Hohmann}} = \pi \sqrt{\frac{a^3}{\mu_e}} = \pi \sqrt{\frac{[(r_p + r_a)/2]^3}{\mu_e}} \qquad (3.64)$$

The calculated time of flight, for our example Hohmann transfer, is then

$$\tau_{\text{Hohmann}} = \pi \sqrt{\frac{[(r_p + r_a)/2]^3}{\mu_e}} = \pi \sqrt{\frac{[6{,}740.5 \text{ km}]^3}{3.986 \times 10^5 \text{ km}^3/\text{s}^2}} = 2{,}754 \text{ s} = 45.9 \text{ minutes}$$
$$(3.64a)$$

This time of flight, 45.9 minutes, is only about 6% longer than the equivalent half-period for a circular parking orbit with an altitude of 115 km.

3.13.5 Direct insertion to a final orbital altitude (without using a parking orbit)

For early Shuttle missions, the total Δv requirement to achieve the desired orbit was assigned entirely to the OMS engines onboard the Orbiter, after it achieved an initial parking orbit of approximately 115 km. But in later Shuttle missions, with the benefit of improved SSME engines and a lighter-weight external tank, NASA determined that it was possible for the SSME engines to deliver a slightly higher burnout velocity at MECO (during what we have been calling the "second stage") than the circular velocity of 7,835 m/s that we previously calculated. Let's call this higher second stage burnout velocity v_{direct}.

If, in fact, v_{direct} achieves exactly the perigee speed, 7,978 m/s according to Equation 3.60, with a zero pitch angle at an altitude of 115 km, then according to our calculations for Δv_1 the Orbiter would be at the correct speed for a *direct transfer* from a 115 km perigee to a 610 km apogee along an elliptical path, making a parking orbit redundant. All it would require to achieve this *"direct insertion"* maneuver (as NASA calls it) would be to increase the second stage burnout velocity by 142.6 m/s above the nominal 7,835 m/s for a circular orbit, and then upon reaching apogee the Orbiter would fire the OMS engines to impart an impulse Δv_2 of 143 m/s to circularize the elliptical orbit at the final altitude of 610 km.

The purpose of writing this orbital mechanics section was to set the burnout velocity requirements for the Shuttle. Henceforth, we assume that launches are capable of *direct insertion* to a higher orbital altitude, and so our second stage burnout velocity requirement will, from Equation 3.54, be

$$v_{\text{burnout, direct}} = \sqrt{\frac{\mu}{(r_e + h_0)}} \sqrt{\left(\frac{2}{[1 + (r_e + h_0)/(r_e + h)]}\right)} \quad (3.65)$$

where: h_0 = Shuttle burnout altitude; h = final outer orbit altitude.

This equation provides us with the expression we need for burnout velocity as a function of outer orbit altitude h; i.e. $v_{\text{burnout}}(h)$.

The additional Δv required to achieve outer-orbit circularization (Equation 3.58), a mission-specific requirement for the OMS engines, will be incorporated into a separate (and assumed launch independent) "OMS velocity budget".

3.14 A BACK-OF-THE-ENVELOPE MODEL TO DETERMINE SHUTTLE PAYLOAD AS A FUNCTION OF ORBIT ALTITUDE

In our Δv budget for each stage of a Shuttle ascent (Table 3.4) we demonstrated that our calculated second stage burnout velocity (with a nominal payload) is more than sufficient to place a given payload into a 115 km parking orbit. We observed that an achievable rocket burnout velocity should, at a minimum, be essentially equal to the velocity required by the laws of orbital mechanics for a given satellite mission. The low-thrust OMS engines and/or the reaction control thrusters are typically used to complete the Shuttle mission by providing low Δv final adjustments to the desired mission orbit. The major fraction of the overall mission Δv is clearly supplied by the first and second stage *ideal* velocity increases during the ascent. The calculations leading to Table 3.4's Δv budget entries and the second stage burnout velocity, were based on an assumed Shuttle payload of 25,000 kg (as prescribed in Table 3.1). Specifically, our calculation for a payload of 25,000 kg yields a burnout velocity v(MECO) of approximately 7.93 km/s (Table 3.4). For *direct* insertion into a 610 km orbit for a Hubble Space Telescope mission, the required burnout velocity is 7.978 km/s. Thus our initial calculation of burnout velocity of 7.93 km/s is about 0.048 km/s or 48 m/s short of that required for a 610 km orbit (Table 3.4).

Sec. 3.14] A Back-of-the-Envelope model to determine Shuttle payload 141

To more precisely estimate the payload that the Shuttle can deliver to a 610 km circular orbit, we need to iterate on the magnitude of that assumed payload. In the next section we present a more complete description of the mathematical foundation of this non-linear payload problem. The goal is to compute an "exact" numerical solution of our analytical payload model as a function of an arbitrary orbital altitude. Such an analysis facilitates determining the payload the Shuttle can safely lift to a given orbital altitude for any given mission. This model shows that the exact numerical solution for payload that can be delivered to a 610 km orbit is 23,430 kg.

Finally, in Section 3.14.2 we develop a simple closed-form linearized solution of this "exact" payload problem. This approximate solution is based on the obvious assumption that payload is a very small fraction of the takeoff mass.

3.14.1 Analytic model for payload as a function of orbital altitude

As previously shown, the Tsiolkovsky "ideal" rocket equation relates the velocity increase of any given rocket stage "i", $\Delta v_{\text{ideal: stage } i}$, to the reduced rocket mass $m_{\text{final},i}$ as a result of consuming the propellant for the "i"th stage of flight. For the "i"th stage it is:

$$\Delta v_{\text{ideal},i} = -C_{\text{avg},i} \ln\left(\frac{m_{\text{final},i}}{m_{\text{initial},i}}\right) = -C_i \ln\left(\frac{m_{\text{initial},i} - m_{\text{propellant},i}}{m_{\text{initial},i}}\right)$$

We assume in this equation that exit velocity C_i can be replaced by a constant average exit velocity for that stage. Note that we defined C_{avg} for the first stage in Section 3.7.2.

To achieve a given orbit, the final burnout velocity for all stages must be equal to that required by the laws of orbital mechanics for a given satellite mission. For example, if we specify the velocity at burnout that is required for direct-insertion transfer to a specified orbit, we can then deduce the maximum possible payload that can be carried. To properly initialize any orbital mechanics calculation, it is necessary to convert our final burnout velocity (measured relative to a rotating earth) into a corresponding velocity measured in an "inertial-space" coordinate system. Hence to calculate velocities in inertial space, a correction to our earth-fixed burnout velocity must be made that accounts for the rotational velocity of the earth. For example, for a due-east launch from the Kennedy Space Center, the final burnout velocity (relative to a rotating earth) must be increased by a velocity increment of 0.408 km/s. This added velocity increment is equal to the earth's rotational velocity at the equator (0.464 km/s) multiplied by the cosine of the latitude of the launch site (28.5°).

As we have shown in our predicted Δv budget for the Shuttle (Table 3.4), the rocket delivered "inertial" burnout velocity is equal to the total ideal velocity change for all stages $\sum_i \Delta v_{\text{ideal},i}$ at burnout after corrections for: (a) the gravity losses $\Delta v_{\text{gravity},i}$ for each stage, (b) the aerodynamic drag losses $\Delta v_{\text{drag},i}$ for each stage, and (c) the launch-geometry rotational velocity of the earth $\Delta v_{\text{earth-rot}}$. In algebraic form, this burnout velocity (defined as Δv_{final} in an inertial coordinate system) is given by

$$\Delta v_{\text{final}} = \sum_i \Delta v_{\text{ideal},i} - \sum_i \Delta v_{\text{gravity},i} - \sum_i \Delta v_{\text{drag},i} + \Delta v_{\text{earth-rot}} \quad (3.66)$$

Let's equate this final rocket-delivered velocity to the *required* orbital velocity for the mission given by the laws of orbital mechanics

$$\sum_i \Delta v_{\text{ideal},i} - \sum_i \Delta v_{\text{gravity},i} - \sum_i \Delta v_{\text{drag},i} + \Delta v_{\text{earth-rot}} = \Delta v_{\text{required}_{\text{orbital-mechanics}}} \quad (3.67)$$

We recall that the ideal velocity gain, $\sum_i \Delta v_{\text{ideal},i}$ is the only term in Equation 3.67 that depends *directly* on the initial loaded payload mass, $m_{\text{p/l}}$. By solving for $\sum_i \Delta v_{\text{ideal},i}$ we obtain a non-linear expression for payload, with Δv_{ideal} directly linked to payload or cargo mass ($m_{\text{p/l}}$) through the initial and final mass terms for each stage:

$$\boxed{\sum_i \Delta v_{\text{ideal},i_{\text{p/l}}} = \Delta v_{\text{required}_{\text{orbital-mechanics}}} + \sum_i \Delta v_{\text{gravity},i} + \sum_i \Delta v_{\text{drag},i} - \Delta v_{\text{earth-rot}}} \quad (3.68)$$

For our Shuttle problem:

$$\sum_i \Delta v_{\text{ideal},i_{\text{p/l}}} = \Delta v_{\text{ideal, 1st stage}_{\text{p/l}}} + \Delta v_{\text{ideal, 2nd stage}_{\text{p/l}}}$$

For the first stage, the Tsiolkovsky rocket equation for Δv_{ideal} (above) can be written as

$$\Delta v_{\text{ideal},1} = C_1 \ln\left(\frac{m_{\text{initial},1}}{m_{\text{initial},1} - m_{\text{propellant},1}}\right)$$

by setting C_{avg} to the equivalent exit velocity C_1 and converting the equation into an alternate form using the well-known logarithmic identity that *the logarithm of the reciprocal is the negative of the logarithm.*

We note that the initial mass in the above equation, $m_{\text{initial},1}$, is the sum of a number of structural, propellant, and payload components; i.e.

$m_{\text{initial},1}$ = structural mass (propellant casing/tanks, engines, Orbiter, etc)

+ propellant mass (liquid, solid propellants) + payload mass; $m_{\text{p/l}}$ (3.69)

Let's rewrite the *initial* first stage mass (Equation 3.69) as the sum of two terms: (i) the *initial* "known" combination of structural and propellant mass $\equiv M_{1a}$, and (ii) the "unknown" payload mass, $m_{\text{p/l}}$.
By this definition,

$$m_{\text{initial},1} = M_{1a} + m_{\text{p/l}} \quad (3.70)$$

Sec. 3.14]   **A Back-of-the-Envelope model to determine Shuttle payload**   143

From Equation 3.3b and Table 3.1 the magnitude of the *initial* combined structural and propellant mass is (subscript "*a*" denotes "initial"):

$$M_{1a} = 2{,}038{,}016 \text{ kg}$$

Similarly, we write the "final" first stage mass in terms of the *final* structural and propellant mass, M_{1b}, and payload as

$$m_{\text{final},1} = M_{1b} + m_{p/l} \tag{3.71}$$

where

$$M_{1b} = M_{1a} - m_{\text{propellant},1} = 2{,}038{,}016 - 1{,}170{,}000 = 868{,}016 \text{ kg}$$

based on Table 3.1 (subscript "b" denotes "final").

Substituting the defined mass components of Equations 3.70 and 3.71 therefore yields

$$\Delta v_{\text{ideal},1} = C_1 \ln\left(\frac{M_{1a} + m_{p/l}}{M_{1b} + m_{p/l}}\right) \tag{3.72}$$

For the second stage we adopt a similar expression for $\Delta v_{\text{ideal},2}$

$$\Delta v_{\text{ideal},2} = C_2 \ln\left(\frac{M_{2a} + m_{p/l}}{M_{2b} + m_{p/l}}\right) \tag{3.73}$$

The initial and final values of the second stage structural and propellant mass (Table 3.1) are:

$$M_{2a} = 702{,}258 \text{ kg}; \quad M_{2b} = 134{,}258 \text{ kg}$$

The first stage and second stage exit velocities (extracted from Table 3.2 and the corresponding specific impulses) are:

$$C_1 = C_{\text{avg}} = 2{,}717 \text{ m/s}; \quad C_2 = 4{,}439 \text{ m/s}$$

Substituting Equations 3.72 and 3.73 into Equation 3.68 and defining $\Delta v_{\text{req}}(h) \equiv$ the right hand side of Equation 3.68 that corrects $\Delta v_{\text{required}_{\text{orbital-mechanics}}}$ for all gravity and drag losses and adjusts for the earth's rotation (with $h =$ the orbital altitude), we obtain the following "Full analytical model equation for payload":

$$\boxed{C_1 \ln\left(\frac{M_{1a} + m_{p/l}}{M_{1b} + m_{p/l}}\right) + C_2 \ln\left(\frac{M_{2a} + m_{p/l}}{M_{2b} + m_{p/l}}\right) = \Delta v_{\text{req}}(h)} \tag{3.74}$$

Given a particular analytical model for $\Delta v_{\text{req}}(h)$, such as that derived for a direct Shuttle insertion to a Hohmann transfer orbit (Equation 3.65), we can solve

Equation 3.74 for the payload $m_{p/l}$ as a function of orbital altitude h. Note that by definition $\Delta v_{req}(h)$ is "offset" from $\Delta v_{required_{orbital-mechanics}}$ by the gravity loss, drag loss, and earth rotation terms of Table 3.4 as spelled out in Equation 3.68. For direct insertion we have the following specific expression

$$\Delta v_{req}(h) = \Delta v_{\text{direct-insertion}} + \sum_i \Delta v_{\text{gravity},i} + \sum_i \Delta v_{\text{drag},i} - \Delta v_{\text{earth-rot}}$$

$$= \Delta v_{\text{direct-insertion}} + (0.892 + 0.477) + (0.120 + 0.0) - (0.408)$$

$$= \sqrt{\frac{\mu}{(r_e + h_0)}} \sqrt{\left(\frac{2}{[1 + (r_e + h_0)/(r_e + h)]}\right)} + 1.081 \text{ km/s} \quad (3.75)$$

Since Equation 3.74 is a non-linear equation for payload mass, $m_{p/l}$ (given a specified Δv_{req}) possessing no analytical closed-form solution, we need to numerically *iterate* upon the unknown dependent variable $m_{p/l}$ to find a specific payload solution of Equation 3.74 for a given altitude h. Note that this numerical operation is equivalent to iteratively finding the root of the equation formed by subtracting the right side of Equation 3.74 from the left side; yielding: (left side − right side) $\equiv f(m_{p/l}) = 0$.

For low orbital altitudes, we take as our first guess $m_{p/l} = 25{,}000$ kg. We then calculate the left side of Equation 3.74 and compare it to the right hand side. If the calculated Δv for the left side is $< \Delta v_{req}$, then the next guess for $m_{p/l}$ must be less than 25,000 kg. If the calculated Δv for the left side is $> \Delta v_{req}$, the next guess for $m_{p/l}$ must be greater than 25,000 kg. This process is repeated until the solution is determined to an accuracy of say 0.01%. This approach is a preliminary version of a more general "bisection method" iteration procedure. It will converge rapidly once it has been established that a specific interval of the independent variable contains some root of the equation. This root-is-present determination is initially made by a numerical trial and error process (see [32, W. Press, "Numerical Recipes", chapter 9]).

After applying this iteration procedure, our "Full analytical model" solution for payload as a function of orbital altitude is presented graphically in Figure 3.19. Our analytical-model payload solution is displayed as the purple curve, where it is also compared to three Shuttle mission payloads at three orbital altitudes; the colored squares [33].

While our basic analytical payload solution agrees well with NASA's nominal mission data for orbital altitudes below 300 km, showing an estimated decline in payload with altitude of approximately 13 kg/km, this solution overestimates by about 7,000 kg the predicted payload for spacecraft servicing missions at altitudes of approximately 600 km.

Sec. 3.14] A Back-of-the-Envelope model to determine Shuttle payload   145

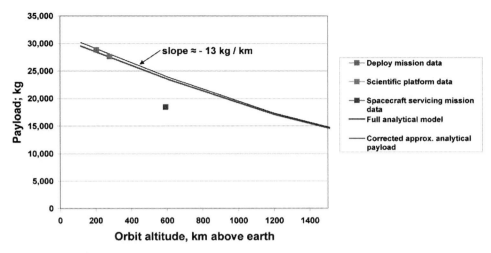

Figure 3.19. Shuttle payload as a function of orbit altitude for eastward launches at Kennedy Space Center. Approximate analytical solution (upper black curve); full numerical evaluation analytic model (lower purple curve). Model solutions of Equations 3.74 and 3.75 compared to nominal Shuttle mission data [33] for: (a) deployment mission, (b) scientific platform mission, and (c) spacecraft servicing mission (e.g. Hubble Space Telescope). Note that the nominal cargo payloads plotted apply to post-1992 flights which incorporated the lighter-weight external tank built under the auspices of the "weight reduction program" and other weight reduction improvements. These payload values average about 7,300 kg higher than pre-1992 flights for similar-type missions.

3.14.2 Approximate linearized solution for payload

While a numerical solution for payload can be obtained after sufficient iterations of Equation 3.74, it is often simpler, and more instructive, to obtain an approximate solution to the payload problem in closed-form.

Such an approximate linearized solution of Equation 3.74 is presented briefly here.

We determine an approximate linearized closed-form solution of the equation by assuming that the unknown payload, x ($= m_{p/l}$), is small compared to the initial Shuttle mass M_{1a}, i.e. $x/M_{1a} \ll 1$. To obtain a linearized solution of the problem we rewrite Equation 3.74 as:

$$\ln\left\{\left(\frac{M_{1a}+x}{M_{1b}+x}\right)^\gamma \left(\frac{M_{2a}+x}{M_{2b}+x}\right)\right\} = \frac{\Delta v_{\text{req}}}{C_2}; \quad \text{where } \gamma = \frac{C_1}{C_2} \qquad (3.76)$$

Taking the inverse logarithm of both sides, we obtain

$$\left(\frac{M_{1a}+x}{M_{1b}+x}\right)^\gamma \left(\frac{M_{2a}+x}{M_{2b}+x}\right) = \exp\left(\frac{\Delta v_{\text{req}}}{C_2}\right); \quad \text{where } \gamma = \frac{C_1}{C_2} \qquad (3.77)$$

Expanding the left side in a Taylor's series to first order using the normalized payload; $\varepsilon = x/M_{1a}$, and dropping terms of order ε^2, we arrive at the following linearized solution for the payload $x = m_{p/l}$

$$m_{p/l} = \frac{M_{2a} - KM_{2b}}{\left\{K - 1 + \gamma\left(K\dfrac{M_{2b}}{M_{1b}} - \dfrac{M_{2a}}{M_{1a}}\right)\right\}}; \quad \text{where } K = \left(\frac{M_{1b}}{M_{1a}}\right)^{\gamma} \exp\left(\frac{\Delta v_{req}}{C_2}\right)$$

(3.78)

The corrected "approximate analytical model" of Equation 3.78 is plotted as the lower thin black curve in Figure 3.19. Equation 3.78 is quite a good approximation to the "Full analytical model" iterated solution for payload given by Equation 3.74. In turn, both the exact and approximate solutions are in reasonable agreement with NASA's nominal mission payload data (the colored squares) but only at the lower orbital altitudes.

3.14.3 Reduction in useful cargo mass due to increases in OMS propellant mass

In this section we develop a simple "payload-reduction model" that corrects our 7,000 kg overestimate of the predicted payload for Shuttle missions at altitudes on the order of 600 km. This correction is related to the mass of the propellant loaded onto the OMS system.

There is an important Shuttle operational mission parameter that increases its impact on the true payload or useful cargo mass with increasing orbital altitude. This is the *OMS propellant mass* (m_p) loaded in different amounts as required to enable the Orbital Maneuvering System to maneuver the Orbiter into its mission-defined orbit and also to initiate the return of the Orbiter to the ground [34]. For missions at lower orbital altitudes (as we will show), the required *propellant* mass is small compared to the Orbiter's baseline 25,000 kg *payload* capability, but the OMS propellant mass becomes a significantly larger fraction at higher orbital altitudes. To offset this added mass there must be a net reduction in *true payload* or *useful cargo* mass that can be delivered to the required higher orbits. Analytical models and empirical representations of the OMS propellant mass are developed in Section 14.4 and used to modify the previous payload solution (Figure 3.19). In Figure 3.20 the OMS propellant-corrected payload estimates are represented by the red and aqua curves for orbital altitudes extending from 220 km to 610 km.

In developing specific analytical and empirical models to calculate the impact of OMS propellant on useful cargo mass, our modeling focuses on the two major maneuvering tasks performed by these engines: (a) *circularization* of the elliptical Hohmann transfer orbit at the apogee of the transfer orbit (requiring an *increase* in the velocity of the Orbiter, supplied by an "OMS-2" burn); and (b) *deorbiting* the vehicle from its near-circular mission orbit by an *OMS deorbit* velocity decrease which will send the Orbiter along a flight path that reenters the atmosphere. To accomplish the deorbit velocity decrease, the Orbiter is "turned around" to point

Sec. 3.14] A Back-of-the-Envelope model to determine Shuttle payload 147

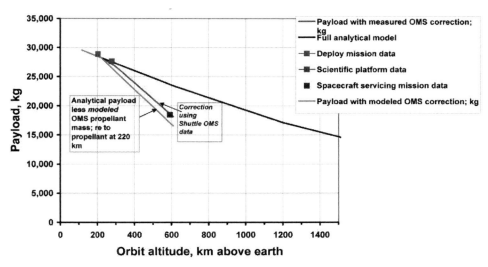

Figure 3.20. Shuttle payload as a function of orbit altitude. The "full analytic model" calculation for payload is compared to nominal Shuttle mission data. In addition, this "analytical payload" curve is corrected (or converted into a "useful payload" curve) by taking into account the added propellant that is loaded before launch into the Shuttle's OMS propellant tanks. This OMS propellant makes it possible for the Orbiter to perform specific maneuvers while in space. This "useful payload" correction is shown in two curves that drop below our "full analytic model" results (Section 3.14.4): (1) a "useful cargo" curve [aqua], based on a simple closed-form OMS orbital/propellant model, and (2) an empirical curve [red], based on the data for the pre-launch loaded OMS propellant mass tabulated by NASA for several of its Shuttle missions (Section 3.14.3).

its nose 180° from the near-circular orbital flight path, after which it is reoriented nose-first for reentry.

Figure 3.20 shows two propellant-mass-corrected "useful cargo" curves; one uses a simplified *analytic* model for the estimated OMS propellant mass and the other *empirically* calculates the payload correction at various altitudes using the actual OMS propellant data for selected missions published by NASA (Table 3.5). Both of our "useful cargo" curves compare well with actual Shuttle payloads for a deploy mission at 200 km, a scientific platform mission at 300 km, and a Hubble Space Telescope servicing mission at 600 to 610 km.

For our *empirical* payload correction model we use tabulated data published by NASA for a number of missions.

Table 3.5 provides NASA-tabulated OMS propellant data in two forms: (a) the total propellant loaded at Shuttle launch in kg, and (b) the total propellant actually used during specific missions (these data are also presented as a percentage of the loaded amount). There is a significant increase in the loaded propellant with mission orbital altitude. It is this loaded amount that determines the net reduction in useful

Table 3.5. OMS propellant mass based on NASA data for selected missions; (a) total propellant *loaded* in kg, (b) total propellant *used* in kg, with the used/loaded percentage [35, "Shuttle flight data and in-flight anomaly list" of April 1996].

Shuttle Missions at the given orbit altitude (STS-mission @ altitude in km)	OMS propellant summary report. JSC/EP2 Propulsion Branch Total propellant *loaded* MMH fuel + N_2O_4 oxidizer; kg	OMS propellant summary report. JSC/EP2 Propulsion Branch Total propellant *used* MMH fuel + N_2O_4 oxidizer; kg and (% of loaded)
STS-59 @ altitude = 224 km	5,957	4,131 (69%)
STS-41 @ altitude = 297 km	6,511	5,797 (89%)
STS-61 @ altitude = 574 km	11,265	10,455 (93%)

cargo. Any propellant "left-over" after performing the actual maneuvers during the mission is returned by the Orbiter to the ground following reentry. In principle, this left-over propellant results in an unnecessary reduction in useful payload that might have been delivered to orbit. It is clear, however, that some fraction of the loaded propellant (of order 10%) must be held in reserve in any mission as a safety margin for unforeseen exigencies.

3.14.4 OMS models for correcting cargo or payload mass

The aqua curve in Figure 3.20 is the propellant correction based on a simple model of the post-Hohmann-transfer circularization maneuver supplemented by an estimate of the Δv required for deorbiting the Shuttle. In order to simplify the calculation, the velocity change for the *deorbit* maneuver is crudely assumed to be equal in magnitude to the Δv for the OMS-2 circularization maneuver, i.e.

$$\Delta v_{\text{deorbit}} \approx \Delta v_2 \equiv \Delta v_{\text{oms-2}} \quad [\text{deorbit } \Delta v \text{ approximate by } \Delta v_2 \text{ of Equation 3.58}]$$

A detailed orbital mechanics analysis of the deorbit maneuver, which takes into account the flight path angle upon reentering the atmosphere, would yield a more accurate estimate for $\Delta v_{\text{deorbit}}$.

The red curve in Figure 3.20 is the corresponding effective payload (or useful cargo) where the payload correction is based on the tabulated OMS propellant mass data "loaded" at launch (Table 3.5). Details of the correction procedure are given below.

Sec. 3.14] **A Back-of-the-Envelope model to determine Shuttle payload** 149

All of the calculated payload corrections, Δ(payload) at the different orbital altitudes h, are (by definition) scaled with respect to the OMS propellant mass at 220 km (the lowest orbital altitude)

$$\Delta \text{ payload}(h) \equiv [\text{OMS propellant}(h) - \text{OMS propellant } (@\ h = 220 \text{ km})] \quad (3.79)$$

It is clear that the payload *correction* increment from Equation 3.79 is identically zero for a mission at the lowest nominal orbital altitude; $h = 220$ km.

One can determine useful cargo mass as a function of orbital altitude by subtracting Δ payload(h) from the full analytical payload model solution shown in Figure 3.20.

The following expression

$$[\text{useful cargo}](h) = [\text{full analytical payload}](h) - [\Delta \text{ payload}](h) \quad (3.80)$$

is used to construct the two different propellant-mass-corrected "useful cargo" curves shown in Figure 3.20. These two solutions differ depending on which of the following methods are employed to calculate the OMS propellant mass as a function of altitude, Δ payload(h): (a) our simple analytical OMS-propellant model (calculated directly from Equation 3.3 with Δv supplied by Equation 3.58], or (b) an empirical function fitted to the NASA-OMS propellant data specifically for the propellant "loaded" onto the Orbiter (Table 3.5).

3.14.4.1 *Figure of merit for useful cargo decrease with altitude*

A good "figure of merit" for useful cargo is the "slope" of the useful cargo curves, since it tells us how rapidly "useful cargo" mass *decreases* with increasing orbital altitude.

As we will show, the estimated average slope of our "modeled OMS-corrected payload or "useful cargo curve" (the aqua curve in Figure 3.20) is approximately

$$\Delta(\text{useful cargo})/\Delta h \approx -30 \text{ kg/km} \quad [\text{OMS model corrected "analytical payload"}] \quad (3.81)$$

The slope of the corresponding "NASA-data-corrected useful cargo curve" (the red curve) is somewhat less steep

$$\Delta(\text{useful cargo})/\Delta h \approx -26 \text{ kg/km} \quad [\text{NASA OMS data corrected "analytical" payload}] \quad (3.82)$$

There is a similar *payload rule of thumb* developed by NASA stating that the "useful" cargo mass drops about *100 lbs per orbital nautical mile increase*. When expressed in metric units, this is a decrease of 24.5 kg in delivered cargo mass per increased orbital kilometer (see [36], in particular the section entitled "Shuttle cargo, capability). Our full analytical model for how payload varies with orbital altitude (Equation 3.81) is within 20% of NASA's rule of thumb ≈ -25 kg/km.

We note from Figures 3.19 and 3.20 that while total Shuttle payload mass delivered to orbit (i.e. cargo plus OMS propellant mass) declines rather *slowly* with altitude (≈ -13 kg/km), the useful cargo mass falls off more *rapidly* (≈ -26 to -30 kg/km). This is so because a significant portion of the delivered payload is, of necessity, replaced with increasingly greater amounts of OMS propellant. This trade-off is particularly significant for those missions carried out at higher orbital altitudes. This loss in useful cargo mass is inevitable, since the additional OMS propellant is required to perform the orbital circularization and deorbit maneuvering tasks that must be carried out to successfully complete those higher altitude Shuttle missions.

3.14.4.2 Limiting orbital altitudes

At an altitude on the order of 700 km (call it $h_{\max}$), the required OMS propellant mass becomes so large that the required propellant volume would exceed the maximum volumetric capacity of the OMS fuel and oxidizer tanks. It is evident, therefore, that no Orbiter mission is possible beyond the orbital altitude $h_{\max}$.

3.14.5 Model for rate of change of "useful cargo" with altitude

The key Shuttle system and orbital variables that determine the OMS propellant requirements (and thus drive the decline in useful cargo mass with altitude) can be readily identified by constructing a simple "closed-form" solution for payload change as a function of altitude. The simple model developed in this section is used to estimate the useful payload rule of thumb (i.e. the "slope" of the useful payload versus orbital altitude curve) and to highlight the key system and orbital parameters that determine the magnitude of that slope. Such a model helps us directly estimate the drop-off in useful cargo with altitude *using only pencil and paper*.

Let's first determine an expression for the OMS propellant mass, $m_{\text{p-OMS}}$ for the circularization maneuver for a Shuttle. This maneuver transitions the Orbiter from the Hohmann transfer ellipse to a circular operating orbit. As illustrated in Figure 3.18, it is the impulsive velocity increase Δv_2 executed at the apogee of the ellipse. Once again we employ the Tsiolkovsky "ideal" rocket equation that relates the velocity increase of a rocket to the finite change in its mass due to consumption of the rocket's propellant

$$\Delta v_{\text{ideal}} = -C \ln \left(\frac{m_{\text{final}}}{m_{\text{initial}}} \right) = -C \ln \left(\frac{m_{\text{initial}} - m_{\text{propellant}}}{m_{\text{initial}}} \right)$$

Once we prescribe the magnitude of the velocity change, $\Delta v_{\text{ideal}} = \Delta v_{\text{OMS-2}}$, in addition to the rocket's initial mass, m_i, and the specific impulse of the OMS engine, we can readily solve this equation and calculate the mass of propellant that must be burned, m_p, in a short OMS engine impulse

$$\frac{m_{\text{propellant}}}{m_i} = 1 - \exp\left(\frac{-\Delta v_{\text{oms-2}}}{C}\right) \tag{3.83}$$

For the OMS engine burning nitrogen tetroxide and monomethyl hydrazine, the vacuum specific impulse for the this oxidizer/fuel combination is found in the rocket fuel literature to be $Isp = 313\,s$ [37]. The corresponding exit velocity, C, is $C = Isp \cdot g = 3{,}070\,m/s$.

3.14.5.1 Sample "exact" useful cargo calculation

For an OMS-2 circularization burn into a 610 km orbit from a Hohmann transfer ellipse the velocity increment, $\Delta v_{\text{oms-2}}$, is about 140 m/s. For an Orbiter mass of 100,000 kg, we see that this burn consumes a propellant mass of

$$m_p = (100{,}000\,\text{kg})\left(1 - \exp\left(\frac{-140.0\,\text{m/s}}{3{,}070\,\text{m/s}}\right)\right) = 4{,}456\,\text{kg}$$

[OMS-2 burn: 610 km circularization]

If we assume that the *deorbit* burn uses the same amount of propellant (our simple estimate) then the total propellant mass for both burns is

$$m_{\text{propellant: total}} = 2(4{,}456) = 8{,}912\,\text{kg} \quad [(\text{OMS-2})_{610\,\text{km}}\text{ burn} + \text{"deorbit" burn}]$$

For a lower baseline orbit of only 220 km the amount of propellant required for circularization of the orbit after a Hohmann transfer from a 115 km parking orbit is

$$m_{\text{propellant: total}} = 2(1{,}012\,\text{kg}) = 2{,}024\,\text{kg}$$

From Equation 3.79 the change in payload (or useful cargo) for a 610 km orbit, relative to 220 km, is: $= [8{,}912 - 2{,}024] = 6{,}888\,\text{kg}$. This 6,888 kg of OMS propellant mass is the reduction in useful cargo compared to the "full analytical payload" of 23,430 kg shown in Figure 3.20 for the 610 km orbit.

The resulting net useful cargo at the *610 km* orbital altitude is

$$[\text{useful cargo}](h) = [\text{full analytical payload}](h) - [\Delta\text{ payload}](h)$$
$$= 23{,}430\,\text{kg} - 6{,}888\,\text{kg} = 16{,}542\,\text{kg}$$

This value of *16,542 kg* is plotted at the lower right end of the "aqua" curve in Figure 3.20 at $h = 610$ km. If we subtract this useful cargo from that useful cargo calculated at *220 km* altitude ($\approx 28{,}256$ kg) then the net change in useful cargo is

$$\Delta(\text{useful cargo}) = 16{,}542 - 28{,}256 = -11{,}714\,\text{kg}.$$

The approximate slope of our cargo payload curve between the altitudes of 220 and 610 km is then

$$\Delta(\text{useful cargo})/\Delta h \approx (-11{,}714\,\text{kg})/(610\,\text{km} - 220\,\text{km}) = -30.0\,\text{kg/km}$$

This is the estimated cargo-reduction slope given above in Equation 3.81.

An even simpler Back-of-the-Envelope cargo loss model and "payload rule of thumb" calculation is presented in the following section.

3.14.6 Approximate analytic model for useful cargo

In Equation 3.83 the ratio $\Delta v_{\text{oms-2}}/C$ is quite small (≈ 0.046). We can then readily expand the exponential in this equation to first order $\exp(-x) \approx 1 - x$ for $x \ll 1$, in order to simplify the resulting mathematics. We thus approximate the propellant mass by

$$\frac{m_{\text{propellant}}}{m_{\text{i}}} \approx \left(\frac{\Delta v_{\text{oms-2}}}{C}\right) \qquad (3.84)$$

Note: Applying this approximate model to the previous OMS-2 610 km orbit circularization calculation problem, we obtain $m_{\text{p}} = 4{,}560$ kg. This differs by only 2.3% from the "exact" result (4,456 kg) calculated from Equation 3.83. Hence we conclude that Equation 3.71 yields a reasonably good linearized approximation for m_{p}.

We previously derived Equation 3.58 for the velocity change Δv_2 for a circularization maneuver performed at the apogee radius r_{a} in the form

$$\Delta v_2 = \sqrt{\frac{\mu}{r_{\text{a}}}}\left(1 - \sqrt{\left(\frac{2}{(1 + r_{\text{a}}/r_{\text{p}})}\right)}\right)$$

The apogee distance from the center of the earth is $r_{\text{a}} = r_{\text{earth}} + h$, where h is the altitude of the outer circular orbit. The corresponding perigee for the transfer ellipse is defined by: $r_{\text{p}} = r_{\text{earth}} + h_0$. Since the magnitude of the change in orbital height between the perigee and apogee of the Hohmann ellipse, $\Delta h = h - h_0$, is fairly small compared to the radius of the earth, $\dfrac{h - h_0}{r_{\text{earth}}} < 0.1$, we can derive an approximate solution for Δv_2 which simplifies the algebra of our mathematical solution. That is, assuming that $\varepsilon \equiv \dfrac{\Delta h}{r_{\text{p}}} \ll 1$, we can expand the solution for Δv_2 in Equation 3.58 in a Taylor's series to order ε and generate the following simple expression

$$\Delta v_2 \approx \sqrt{\frac{\mu}{r_{\text{p}}}}(\varepsilon/4) = \sqrt{\frac{\mu}{r_{\text{p}}}}\left(\frac{\Delta h}{4r_{\text{p}}}\right) = (v_{\text{circular@perigee}})\left(\frac{\Delta h}{4r_{\text{p}}}\right) \qquad (3.85)$$

The derivation of Equation 3.85 from Equation 3.58 is left as an exercise for the reader.

Substituting the circularization velocity change of Equation 3.85 into Equation 3.84 yields the following approximate solution for the required OMS-2 propellant mass

$$m_{\text{propellant:OMS-2}} = m_{\text{i}}\left(\frac{v_{\text{circular@perigee}}}{C}\right)\left(\frac{(h - h_0)}{4r_{\text{p}}}\right) \qquad (3.86)$$

We now crudely assume that the propellant mass required to deorbit the Orbiter is *equal* to the propellant mass that we calculated for the OMS-2 circularization burn (Equation 3.86). As a result of this estimate, the total propellant mass as a function of orbital altitude is equal to *twice* the mass given in Equation 3.86; i.e.

$$m_{\text{propellant:total}} \approx 2 \cdot m_{\text{i}} \left(\frac{v_{\text{circular@perigee}}}{C} \right) \left(\frac{(h - h_0)}{4 r_{\text{p}}} \right) \quad (3.87)$$

We now assume that the actual payload at each orbital altitude is reduced by the OMS propellant mass required to circularize the orbit at that altitude and subsequently deorbit. We now use the following expression (derived from Equation 3.80) in order to calculate the "useful cargo" as a function of orbital altitude

$$[\text{useful cargo}](h) = [\text{full analytical p/l}](h)$$
$$- \{[m_{\text{propellant}}](h) - [m_{\text{propellant}}](220 \text{ km})\} \quad (3.88)$$

Equation 3.88 has been used to construct the two different propellant-mass-corrected "useful cargo" curves shown in Figure 3.20. Note that $m_{\text{propellant}}(220\text{km})$ is a constant reference mass in this equation.

To obtain our "useful cargo" rule of thumb slope estimate, we first calculate the derivative $dm_{\text{propellant:total}}/dh$ by differentiating Equation 3.87, viz.

$$\frac{dm_{\text{propellant:total}}}{dh} = \left(\frac{v_{\text{circular@perigee}}}{2(\text{Isp}_{\text{oms}})g} \right) \left(\frac{m_{\text{i}}}{r_{\text{p}}} \right) \quad (3.89)$$

where $v_{\text{circular@perigee}} = \sqrt{\frac{\mu}{r_{\text{p}}}}$.

We see that the first-order slope expression in Equation 3.89 is independent of orbital altitude h and that the three parameters that set the slope's magnitude are the Orbiter mass m_{i}, the specific impulse for the OMS engines, Isp_{oms}, and the burnout altitude h_0. We have assumed a fixed Shuttle burnout altitude, $h_0 = 115$ km, corresponding to $r_{\text{p}} = 6{,}493$ km.

Setting the Orbiter mass and OMS specific impulse:

$$\text{Orbiter mass:} \quad m_{\text{i}} \approx 100{,}000 \text{ kg}$$

$$\text{OMS specific impulse:} \quad \text{Isp}_{\text{oms}} = 313 \text{ s } [37]; \text{ exit velocity } C = 3{,}070 \text{ m/s},$$

along with

$$v_{\text{circular@perigee}} = \sqrt{\frac{3.986 \times 10^5}{6{,}493}} = 7.835 \text{ km/s} = 7{,}835 \text{ m/s}$$

into Equation 3.89, we obtain the following numerical value for the rate of change of propellant mass with orbital altitude

$$\frac{dm_{\text{propellant:total}}}{dh} = \left(\frac{7{,}835 \text{ m/s}}{2(313 \text{ s})(9.81 \text{ m/s}^2)} \right) \left(\frac{10^5 \text{ kg}}{6{,}493 \text{ km}} \right) = 19.6 \text{ kg/km}$$

Differentiating Equation 3.88 with respect to h yields the relationship between the "useful cargo" derivative and the propellant mass derivative

$$\frac{d[\text{useful cargo}]}{dh} = \frac{d[\text{full analytical payload}]}{dh} - \frac{d\lfloor m_{\text{propellant:total}} \rfloor}{dh} \quad (3.90)$$

Since we previously determined that the rate of change of the "full analytical payload with orbital altitude" is approximately $-13\,\text{kg/km}$ (Figure 3.19) we obtain the final cargo rule of thumb estimate

$$\boxed{\frac{d[\text{useful cargo}]}{dh} = -13\,\text{kg/km} - 19.6\,\text{kg/km} = -32.6\,\text{kg/km} \quad (3.91)}$$

In comparison, the "exact" mathematical solution of Equations 3.3 and 3.58 that did not involve making any small parameter approximations is numerically given in Equation 3.68 and evaluated in Section 3.14.5.1 as

$d(\text{useful cargo})/dh \approx -30\,\text{kg/km}$ [OMS model corrected "analytical payload"]

A straight line "useful cargo" curve having this slope of $-30\,\text{kg/km}$ is shown in Figure 3.20. As previously noted, our analytical BotE model for payload change with orbital altitude is within 20% of NASA's payload rule of thumb $\approx -25\,\text{kg/km}$ [36]. Our small-parameter linearized-approximation (Equations 3.89 and 3.91) differs by about 30% from the NASA rule of thumb. Nevertheless, it is certainly "within the ball park" of this complex OMS maneuvering process.

In conclusion, we have demonstrated that a Back-of-the-Envelope model can explain and quantitatively estimate the rapid falloff of "useful" cargo with orbital altitude and can be used to estimate the payload that the Shuttle can deliver to any typical mission orbit.

3.14.7 Modeling missions to the International Space Station

In the preceding analysis we have, for simplicity, not considered missions to the International Space Station that require a Shuttle launch trajectory which is not directly due-east from the Kennedy Space Center; i.e. for launch azimuths $\neq 90°$, as measured clockwise from north. A *non-easterly* correction to our useful cargo mass as a function of orbital altitude can also be made using our BotE methods. One must first consider the fact that the ISS orbit is inclined at $51.6°$ [38], which differs in inclination from the due-east launches of with orbits inclined at $28.5°$ (the inclination used in our calculations in Section 10.4 and carried forward in our studies).

A modification of the burnout velocity magnitude that takes into account the smaller "earth-spin velocity component" associated with launches to inclinations not set by the takeoff latitude needs to be incorporated into a more generalized predictive model for cargo mass. This is left as a difficult, but certainly doable, exercise for the reader (see [26, pp. 55–58] for help with the appropriate spherical-geometry expressions that link burnout velocity to the orbital velocity required to achieve a given elliptical orbit, for a launch from a given latitude and azimuth angle).

3.15 TABULATED SUMMARY OF BACK-OF-THE-ENVELOPE EQUATIONS AND NUMERICAL RESULTS

The accompanying four tables summarize the key BotE equations in the preceding sections. Numerical results are given for important Shuttle parameters.

Table 3.6 presents the key BotE results for the first stage ideal burnout velocity of the Shuttle without the effects of gravity. Table 3.7 presents the first stage ideal burnout velocity taking into account gravity loss and drag loss. Table 3.8 gives the second stage ideal burnout velocity without and with the effects of gravity. And Table 3.9 presents the orbital velocity requirements based on a "direct insertion" maneuver for the Shuttle, and an estimate of payload for given orbital altitude. The calculations for the payload decrease at the higher orbital altitudes due to the use of additional OMS propellants are not included in Table 3.9. See instead Section 3.14.4.

Table 3.6. BotE input parameters and summary of calculation of first stage burnout velocity and height at $t = 110$ s *without the effects of gravity*.

			First stage ideal velocity estimates (Eqs. 3.1c to 3.17)	
		Given	Data source	
$m_{\text{initial},1} = m_{\text{total@takeoff}}$ (kg)		2,063,016	Table 3.1	
$m_{\text{propellant2SRBs}}$		1,008,000	Figure 3.3 schematic	
SRB effective burn time (s)		110.0	[16, "Solid Propellant Rocket Motor Fundamentals"]	
SSME total sea-level thrust (N)		5.25×10^6	Table 3.2	
Isp—LOX/LH2 engine (s)		363.0	Table 3.2	
			Calculation	
Isp solid avg. between sea level(242 s) & vacuum(268 s)			$\text{Isp}_{S,\text{avg}} \approx \dfrac{(242\text{ s} + 268\text{ s})}{2} = 255$ s [Eq. 3.16a/Table 3.2]	
Isp liquid avg. between sea level(363 s) & vacuum(453 s)			$\text{Isp}_{L,\text{avg}} \approx \dfrac{(363\text{ s} + 453\text{ s})}{2} = 408$ s [Eq. 3.16a/Table 3.2]	
Avg. solid propellant SRB flow rate: $\dot{m}_S$ (kg/s); $= m_{\text{propellant2SRBs}}/t_{b,\text{eff}}$			$\dot{m}_S = \dfrac{1.008 \times 10^6 \text{ kg}}{110 \text{ s}} = 9164$ kg/s (Eq. 3.7b)	
Total SSME flow rate (kg/s) based on sea level thrust, $T_{3\text{SSME}}$ and $\text{Isp}_{\text{H2/O2}}$ (Equation 3.1a)			$\dot{m}_L = \dfrac{T_{3\text{SSME}}}{(\text{Isp}_{\text{H2/O2}})g} = \dfrac{5.25 \times 10^6}{(363)(9.81)} = 1{,}474$ kg/s (Section 3.7.1)	
Liquid propellant mass consumed during the first stage (kg)			$m_{\text{liquid propellant burn, 1st stage}} = (\dot{m}_L)_{b,\text{eff}} = (1{,}474 \text{ kg/s})(110 \text{ s})$ $= 162{,}000$ kg (Section 3.7.1)	

Sec. 3.15] Tabulated summary of Back-of-the-Envelope equations 157

Description	Equation
Final first stage burnout mass	$m_{\text{final},1} = m_{\text{initial},1} - m_{\text{solid propellant},1} - m_{\text{liquid propellant},1}$ $m_{\text{final},1} = m_{\text{initial},1} - m_{\text{propellant},1}$ $= 2{,}063{,}016 - 1{,}008{,}000 - 162{,}000 = 893{,}016 \text{ kg}$ (Eq. 3.3b)
α, solid/liquid mass ratio constant for "parallel" (solid + liquid) first stage burn (non-dimensional)	$\alpha = \dfrac{\dot{m}_S}{(\dot{m}_S + \dot{m}_L)} = \dfrac{9{,}164}{(9{,}164 + 1{,}474)} = 0.86$ (Eq. 3.13)
Isp avg. both engines together (s) $\text{Isp}_{\text{avg}} = \text{Isp}_S\,\alpha + \text{Isp}_L(1-\alpha)$	$\text{Isp}_{\text{avg, both engines}} = (\text{Isp}_{S,\text{avg}})(0.861) + (\text{Isp}_{L,\text{avg}})(0.139)$ $= (255)(0.86) + (408)(0.14) \approx 277 \text{ s}$ (Eq. 3.14)
C avg. both engines together (m/s)	$C_{\text{avg, both engines}} = (\text{Isp}_{\text{avg, both engines}})g = (277)(9.81)$ $= 2{,}717 \text{ m/s} = 2.72 \text{ km/s}$ (Eq. 3.16b)
$\Delta v_{i,\text{ideal}}$ estimate for Shuttle first stage	$\Delta v_{1,\text{ideal}} = -C_{\text{avg}} \ln\left(\dfrac{m_{\text{final},1}}{m_{\text{initial},1}}\right) = -(2{,}717)\ln\left(\dfrac{893{,}016}{2{,}063{,}016}\right) = 2{,}275 \text{ m/s}$ (Eq. 3.17)
Ideal height at Shuttle first stage burnout (km) without gravity. Using inverse time constant based on Total mass flow rate $\dot{M}_e$ over initial mass m_i $\left(\dfrac{\dot{M}_e}{m_i}\right) = \left(\dfrac{9164 \text{ kg/s} + 1474 \text{ kg/s}}{2{,}063{,}016 \text{ kg}}\right) = 5.156 \times 10^{-3} \text{ s}^{-1}$	$h(t_b) = \dfrac{C_{\text{avg}}}{(\dot{M}_e/m_i)}\left[1 - \left(\dfrac{\dot{M}_e}{m_i}\right)t_b\right]\cdot\ln\left[1 - \left(\dfrac{\dot{M}_e}{m_i}\right)t_b\right] + (C_{\text{avg}})t_b$ $= \dfrac{2.717}{(5.156\times 10^{-3})}[1 - (5.156\times 10^{-3})110]$ $\times \ln[1 - (5.156\times 10^{-3})110] + (2.717)110$ $h(t_b) = $ Ideal height at 110 s burnout $= \underline{108 \text{ km}}$ (Eq. 3.9)

158   Estimating Shuttle launch, orbit, and payload magnitudes   [Ch. 3

Table 3.7. BotE input parameters and summary of calculation for first stage burnout velocity at $t = 110\,\text{s}$ including the effects of gravity loss and drag loss.

	First stage velocity—gravity—drag losses (Eqs. 3.19 to 3.33)	
	Given	Data source
Initial pitch angle, β_{initial}, at time of launch (radians)	0 radians (or 0°)	Section 3.8.2
Final pitch angle, β_{final}, at first stage burnout (radians)	$\pi/6$ (or 30°)	Section 3.8.2
Drag coefficient	$C_d = 1.0$	Section 3.9.2
Drag Front Area (m²)	$A = 126.7\,\text{m}^2$	Figure 3.11, Section 3.9.2
Time (s) to peak q, i.e. peak dynamic pressure	5 ≈ 59 s	Figure 3.10, Section 3.9.2
Shuttle velocity (m/s), gravity corrected, at $t = 59\,\text{s}$	$v \approx 437\,\text{m/s}$	Figure 3.6, Section 3.8.3. See also Section 3.9.2
Shuttle altitude (km), gravity corrected, at $t = 59\,\text{s}$	$h \approx 9.8\,\text{km}$	Figure 3.7, Figure 3.8, Section 3.8.4. See also Section 3.9.2
$\Delta t_{\text{(significant drag duration)}}$ based on dynamic pressure vs time	$\Delta t_{\text{(significant drag duration)}} \approx 50\,\text{s}$	Figure 3.10, Section 3.9.2
		Calculation
Gravity loss, first stage Δv_{loss}, m/s		$\Delta v_{\text{loss}} = \left[\dfrac{\sin(\pi/2 - \beta_{\text{final}})}{(\pi/2 - \beta_{\text{final}})}\right] g t_b = \left[\dfrac{\sin(\pi/3)}{\pi/3}\right](9.81)(110) = 892\,\text{m/s}$ (Eq. 3.23), Section 3.8.2

$\Delta v_{ideal,1}$ estimate for Shuttle *first stage* with gravity loss; m/s	$\Delta v_{\text{1st stage with gravity loss}} = -C_{avg} \ln\left(\dfrac{m_{final,1}}{m_{initial,1}}\right) - \Delta v_{\text{gravity loss}}$ $= 2275 - 892 = 1383$ m/s (Eq. 3.19 and 3.26); Sections 3.8.1 and 3.8.8
Gravity-corrected height: first stage burnout ($t = 110$ s) Calculated using numerical quadrature of the height integral	$h(t) = \displaystyle\int_0^t v(t) \sin \beta(t)\, dt$ $= -C \displaystyle\int_0^t \ln\left[1 - \left(\dfrac{(T/W)_i}{I_{sp}}\right) t\right] \cos\left(a \dfrac{t}{t_{max}}\right) dt$ $- \dfrac{g t_{max}}{a} \displaystyle\int_0^t \sin\left(a \dfrac{t}{t_{max}}\right) \cos\left(a \dfrac{t}{t_{max}}\right) dt$ where $a \equiv \left(\dfrac{\pi}{2} - \beta_{final}\right)$ $h(t = 110$ s$) \approx 38.0$ km (Eq. 3.25), Section 3.8.4
Shuttle's mass at max q ($t = 59$ s)	$m_{\text{propellant burn}} = m_i - (\dot{M}_e)(t_{q\,max})$ $= 2{,}063{,}016$ kg $- (10{,}638$ kg/s$)(59$ s$) = 1.43 \times 10^6$ kg (Eq. 3.33), Section 3.9.2
Air density ρ (kg/m^3) at max q where $h \approx 9.8$ km For exponential ρ model: Ref. air density $\rho_0 = 1.23$ kg/m^3; scale height $H = 6.73$ km	$\rho = \rho_0 \exp(-h/H) = 1.23 \exp(-9.8 \text{ km}/6.73 \text{ km}) = 0.29$ kg/m^3 (Eq. 3.30), Section 3.9.2

(*continued*)

Table 3.7 (*cont.*)

First stage velocity—gravity—drag losses (Eqs. 3.19 to 3.33)

	Given	Data source
Drag, D, at max q, first stage flight, Newtons		$D = (\frac{1}{2}\rho v^2) C_d A = (0.5)(0.29)(437.)^2(1.0)(126.7) = 3.50 \times 10^6$ N (Eq. 3.31), Section 3.9.2
Drag Loss, $\Delta v_{\text{drag loss}}$, m/s		$\Delta v_{\text{drag loss}} = \int_0^t \dfrac{D(\tau)}{m(\tau)} d\tau \approx \dfrac{D(\Delta t_{\text{drag duration}})}{m_{\max q}}$ $= \dfrac{(3.50 \times 10^6)(50)}{1.43 \times 10^6} = 122$ m/s (Eqs. 3.35, 3.36), Section 3.9.2
Δv_{ideal} estimate for Shuttle first stage with gravity loss and drag loss; m/s		$\Delta v_{\text{1st stage with gravity+drag loss}} = -C_{\text{avg}} \ln\left(\dfrac{m_{\text{final},1}}{m_{\text{initial},1}}\right) - \Delta v_{\text{gravity loss}}$ $\quad - \Delta v_{\text{drag loss}}$ $= 2{,}275 - 892 - 122 = 1{,}261$ m/s ≈ 1.26 km/s (Eq. 3.37), Sections 3.9.3

Table 3.8. BotE input parameters and summary of calculation for second stage burnout velocity at $t = 495$ s including the effects of gravity loss and drag loss. Final burnout velocity in inertial coordinates includes the Δv due to the earth's rotation.

	Second stage burnout velocity estimates	
	Given	Data source and/or Calculation
$m_{\text{SRBcasing}}$ (kg) for 2 SRBs	165,758	Figure 3.3
$m_{\text{initial},2} = m_{\text{final},1} - m_{\text{SRBcasing}}$ (kg)		$m_{\text{i2a}} = 893{,}016 - 165{,}758 = 727{,}258$ kg (Eq. 3.41c)
Liquid propellant mass, second stage, $m_{\text{propellant},2}$ (kg)	568,000	Figure 3.3
The time to MECO, t_{MECO} (s)		$t_{\text{MECO}} = \dfrac{m_{\text{propellant},2}}{\dot{m}_L} + t_{\text{1st stage burnout}} = \dfrac{568{,}000 \text{ kg}}{1{,}474 \text{ kg}} + 110 \text{ s} = 495.3$ s Section 3.10.3
Final burnout mass, second stage, $m_{\text{final},2}$		$m_{\text{final},2} = m_{\text{initial},2} - m_{\text{propellant},2} = 727{,}258 - 568{,}000 = 159{,}258$ kg (Eq. 3.3b, using second stage masses)
Isp liquid propellants in vacuum, $\text{Isp}_{L,\text{vac}}$ (s)	452.5	Table 3.2
Δv_{ideal} estimate for Shuttle second stage (m/s)		$\Delta v_{2,\text{ideal}} = -(\text{Isp}_{L,\text{vac}} \cdot g) \ln\left(\dfrac{m_{\text{final},1}}{m_{\text{initial},1}}\right)$ $= -(452.5) \cdot (9.81) \ln\left(\dfrac{159{,}258}{727{,}258}\right) = 6{,}742$ m/s (Eq. 3.3, using second stage masses); also Table 3.4

(continued)

Table 3.8 (*cont.*)

Second stage burnout velocity estimates

Given	Data source and/or Calculation
Gravity loss, second stage $\Delta v_{g\text{-loss},2}$, m/s Pitch angle time scale for second stage (region 2a) $\Delta t_2 = (300 - 110) = 190$ s	$\Delta v_{g\text{-loss},2} = \left(\dfrac{1}{\pi/6}\right)\left\{1 - \cos\left[\dfrac{\pi}{6}\right]\right\}(g\,\Delta t_2)$ $= 1.91(0.134))(9.81 \cdot 190) = 477$ m/s (Eq. 3.39), Section 3.10.2 with pitch $= 0$ at $t = 300$ s. See also Table 3.4
$\Delta v_{\text{ideal},2}$ estimate for Shuttle *second stage* with second stage gravity loss; m/s	$\Delta v_{\text{2nd stage with gravity loss}} = -C_{\text{vac}}\ln\left(\dfrac{m_{\text{final},2}}{m_{\text{initial},2}}\right) - \Delta v_{g\text{-loss},2}$ $= 6{,}742 - 477 = 6{,}265$ m/s Table 3.4
$\Delta v_{\text{total with gravity}} =$ combined first stage and second stage Δv with gravity loss; m/s	$v_{\text{total with gravity}} = (\Delta v_{\text{1st stage}}) + (\Delta v_{\text{2nd stage}})$ $= 1{,}383 + 6{,}265 = 7{,}648$ m/s Section 3.10.3 and Table 3.4
Final second stage burnout or MECO velocity with drag loss subtracted; m/s	$v_{\text{MECO}} = (\Delta v_{\text{total with gravity}}) - (\Delta v_{\text{1st stage drag}}) = 7{,}648$ m/s $- 122$ m/s $= 7{,}526$ m/s ≈ 7.53 km/s Section 3.10.4 and Table 3.4
Added Δv due to rotation of earth at 28.5° latitude, eastward launch, m/s Earth's radius: $r_e = 6493$ km	$\Delta v_{\text{rotational}} = \dfrac{2\pi r_e}{(24\,\text{hr})(3{,}600\,\text{s/hr})}\cos(\text{latitude})$ $= (0.464\,\text{km/s})\cos(28.5°) = 0.408$ km/s Section 3.11 and Table 3.4

Sec. 3.15] **Tabulated summary of Back-of-the-Envelope equations** 163

Final second stage MECO velocity in inertial framework with eastward launch rotational velocity added; m/s Cargo Payload = 25,000 kg	$v_{\text{MECO, inertial}} = (\Delta v_{\text{MECO}}) + (\Delta v_{\text{rotational}}) = 7{,}526 \text{ m/s} + 408 \text{ m/s}$ $= 7{,}934 \text{ m/s} \approx 7.93 \text{ km/s}$ Section 3.11 and Table 3.4
Gravity-corrected height: second stage max altitude ($t = 300$ s where pitch $\beta \to 0$ deg.) Height calculated using numerical quadrature of the height integral + (first stage burnout height; 38 km)	$h(t = 300 \text{ a}) = \dfrac{1}{1000} \displaystyle\int_{110\,\text{s}}^{300\,\text{s}} v_{2a}(t) \sin \beta(t)\, dt + 38.0 \text{ km}$ where pitch angle and velocity for region 2a is $\beta(t) = \left(\dfrac{\pi}{6}\right)\left[1 - \left(\dfrac{t - 110}{190}\right)\right]$ $v_{2a}(t) = (1{,}382 \text{ m/s}) - (4{,}439 \text{ m/s}) \ln\!\left[1 - (2.03 \times 10^{-3}\text{ s}^{-1})(t - 110 \text{ s})\right]$ $\quad - \left(\dfrac{1}{\pi/6}\right)\!\left\{\cos\!\left[\dfrac{\pi}{6}\!\left(1 - \dfrac{(t - 110\text{ s})}{190\text{ s}}\right)\right] - \cos\!\left[\dfrac{\pi}{6}\right]\right\}(g \cdot 190\text{ s})$ $\boxed{h(t = 300\text{ s}) \approx 125 \text{ km}}$ Section 3.10.2 and Figure 3.14, Section 3.12.2

Table 3.9. BotE results for: (a) "required orbital velocity" for a "direct insertion" maneuver to a height of 610 km with corrections for gravity loss, drag loss, and the earth's rotation (b) an example of the calculation of Shuttle payload delivered to a 610 km orbital height.

	Orbital velocity requirements and payload estimate	
	Given	*Data source and/or Calculation*
Altitude of parking orbit, h_0, (km)	115 km	
Altitude of higher orbit, h, after Hohmann transfer (km)	610 km	
Universal Gravitation parameter, μ for earth	3.986×10^5 km^3/s^2	
Required direct-insertion velocity to height $h = 610$ km, $\Delta v_{\text{req}}(h)$; m/s		(Eq. 3.44), Section 3.13.1 $\Delta v_{\text{req}}(h) = \Delta v_{\text{direct-insertion}} + \sum_i \Delta v_{\text{gravity},i} + \sum_i \Delta v_{\text{drag},i} - \Delta v_{\text{earth-rot}}$ $= \Delta v_{\text{direct-insertion}} + (0.892 + 0.477) + (0.120 + 0.0) - (0.408)$ $= \sqrt{\dfrac{\mu}{(r_e + h_0)}\left(\left[1 + (r_e + h_0)/(r_e + h)\right]\dfrac{2}{}\right)} + 1{,}081$ m/s $\Delta v_{\text{req}}(610 \text{ km}) = 7{,}978 + 1{,}081 = 9{,}059$ m/s (Eq. 3.60), Section 3.13.3; (Eq. 3.65), Section 3.13.5; and (Eq. 3.75), Section 3.14.1
Payload estimate model Mass w/o payload input M_{1a}, M_{1b}, M_{2a}, M_{2b} (kg) and C_1, Avg exit velocity (stage 1) C_2 Exit velocity (stage 2); (m/s)		$M_{1a} = m_{\text{initial},1} - 25{,}000$ kg $= 2{,}038{,}016$ $M_{1b} = m_{\text{final},1} - 25{,}000$ kg $= 868{,}016$ $M_{2a} = m_{\text{initial},2} - 25{,}000$ kg $= 702{,}258$ $M_{2b} = m_{\text{final},2} - 25{,}000$ kg $= 134{,}258$ $C_1 = 2{,}712$ m/s; $C_2 = 4{,}439$ m/s (Eq. 3.74, 3.75), Section 3.14.2

Sec. 3.15] **Tabulated summary of Back-of-the-Envelope equations** 165

Payload mass $m_{p/l}$ kg for direct insertion to orbit height $h = 610$ km; calculated using approx. linear solution Payload based on the *approximate payload model*, Eq. 3.78, Section 3.14.2 Input parameters given above	$$m_{p/l} = \frac{M_{2a} - KM_{2b}}{\left\{ K - 1 + \gamma \left(K \frac{M_{2b}}{M_{1b}} - \frac{M_{2a}}{M_{1a}} \right) \right\}};$$ with $K = \left(\frac{M_{1b}}{M_{1a}}\right)^\gamma \exp\left(\frac{\Delta v_{req}}{C_2}\right) = 4.565$ where $\gamma = \frac{C_1}{C_2} = \frac{\text{Isp}_{avg,1}}{\text{Isp}_2} = \frac{276.4}{452.5} = 0.611$ $m_{p/l} \approx 23{,}600$ kg ($h = 610$ km) (Eq. 3.78), Section 3.14.2; Figure 3.19 (approx analytical solution)

3.16 REFERENCES

[1] Albaugh, James, "Spacecraft: The art of designing and crafting space-bound platforms", speech given at Columbia University Graduate School of Engineering, New York, June 01, 2002.

[2] Jenkins, D. R., *The History of the National Space Transportation System—The First 100 Missions*, Cape Canaveral Florida: Specialty Press, 2001.

[3] Heppenheimer, T. A., *The Space Shuttle Decision*, NASA SP-4221, Chapter 6, "Why People Believed in Low-Cost Space Flight", Complete Online Book, NASA History Series. http://www.nss.org/resources/library/shuttledecision/index.htm

[4] Futron Corporation, Space Transportation Costs, Trends in price per pound to Orbit 1990-2000, September 2002. http://www.futron.com

[5] Gehman, H.W. et al., "Columbia Accident Investigation Board Report," Volume 1, The Evolution of the Space Shuttle Program, U.S. Government Printing Office, Washington DC, August, 2003.

[6] NASA Spaceflight web site, Space Shuttle Basics, second stage ascent. http://www.spaceflight.nasa.gov/shuttle/reference/basics/ascent.html

[7] Space Shuttle, Wikipedia Space Shuttle entry, see STS mission profile graphic. http://en.wikipedia.org/wiki/Space_Shuttle

[8] Braeunig, R.A., "Rocket and Space Technology" web site: see Space Shuttle Specifications for dimensions, mass, thrust of various components. http://www.braeunig.us/space/

[9] Information summary, Kennedy Space Center, "NASA Space Shuttle & Facilities", IS-2005-06-017-KSC. http://www-pao.ksc.nasa.gov/kscpao/nasafact/pdf/Countdown 2005.pdf

[10] Wertz, J.R. and W.J. Larson, *Space Mission Analysis and Design*, 3rd edition, published by Space Technology Library, 1999.

[11] Fortescue, P. and J. Stark editors, *Spacecraft Systems Engineering*, Wiley, 1992. (See page 160, Table 7.2 for Ariane 1 data).

[12] *Top Chef*, Television show, Bravo Network, Season 1, March 2006.

[13] READING: The External Tank: educational reading materials and graphics developed by NASA KSC and funded by NASA SOMD. http://www.alivetek.com/klass/Curriculum/Get_Training%201/ET/RDG_ET.pdf

[14] Hutchinson, V.L. and J. Olds, "Estimation of Launch Vehicle Propellant Tank Structural Weight Using Simplified Beam Approximation", AIAA 2004-3661, 40th IAA/ASME/SAE/ASEE Joint AIAA 2004-3661Propulsion Conference, July 11–14, 2004.

[15] Sutton, George P., *Rocket Propulsion Elements*, 6th edition, New York, John Wiley and Sons, 1992.

[16] Boardman, T., "Solid Propellant Rocket Motor Fundamentals", Design and Manufacturing baseline course briefing, Reusable Solid Rocket Motor Engineering, ATK Thiokol, Feb. 2005.

[17] National Aeronautics and Space Administration. *Shuttle Systems Design Criteria. Volume I: Shuttle Performance Assessment Databook*. NSTS 08209, Volume I, Revision B. March 16, 1999, see Table 5.18, Predicted Performance Time History for RSRM (TP-R074-99), Total flow-rate vs time.

[18] Ryan, R.S. and J.S. Townsend, "Fundamentals and Issues in Launch Vehicle Design", *Journal of Spacecraft and Rockets*, Vol. 34, No. 2, pp. 192–198, March 1997.

[19] Martin, John J., *Atmospheric Reentry*, Prentice-Hall, Inc., Englewood Cliffs, New Jersey, (see Equation 3-59), 1966.
[20] "SinIntegral" and "CosIntegral" functions; see http://functions.Wolfram.com where: Si(z): http://functions.wolfram.com/06.37.02.0001.01   Ci(z): http://functions.wolfram.com/06.38.02.0001.01
[21] Arrington, J. and J. Jones, "Shuttle Performance: Lessons Learned", NASA Conference Publication 2283, [see ambient pressure and Mach number vs. time plots on p. 116 of this publication] March, 1983.
[22] 1976 "U.S. standard atmosphere" table of speed of sound as a function of altitude. http://www.digitaldutch.com/atmoscalc/tableoptions1.htm
[23] Camp, D.W. (Operations Engineering Office; Space Shuttle Program Office NASA Houston), "Shuttle flight data and in-flight anomaly list (STS-1 to STS-74)", (see Ascent and Orbit Insertion Data, pp 1-78 to 1-88 for Shuttle velocity and altitude at SRB separation), May 1996.
[24] Fleming, F.W. and V. E. Kemp, "Computer Efficient Determination of Ascent Trajectories", *Journal of the Astronautical Sciences*, Vol. 30, No.1, pp. 85–92, Jan–March, 1982.
[25] Griffin, M. D. and J. R. French, *Space Vehicle Design*, 2nd edition, AIAA education series, 2004.
[26] Humble, R.W., G.N. Henry, and W.J. Larson, *Space Propulsion Analysis and Design*, Space Technology Series, McGraw Hill, page 66, 1995.
[27] Allen, H.J. and A.J. Eggers, "A study of the motion and aerodynamic heating of ballistic missiles entering the earth's atmosphere at high supersonic speed," NACA report 1381, page 2 for density scale height, 1958.
[28] Science Made Simple web-site, "Why is the sky blue? What is the atmosphere?", Copyright 1997. http://www.sciencemadesimple.com/sky_blue.html
[29] Edinger, L.D. "The Space Shuttle Ascent Flight Control System", proceedings of the Guidance and Control Conference, San Diego, CA, AIAA paper #1976-1942, pp. 225–235, August 16–18, 1976
[30] Allaway, Howard, "The Space Shuttle at Work", SP-432/EP-156, Cover page graphic, Scientific and Technical Information Branch and Division of Public Affairs, NASA History Office, Washington, D.C. 1979, on-line update August 2004. http://history.nasa.gov/SP-432/sp432.htm
[31] Harwood, William, space news reporter affiliated with CBS News. The NASA predicted flight data for STS-119, STS-119 SpaceCalcPC, is posted as an Excel spreadsheet on http://www.cbsnews.com/network/news/space/downloads.html. Note that *SpaceCalc* is an Excel workbook assembled by Harwood. It is loaded with Shuttle and space station flight, timeline, and statistical data for a number of Shuttle missions. Also see The CBS News Space Reporter's Handbook Mission Supplement, Shuttle Mission STS-119: Space Station Servicing Mission ULF-2, March 9, 2009 available at http://www.spaceflightnow.com/.
[32] Press, William, B. Flannery, S. Teukolsky, and W. Vetterling, *Numerical Recipes*, Chapter 9, Cambridge University Press, 1986.
[33] http://shuttlepayloads.jsc.nasa.gov/flying/capabilities/capabilities.htm is the original (accessed 9/28/06) for the Shuttle mission data plotted in Figure 3.18 (This link is no longer viable). Currently available information on Shuttle payloads are posted at http://science.ksc.nasa.gov/shuttle/technology/sts-newsref/stsover-prep.html
[34] "Orbital Maneuvering System" description is given at the following URL. http://science.ksc.nasa.gov/shuttle/technology/sts-newsref/sts-oms.html #sts-oms

[35] See Orbital Maneuvering System- Propellant Usage Summary in Ref. [23], "Shuttle flight data and in-flight anomaly list", Johnson Space Center/EP2 propulsion branch summary (report JSC 19413, Rev. V, April 1996), page 1-192 to 1-193.
[36] Sutton, George P. and Oscar Biblarz, *Rocket Propulsion Elements*, 7th edition, New York, John Wiley and Sons, 2001. See chapter entitled "Flight performance", page 148 for Shuttle payload rule of thumb that states that the "payload decreases about 100 lbs for every nautical mile increase in altitude".
[37] See Reference [36], page 210, Table 6-3, "Characteristics of the Orbital Maneuvering System (OMS) and the Reaction Control System of the Space Shuttle in One of the Aft Side Pods.
[38] See the International Space Station article http://en.wikipedia.org/wiki/International_Space_Station

4

Columbia Shuttle accident analysis with Back-of-the-Envelope methods

4.1 THE COLUMBIA ACCIDENT AND BACK-OF-THE-ENVELOPE ANALYSIS

On February 1, 2003, the Columbia Shuttle Orbiter and its brave seven member crew perished during reentry into the earth's atmosphere at the conclusion of its 28th flight, STS-107, a multi-disciplinary science mission.

Since its inaugural flight in 1981, Columbia and its crews had distinguished themselves in a wide range of missions, including the deployment of the Chandra X-ray Observatory in July 1999 and a flight to service the Hubble Space Telescope's optical system in March 2002.

The loss of Columbia proved to have been triggered by an accidental breach of the leading edge of the Orbiter's left wing at the beginning of the mission. This finding was made after seven months of investigation by the Columbia Accident Investigation Board (CAIB) and its staff of 120 experts (supported by 400 NASA engineers). The breach occurred about 82 seconds after launch on January 16, 2003 when a piece of insulation foam separated from the external tank and struck the left wing in the vicinity of leading-edge panel #8 (Figure 4.1). The Orbiter nose and the leading edges of the wings were made of reinforced carbon-carbon (RCC) and, as part of the thermal protection system (TPS), were designed to withstand the highest heating loads during reentry.

The CAIB final report stated: "... during reentry, the breach in the TPS allowed super-heated air to penetrate the leading edge insulation and progressively melt the aluminum structure of the left wing, resulting in a weakening of the structure until increasing aerodynamic forces caused loss of control, failure of the wing and breakup of the Orbiter, just 16 minutes before its scheduled touchdown" [1, p. 49].

Although foam strikes had been observed on many previous flights, and an older empirical penetration model had suggested the possibility of damage from a large piece of foam impacting the TPS at certain angles, engineers at NASA chose to

---

I. E. Alber, *Aerospace Engineering on the Back of an Envelope.*
© Springer-Verlag Berlin Heidelberg 2012.

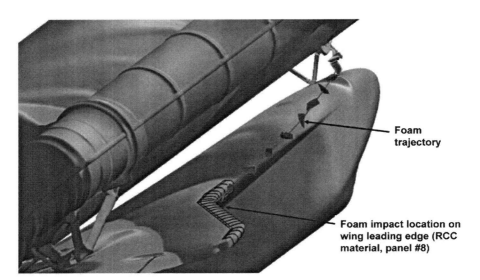

Figure 4.1. A 1.67 pound slab of insulating foam, falling off Columbia's external tank and hitting the left wing of the orbiter at panel #8, is graphically depicted. This image, included in the CAIB final report [1], was created by analysts using the graphical code *Tecplot* to visualize the output of NASA's Cart3D aerodynamic trajectory simulation software. (This image was originally developed by NASA Ames' Supercomputing Division using their computer software and hardware in 2003).

downgrade the significance of this information both prior to and during STS-107's flight. Some of the reasons given for ignoring the safety threat were that no mission-critical damage, such as wing burn-throughs or large scale tile ruptures, had ever occurred. Also, the operations team was under considerable pressure to fulfill the Shuttle launch schedule. The CAIB stated, "Assessments of foam shedding and strikes were not thoroughly substantiated by *engineering analysis*" [1, p. 130]. To follow up this comment, an engineering team supporting the CAIB made a series of on-the-ground impact tests in order to determine the extent of the damage that might be caused by a foam strike on either the RCC leading edge or the nearby insulating tiles of the TPS. The impact test program, augmented by a number of computer-simulation impact studies, demonstrated that an approximately 1 to 2 lbm piece of foam could, *if impact occurred at the right speed and impact angle*, cause a wide range of impact damage, ranging from cracks to a sizable hole in the wing's RCC leading edge [2]. Experimental and numerical modeling studies conducted by the Southwest Research Institute [2] later established that impact damage to the *insulating tiles* was either quite small or non-existent. This lack of tile damage, for a given impact speed, was found to be related to the *orientation* of the tiles relative to the foam trajectory rather than to any specific tile-material properties.

The impact angles of the foam trajectories relative to the tile surfaces (defined in the sketch below) were estimated to be considerably smaller than those for the foam trajectories against the wing leading edge. Smaller normal velocity magnitudes for the impacted tiles, in turn, lead (as will be shown in Section 4.4.7) to reduced impact loads at sub-crush levels.

In Section 4.5.2 of our BotE analysis, a simple sine function scaling, *sin* (impact angle), is used to characterize the dependence of the impact load on impact angle.

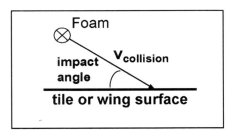

4.1.1 BotE modeling goals for the Columbia accident

Based on the Columbia accident, we pose the question: Can a physics-based BotE model provide a credible engineering evaluation of: (a) the impact velocity of a piece of foam onto the Orbiter wing, (b) the resultant impact pressure or load on the wing, and can it (c) estimate whether serious damage to the wing is likely?

As will be shown, the answer is yes! We can readily estimate the impact velocity, calculate the resultant impact load, and develop a simple criterion for possible wing damage. We will also generate simple plots of these quantities as a function of the driving engineering parameters, such as foam mass and angle of incidence, over their estimated ranges to determine the approximate magnitudes of the speeds and forces that were produced in this critical impact problem.

Initial BotE calculations were first applied to the Columbia foam impact by Douglas Osheroff, Nobel Laureate, shortly after the Columbia tragedy [3]. A prominent member of the CAIB panel, he estimated the foam mass and impact velocity using a simple constant acceleration model supported by data gleaned from movies of the impact event. His numerical results were published in the transcripts of the CAIB deliberations. The physics and mathematics underlying these first-order estimates will be presented in our Quick-Fire example and in Section 4.3 of this chapter.

Robert Kennedy, P.E. quickly followed Osheroff's analysis with a web-based article entitled, "It ain't rocket science, folks!" [4]. In this note, Kennedy used BotE methods to calculate the impact velocity using an approximate estimate of the foam mass as input. The results were published on his web site in February 2003 and revised in July 2005 [www.ultimax.com]. Kennedy's calculation of the foam impact velocity was in the "ball park" of the final CAIB reported estimates, but his mathematical model for this velocity contains a factor of 2 error which is clearly apparent when one compares his model and numerical results with the exact and approximate analytical solutions of Newton's second law of motion for a point mass (Section 4.3). Nevertheless, his straightforward analysis was quite valuable in that it further demonstrated that the BotE approach could generate simple and meaningful real-world engineering results.

To obtain the BotE results presented here, we require certain data gleaned from the real-time movies of the foam impact event, and also some of the basic physical properties of the Shuttle foam and wing materials (e.g. material density, and elastic/plastic wave speeds). We will utilize these data to approximate not only foam/wing impact speeds, but also the impact loads on the leading edge of the wing and the resultant stresses within the leading-edge RCC material. We will further determine the dependence of these velocity and stress quantities on a range of system parameters.

4.1.2 Quick estimation vs accurate estimation

The level of detail and the sophistication of the model that an engineer employs in tackling a real-world problem, such as the foam impact problem, will depend primarily on two key constraints: (1) the deadline set by his customer or manager (e.g. "I need a rough answer by noon today!") and (2) the accuracy demanded of the result ("I need to know that velocity with better than 10% accuracy"). Usually one of these constraints dominates over the other and often cost or manpower is an important factor as well ("I can only let one person work on this task because of our limited budget").

To obtain a quick *first-order cut* at estimating the approximate velocity of the foam impact, we will again use the BotE "Quick-Fire" [5] methodology that we have shown in Chapter 1 can *quickly* produce a result using a minimum amount of technical information and a simple modeling effort.

As previously, in the Quick-Fire approach to a BotE problem, we will develop our estimates by systematically following, in sequence, the following five steps:

(a) Define and/or conceptualize the problem using a short sketch or brief mathematical description.
(b) Select a model or approach, either mathematical or empirical that describes the basic physics of the problem.
(c) Determine the input data parameters and their magnitudes as required to solve the problem, either from data sources or by scaling values by analogy, or simply by using an educated guess.
(d) Substitute the input data into the model and compute a value or range of values for the estimate.
(e) Present the results in a simple form and provide a brief interpretation of their meaning.

4.2 QUICK-FIRE MODELING OF THE IMPACT VELOCITY OF A PIECE OF FOAM STRIKING THE ORBITER WING

After reviewing video of a piece of foam (about 1 foot in scale) being shed from the external fuel tank, the Columbia Accident Investigation Board asked the following question:

- How could a lightweight piece of foam travel so fast and hit the wing of the Orbiter at over 500 miles/hour ($v = 223$ m/s) in under 0.2 seconds?

Here is our Quick-Fire BotE response to that question.

Problem definition

- Estimate the impact velocity for a piece of foam, 1 foot on a side, which suddenly breaks away from the Shuttle's external tank and hits the wing of the Orbiter. We are told by the CAIB that this event occurred less than 100 seconds after launch [1].
- Estimate the time it takes for the foam to travel about 60 feet before hitting the Orbiter wing.

The sketch

The following sketch shows the foam impact event in a frame of reference that *moves with the Shuttle*.

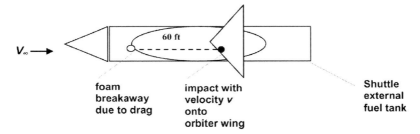

The model

Newton's conservation of energy for an object of mass m subject to a constant aerodynamic drag force F_d is

$$\Delta(\text{Kinetic Energy of foam}) = \text{Work of drag force } F_d \text{ acting over a distance } s$$

$$\Delta(\tfrac{1}{2}mv^2) = F_d s = m a_0 s$$

$$\Rightarrow \quad v(s) = \sqrt{2 a_0 s}$$

where $s = 60$ ft $= 18.3$ m

The force F_d is calculated from the semi-empirical fluid dynamics drag formula that is taught in freshman physics (the origin of the formula has been attributed to Newton)

$$\text{Drag} = (\text{drag coefficient})(\text{dynamic pressure})(\text{foam frontal area})$$

$$F_d = C_d (\tfrac{1}{2} \rho_{\text{air}} V_\infty^2) A$$

Acceleration of foam mass due to drag (Newton's 2nd law), $a_0 = F_d / m$

Physical parameters and data

The estimated foam mass, m, is simply the foam density times the foam volume. Foam is very light. It easily floats in water. Our guess for its density is about 5% that of water $= 0.05 \, (1{,}000 \, \text{kg/m}^3) = 50 \, \text{kg/m}^3$. Let's assume that the piece of foam is a cube. If each side $\approx 1 \, \text{ft} = 0.3 \, \text{m}$, then its volume is $\approx (0.3 \, \text{m})^3 = 0.027 \, \text{m}^3$. Hence the estimated mass is

$$m = (50 \, \text{kg/m}^3)(0.027 \, \text{m}^3) = 1.35 \, \text{kg}$$

Foam frontal area (acted on by the Shuttle "wind"), A, is assumed to be equal to the area of a full cube face. Thus

$$A \approx (0.3 \, \text{m})^2 = 0.09 \, \text{m}^2$$

Let's consider that we are given *no* specific information about the Shuttle's velocity, V_∞, 100 seconds after launch. Nevertheless, V_∞ can be estimated if we make an educated guess about its average vertical acceleration, a_{shuttle}, during those first 100 seconds because $V_\infty \approx (a_{\text{shuttle}})(\text{time})$. (We do not account for the Shuttle slowly pitching away from the vertical over those first 100 seconds.)

We know from the movies we've all seen of Shuttle launches that the rocket initially rises slowly off the launch pad and then rapidly increases its speed with time. From a simple force balance we also know that the initial Shuttle thrust has to exceed the total weight of the vehicle by some reasonable amount, otherwise it would not get off the ground. Let's assume that the initial thrust of the full Shuttle is 50% greater than its initial weight (mg) where g is the strength of gravity at the earth's surface. That means that the net initial rocket acceleration (thrust $- \, mg)/m$ would be $0.5g$.

So our guess for the rocket's average vertical acceleration is

$$a_{\text{shuttle}} = 0.5(9.8 \, \text{m/s}^2) \approx 5 \, \text{m/s}^2$$

The corresponding vehicle velocity 100 seconds after launch is

$$V_\infty \approx (5 \, \text{m/s}^2)(100 \, \text{s}) = 500 \, \text{m/s}$$

Of course, this velocity is probably on the low side because we do not account for the loss of propellant up to that point; but we only need a ball-park estimate. The density of air 100 seconds after launch, ρ_{air}, can be estimated once we know the approximate Shuttle altitude at that time. For the initial acceleration, the altitude, z, can be obtained from the well-known solution for a point mass undergoing uniform acceleration

$$z = \tfrac{1}{2}(a_{\text{shuttle}})t^2 = 0.5(5 \, \text{m/s}^2)(100 \, \text{s})^2 = 25 \, \text{km} \approx 80{,}000 \, \text{ft} \approx 15 \, \text{miles}$$

A rule of thumb for air density is that density drops off with altitude by about a factor of 10 for every 10 miles of atmosphere [4]. Hence the air density at an altitude of 15 miles will be about 1/15 of that the earth's surface. A simple rule of thumb for the air density at sea level is about one-thousandth of that of water, or

$$\rho_{\text{ground}} \approx (\text{density of water})/1{,}000 = 10^{-3} \text{ g/cm}^3 = 1 \text{ kg/m}^3.$$

Thus our estimate for the air density at the time of foam break-away is

$$\boxed{\rho_{\text{air}} = \rho_{\text{ground}}/15 \approx 0.07 \text{ kg/m}^3}$$

The remaining parameter to be estimated is the drag coefficient, C_d. Here we require an elementary knowledge of the magnitude of some basic fluid dynamics quantities. In particular, it is known from wind tunnel or water tunnel experiments on spheres, plates, and circular disks at high speeds (i.e. at high Reynolds numbers) that the measured drag or resistance coefficients for these non-streamlined bodies are approximately of order unity [6]. In this case, a reasonable estimate for the full drag is equal to the dynamic pressure multiplied by the frontal area A.

Thus we take the drag coefficient to be

$$\boxed{C_d = 1}$$

Numerical calculation of results

Inserting the parameters that we have just estimated into our model for drag on the foam debris, we can now calculate the acceleration of a small foam object by the Shuttle "wind". In a coordinate system moving with the Shuttle, it appears that an external wind moves past the Shuttle with speed V_∞ producing a drag force F_d on our hypothetical foam cube acting along the longitudinal principal axis of the vehicle.

Acceleration of foam due to drag, $a_0 = F_d/m$ is then given by

$$a_0 = \left(\frac{C_d A}{m}\right)\left(\tfrac{1}{2}\rho_{\text{air}} V_\infty^2\right)$$

$$= \left(\frac{(1.0)(0.09 \text{ m}^2)}{1.35 \text{ kg}}\right)\left(\tfrac{1}{2}(0.07 \text{ kg/m}^3)(500 \text{ m/s})^2\right) = 583 \text{ m/s}^2$$

The impact velocity (i.e. the velocity of the foam relative to the wing) can be calculated from our initial conservation of energy relationship. The numerical estimate of this velocity is

$$\boxed{v = \sqrt{2a_0 s} = \sqrt{2(583 \text{ m/s}^2)(18.3 \text{ m})} = 146 \text{ m/s}}$$

The time to foam impact on the wing, t, is calculated using the same physics relationship for distance as a function of time that was employed in our estimate for the height of the Shuttle after launch. The equation and numerical result is then

$$s = \tfrac{1}{2} a_0 t^2$$

Solving for impact time t gives

$$t_{\text{impact}} = \sqrt{\frac{2s}{a_0}} = \sqrt{\frac{2(18.3\text{ m})}{583\text{ m/s}^2}} = 0.25\text{ sec}$$

4.2.1 Interpretation of Quick-Fire results

Our estimate for the foam impact velocity, $v = 146$ m/s, is quite high. In English units it is over 325 miles per hour; considerably faster than the speed of a "bullet train". As we will show later, a foam/wing speed of this order may lead, in certain circumstances, to serious damage to the wing of the Orbiter. As a benchmark for these calculations, we note that the CAIB team of engineers, working extensively with complex computer algorithms to simulate and track three-dimensional foam trajectories, estimated an impact velocity of about 500 miles/hour and the time to impact as approximately 0.17 seconds [1, 7].

It appears that our Quick-Fire results are within the "ball park" of these published CAIB findings; however, we differ by about 40 to 50%. By using more detailed information from the flight and from the movies of the foam break-off event, our recalculated BotE velocities and impact times improve, and move closer to the computer-algorithm-processed tracking results of the official study. Indeed, as will be shown in the calculations in Section 4.3, when we input better estimates for the Shuttle flight velocity, foam mass, drag coefficient, and foam dimensional scales, our computed impact velocity differs by less than 20% from the published CAIB findings.

4.2.2 The bridge to more accurate BotE results

Sometimes our engineering customer or manager may demand a higher level of accuracy than we would normally expect to achieve by a BotE analysis conducted in a Quick-Fire time frame of less than an hour.

Let's say that because of a particular accuracy requirement we are given a somewhat longer project time than for a Quick-Fire estimation effort in order to complete our estimation task; perhaps an assignment lasting several days. This is still quite brief in comparison to typical time scales allotted for most complex engineering projects. However, the additional time allows us the opportunity to gather more accurate information to serve as input to our baseline model, or to develop an improved, yet still simple, BotE physical model of the situation. This additional estimation time does not guarantee that we will find a more accurate solution, but it certainly improves our chances for success while still utilizing a minimum amount of engineering resources.

The presentation in Section 4.3 for the foam impact problem provides us with an example of a BotE analysis that is more accurate than is practicable for the short Quick-Fire time frame. We will not only use better data, we will "upgrade" our simple constant-drag-force model by mathematically deriving a closed-form solution of Newton's second law of motion that will allow for more realistic non-constant acceleration of the foam. This modest increase in model complexity will significantly improve the accuracy of our estimated foam impact velocity.

In Section 4.3 we will also give a fuller description of the fundamentals of the underlying dynamics and the rationale for the supporting data approximations used in our BotE calculations of foam velocity and impact time. We present a detailed derivation of the equations of motion for the constant and non-constant acceleration of foam particles due to drag. Comparisons of our detailed results with the published CAIB findings are fully discussed. The sensitivities of our detailed BotE results to different model approximations and to a range of input data uncertainties are presented in graphical form.

We follow our Section 4.3 treatment of the foam velocity estimation problem with an even more difficult BotE estimation project, one in which we seek to answer the following questions that are central to determining the physical cause of the Columbia accident:

- Can a lightweight piece of foam impacting the wing of the Orbiter at high speed produce significant damage to the wing?
- What impact loads on the wing can be expected? What is the estimated maximum stress possible within the wing leading-edge?

Detailed models for, and calculations of, the resultant loads on the Orbiter's wing caused by the high speed impact, and the models and calculations of the resulting high stress levels that develop within the vulnerable wing leading-edge panel are presented in Sections 4.4 and 4.5, respectively. Criteria are presented for assessing damage to the wing based on the calculated impact velocity, angle of incidence of the foam trajectory to the wing's surface, and the maximum impact stress possible in the RCC wing panel.

We then summarize all our detailed BotE equations and numerical results in tabular format in Section 4.6.

4.3 MODELING THE IMPACT VELOCITY OF A PIECE OF FOAM DEBRIS RELATIVE TO THE ORBITER WING; ESTIMATIONS BEYOND THE QUICK-FIRE RESULTS

The Columbia Accident Investigation Board asked the following question during its investigation after reviewing the movies of the flight [1, p. 60]:

- *How could a lightweight piece of foam travel so fast and hit the wing of the Orbiter at over 500 miles/hour ($v = 223 \, m/s$) in under 0.2 seconds?*

In this section we again use a one-dimensional BotE analysis to estimate the range of possible foam impact speeds for a hypothetical piece of foam that has broken away from the external tank at a critical time during the Shuttle's ascent.

We use a straightforward solution of Newton's second law of motion for a point mass undergoing deceleration due to aerodynamic drag to estimate the impact velocity and other key impact scales.

4.3.1 Looking at the collision from an earth-fixed or moving Shuttle coordinate system

According to the CAIB report, "just prior to separating from the Shuttle's external fuel tank, the foam was traveling with the entire Shuttle stack (or assembly) at a velocity of about 1,600 miles/hr (715 m/s) relative to a fixed earth. In just under 0.2 seconds, visual evidence indicated that the foam velocity dropped to about 1,000 mph (447 m/s). Since the foam slowed down while the Orbiter kept moving at 1,600 mph, as viewed from an observer on the ground, the Orbiter wing ran into the foam [1, p. 60]. The relative velocity between the foam and the wing at the moment of impact was therefore 600 mph (268 m/s)".

In the following analysis we will show that this relative velocity magnitude is consistent with simple Newtonian physics, provided that we make reasonable estimates of the mass of the foam, the distance between the foam breakaway point on the external tank and the wing (Figure 4.1), and the aerodynamic drag force on the foam. A straightforward solution of the equations of motion with a prescribed force can either be carried out in a ground-fixed coordinate system (the wing runs into the decelerated foam) or in a coordinate system moving with the Shuttle velocity at the time of the foam breakaway; a so-called "wind tunnel" or moving-coordinate system in which it appears that the wind accelerates the foam towards a stationary wing located downstream at some distance Δx. The two coordinate systems are depicted in Figure 4.2.

The mathematical treatment is more readily handled in a moving coordinate system, so we define our initial conditions and drag model in this way. For example, in this moving framework (Figure 4.2, bottom) the attached foam is stationary at the release time ($t = 0$). Once the foam has "broken free" from the Shuttle ($t > 0$), it is accelerated by an aerodynamic drag force in the positive x direction to a speed of several hundred meters per second as it is transported over a distance of about 58 feet (17.7 m). The drag force on the foam is created by an oncoming "wind tunnel" flow with a constant speed V_∞ of about 1,600 mph (715 m/s), that flows over the foam body.

For this one-dimensional *Shuttle-fixed* coordinate system, Newton's second law for a point object with mass m moving at velocity $v(t)$ and subject to a drag force F_d acting in the $+x$ direction is given by

$$m\frac{dv}{dt} = F_d \tag{4.1}$$

From basic fluid mechanics we know that "drag" is the component of the total force exerted by a fluid stream on a body (with combined pressure and shear induced contributions) that is *parallel* to the undisturbed velocity vector $\vec{V_\infty}$ and acts in the direction of that vector. The lift is the component of force on the body that is

Sec. 4.3]   Modeling the impact velocity of a piece of foam debris   179

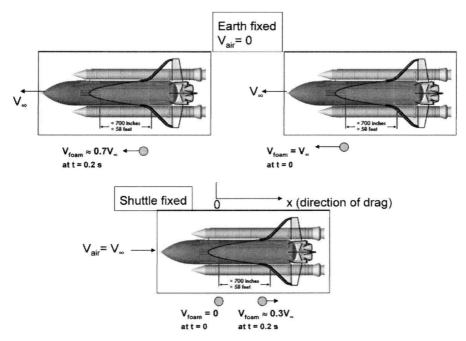

Figure 4.2. Point foam mass, designated above by stationary and moving blue circles, acted upon by a drag force in the $+x$ direction for: (top) Earth fixed coordinate system; (bottom) Shuttle-fixed or "wind tunnel" coordinate system.

perpendicular to the free stream velocity vector. We will assume that the effects of lift on the ideal foam debris trajectory are small relative to drag in our simplified dynamics model, although this is not correct if tumbling or dynamic rotation of the foam is significant. From the bottom sketch of Figure 4.2 we note that the undisturbed (or free-stream) velocity magnitude V_{air} in our "wind tunnel" coordinate system is equal to the speed of the Shuttle at the time that the foam breaks away. We will designate this speed by the symbol V_∞. The direction of the local free-stream velocity vector is assumed, for simplicity, to be parallel to the principal axis of the Shuttle.

From dimensional analysis, the pressure and shear stresses on a body fully immersed in a fluid stream scale directly with the free-stream dynamic pressure, q, a fluid mechanics measure proportional to the square of the *relative* velocity between air and body, i.e.

$$\text{dynamic pressure} \equiv q = \tfrac{1}{2}\rho_{air}V_{rel}^2$$

where ρ_{air} is the free stream air density.

Note that the units for the dynamic pressure are N/m^2 or pascals, Pa. The overall drag force on the body is proportional to q multiplied by a characteristic area A of the body, usually taken to be the frontal area exposed to the flow direction. The non-dimensional constant of proportionality is the drag coefficient C_d that depends, in turn, on the local Reynolds and Mach numbers (R_∞, M_∞) and on the particular body shape. For simplicity, based on published experimental data for bluff or blunt bodies, we shall assume C_d to be a constant of order 1 [6].

Using the above scaling arguments, the drag force is written in standard form as

$$F_d = (\tfrac{1}{2}\rho_{air} V_{rel}^2) C_d A \tag{4.2}$$

For our wind-tunnel-equivalent problem, the velocity of the oncoming flow *relative* to our foam "body" moving at speed v is given by

$$V_{rel} = V_\infty - v \tag{4.3}$$

Inserting Equations 4.2 and 4.3 into Equation 4.1 yields the one-dimensional form of Newton's second law for a point mass "body" acted upon by a velocity-dependent drag force

$$m\frac{dv}{dt} = (\tfrac{1}{2}\rho_{air}(V_\infty - v)^2) C_d A \tag{4.4}$$

At time t, the straight-line displacement of the "body" from its fixed breakaway location (Figure 4.2) is defined to be $s(t)$. The body displacement is positive along the x-axis for the free-stream velocity vector.

Since the velocity $v(t)$ is the time rate of change of its displacement

$$v = \frac{ds}{dt}$$

the acceleration can be rewritten as the rate of change of velocity with respect to displacement, dv/ds, multiplied by the velocity v, since

$$\frac{dv}{dt} = \frac{ds}{dt}\frac{dv}{ds} = v\frac{dv}{ds} \tag{4.5}$$

Using Equation 4.5, another way of writing Newton's second law is

$$mv\frac{dv}{ds} = (\tfrac{1}{2}\rho_{air}(V_\infty - v)^2) C_d A \tag{4.6}$$

Being a first-order differential equation, Equation 4.6 can be integrated to solve for v as a function of the independent distance variable s, subject to the initial condition

$$v = 0 \quad \text{at } s = 0 \tag{4.7}$$

Recall that our foam "body" is stationary at $t = 0$ in the wind-tunnel coordinate system, i.e. in which the foam displacement s is equal to zero.

We now seek to obtain an approximate solution of Equation 4.6 that is valid for small s (or small time t), i.e. for short travel distances when the foam velocity v is considerably smaller than V_∞. Under this assumption, the drag force on the right side of Equation 4.6 can be taken to be approximately constant. Later, we will find the exact solution to Equation 4.6 when $v(s)$ is not small compared to V_∞. For this more general case, the drag force varies with s (or v) in our model differential equation.

4.3.2 The constant drag (small time) approximation

Dividing both sides of Equation 4.6 by the mass m, then setting the acceleration F_d/m equal to a constant $= F_d(t=0)/m = a_0$, leads us to the following differential equation for v in the constant drag approximation

$$v\frac{dv}{ds} = \frac{d}{ds}\left(\frac{v^2}{2}\right) = a_0 \quad (4.8)$$

Using the drag relationship, Equation 4.2 evaluated at $s=0$ and $t=0$, the constant acceleration a_0 is

$$a_0 = \frac{1}{2m}\rho_{\text{air}}(V_\infty)^2 C_d A \quad (4.9)$$

Integrating both sides of Equation 4.8 subject to the initial condition specified in Equation 4.7, yields the solution for a point mass subject to a constant acceleration a_0 that we all learned in our first physics course

$$v^2 = 2a_0 s \quad (4.10)$$

Note that the above equation is also a form of the law of the conservation of energy for a point mass subject to a constant force F_d,

$\Delta(\text{Kinetic Energy}) = $ Work done by drag force acting over a distance s, i.e.

$$\Delta(\tfrac{1}{2}mv^2) = F_d s = m a_0 s$$

For a particle that moves a distance $s = x$, the velocity from Equation 4.10 is

$$v(x) = \sqrt{2a_0 x} \quad (4.11)$$

We can determine the time t that it takes to move a point mass a distance x by integrating the differential equation for the reciprocal of v, $\frac{dt}{ds} = \frac{1}{v}$, to yield

$$\left.\begin{array}{c} t = \displaystyle\int_0^x \frac{ds}{v(s)} = \int_0^x \frac{ds}{\sqrt{2a_0 s}} = \frac{1}{\sqrt{2a_0}}(2\sqrt{x}) \\[2mm] \Rightarrow \quad t = \sqrt{\dfrac{2x}{a_0}} \end{array}\right\} \quad (4.12)$$

Using Equation 4.11, we now calculate a first-order quantitative estimate for the foam velocity v (in m/s) as a function of the distance x (in meters). To do this we will need good engineering estimates for the mass m of the foam debris, the free-stream air density ρ_{air}, the free stream velocity V_∞, the drag coefficient C_d, and the frontal area A of the foam debris.

The foam/wing *collision* velocity, v_c, can then calculated from Equation 4.11 by setting x equal to the distance, D, as measured from the foam release point to the wing. This is a distance of about 58 feet or 17.7 m (as depicted in Figure 4.2, which is based on [7, Fig. 2-1]. The time t_c it takes for the foam to travel that distance can be calculated from Equation 4.12 or from the following simple equation that combines Equations 4.11 and 4.12.

$$t_c = \sqrt{\frac{2D}{a_0}} = \sqrt{\frac{2D}{v_c^2/2D}} = \frac{2D}{v_c} \qquad (4.13)$$

Observe that t_c does not depend explicitly on a_0 in the expression $t_c = 2D/v_c$, but varies inversely with the collision velocity v_c. If we are given v_c empirically or if we can determine an estimate for it, for example from Equation 4.11 for a given acceleration a_0, then we can calculate t_c for any distance D. Conversely, given an estimate for t_c, we can calculate the impact velocity v_c without specifically knowing the magnitude of a_0.

4.3.3 Analytically solving for the impact velocity and mass, given the time to impact

Solving Equation 4.13 for v_c yields a reciprocal relationship for the impact velocity as a function of the time to impact t_c

$$v_c = \frac{2D}{t_c} \qquad (4.14)$$

If we are given the time to impact t_c, for example based on the video footage of the actual foam impact event, then from Equation 4.14 we have a quick way to estimate the impact velocity v_c without knowing anything specific about the foam mass and the drag. As noted earlier, a simple BotE model was first proposed by Douglas Osheroff, a prominent member of the CAIB panel [3, 4] who obtained an estimate for v_c directly from Equation 4.14. We will present his calculation in Section 4.3.6, the part of our BotE analysis entitled "the inverse problem".

In addition to Equation 4.14, the initial acceleration due to drag, a_0, can readily be calculated from a similar expression

$$a_0 \approx \frac{v_c}{t_c} = \frac{2D}{(t_c)^2} \qquad (4.15)$$

that is valid for small values of time, t_c.

Using Equation 4.9, which defines a_0 in terms of the drag force divided by the mass, the unknown foam mass m can then be quickly estimated from

$$m = \frac{1}{2a_0} \rho_{air}(V_\infty)^2 C_d A = \frac{(t_c)^2}{4D} \rho_{air}(V_\infty)^2 C_d A \qquad (4.16)$$

provided that we also prescribe values for the drag-area product parameters ($C_d A$) and the dynamic pressure, $\frac{1}{2}\rho_{air}(V_\infty)^2$.

In the next section, we will proceed in two steps. In Step 1, we first estimate the derived impact quantities v_c and t_c assuming that m and $C_d A$ are given. The ratio of object weight, mg, to the drag parameters $C_d A$ that appear in Equation 4.9 is typically defined by a single parameter called the ballistic coefficient, BN. This quantity is often used in the scaling of both simple and complex trajectories. We will discuss the importance of BN and estimate its magnitude for our problem as part of our Step 1 analysis. We then follow up this "direct" Step 1 problem, and *invert* the solution in Step 2 solving for mass m when t_c is given. We generate a Step 2 estimate for the foam mass m using Equation 4.16 and an estimate for the impact velocity v_c based on Equation 4.14, assuming that the impact time t_c is a known quantity. We will use the approximate value for t_c deduced by the CAIB team based on movies of the foam impact event.

4.3.3.1 An estimate of impact velocity and time to impact, given the mass and drag properties ("the direct problem")

Let's assume that the input data are given by the values in the final CAIB report (Table 4.1). For the constant-drag approximation, the initial acceleration parameter can be calculated by using Equation 4.9 once we specify the product of the drag coefficient C_d and the frontal area A.

4.3.3.2 *Estimating the drag coefficient*, C_d

Basic wind tunnel measurements show that the drag coefficient for a bluff body like a foam wedge or cube has a value on the order of 1.0. Particular C_d values for a number of different geometries and Mach numbers can be found in several

Table 4.1. STS-107 flight data [7, CAIB report, Vol. II, App D08, Oct. 2003].

Trajectory parameter	Tabulated values for STS-107 CAIB report, Volume II, App. D08	
Time after launch for foam break-off	81.7 s	
Shuttle velocity, V_∞	708 m/s	(1,585 mph)
Shuttle altitude, z	19.9 km	(65,280 ft)
Air density, $\rho_{air}(z)$	0.092 kg/m^3	(7.1% of sea-level ρ_{air})
Estimated foam mass, m (in kg)	0.54–0.76 kg	(CAIB trajectory analysis)
Estimated foam dimensions, triangular-pyramid shape (Figure 4.3)	Base = 0.48 m Height = 0.29 m Thickness = 0.14 m	(19.0″) (11.5″) (5.5″)
Distance from foam break-away to wing impact location, D (meters)	17.7 m	

aerodynamic references, e.g. "Fluid-dynamic drag" by S.F. Hoerner [8]. The values typically fall in the range

$$1.0 < C_d < 1.7$$

For our sample calculation, we arbitrarily select a value of C_d at the mid-point of the bluff-body range

$$C_d = 1.35$$

4.3.3.3 Estimating the reference frontal cross-sectional area, A

By definition the reference area for the foam, A, is the frontal area exposed to the free stream. The magnitude of A is difficult to determine since the foam shape is not well-known and since the foam tumbles considerably during its short lifetime. After extensive analysis and testing, CAIB trajectory image analysts estimated the rate of rotation or tumbling to be about 18 revolutions per second [7].

We will start our simplified estimate for A by assuming that the foam object is a triangular wedge shape (Figure 4.3) with the dimensions given in Table 4.1. If the largest face of the foam pyramid is always perpendicular to the one-dimensional wind-tunnel velocity vector, then an estimate for the *maximum* frontal area would be

$$A_{\max} = \tfrac{1}{2}(\text{base} \times \text{height}) = \tfrac{1}{2}(0.48\,\text{m} \times 0.29\,\text{m}) = 0.07\,\text{m}^2$$

If the smallest triangular face of the pyramid is always normal to the flow, the *minimum* area would be approximately

$$A_{\min} = \tfrac{1}{2}(\text{thickness} \times \text{height}) = \tfrac{1}{2}(0.14\,\text{m} \times 0.29\,\text{m}) = 0.02\,\text{m}^2$$

For our sample calculation, we choose a value for A at the mid-point of this range, $A = 0.045\,\text{m}^2$, because the high rate of tumbling will cause the foam to present a number of different faces to the wind over a short period of time and any foam orientation is possible.

4.3.3.4 Estimating the mass of the foam debris, m

The foam mass estimates deduced by CAIB analysts from numerical trajectory studies based on movies showing the breakaway process (Table 4.1) lie in the range

$$0.54 < m < 0.76\,\text{kg}$$

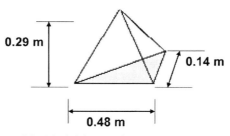

Figure 4.3. Model foam triangular-pyramid shape.

The mid-point of that range, $m = 0.65$ kg, is consistent with a foam volume of 0.017 m^3 if we assume a nominal foam density of 38.4 kg/m^3 [2, 7]. This volume corresponds to that of a cube with each side 0.26 m in length.

4.3.3.5 Impact velocity estimate

Using the above estimated values for C_d, A, and m for a triangular-pyramid of foam, plus the flight data listed in Table 4.1 as input, we use Equation 4.9 to calculate the following magnitude for the initial foam acceleration

$$a_0 = \frac{1}{2m}\rho_{air}(V_\infty)^2 C_d A = \frac{1}{2(0.65)}(0.092)(708)^2(1.35)(0.045) = 2.16 \times 10^3 \text{ m/s}^2$$

Observe that a_0 is about 220 times the acceleration of gravity at the earth's surface, $g = 9.81$ m/s^2. This is a very large acceleration!

Evaluating Equation 4.11 we see that the impact velocity for a piece of foam debris undergoing this degree of acceleration over a distance $D = 17.7$ m is

$$v_c = \sqrt{2a_0 D} = \sqrt{2(2.16 \times 10^3)17.7} = 276 \text{ m/s} \quad \text{[BotE solution]} \quad (4.17)$$

Using the available visual evidence and the supercomputer trajectory analysis conducted by the CAIB study teams, the best reported impact speed estimate calculated after the STS-107 mission was 528 miles/hr [7], or

$$v_c = 236 \text{ m/s} \quad \text{[CAIB best-estimated impact speed]} \quad (4.18)$$

Our first-order estimate of 276 m/s for the impact speed, based on the one-dimensional drag-force equation of motion, is within 17% of the CAIB estimate. Of course, we had some important help. We input into our BotE solution *a-priori* information for the average values of m and A based on the CAIB estimates for mass and foam dimensions. However, and this is the point, we did not rely on any large-scale computer calculations to estimate the foam impact speed.

4.3.3.6 Time to impact estimate ("the direct problem", *continued*)

Using Equation 4.13, we calculate the corresponding time to impact, t_c, using our BotE result for v_c

$$t_c = \sqrt{\frac{2D}{a_0}} = \frac{2D}{v_c} = \frac{2(17.7)}{276} = 0.13 \text{ s} \quad \text{[BotE solution]} \quad (4.19)$$

Again this number is consistent with the CAIB best estimate for the impact time [7] of

$$t_c = 0.17 \text{ s} \quad \text{[CAIB best estimated time to impact]} \quad (4.20)$$

Our BotE solution is within 24% of the CAIB value for impact time.

In addition to the calculated estimates presented here, we will show, using our simple model, the direct dependence of our calculated impact speed and time on the mass, geometry, and drag coefficient parameters. Furthermore, by grouping the three input parameters of m, A, and C_d into a composite quantity called the "ballistic number", or BN, we can estimate the range of uncertainty for calculated impact velocities and impact times solely from the uncertainty in BN. This gives us a simple means of relating the accuracy of our impact calculations to the uncertainty of the input data.

4.3.3.7 The ballistic number, BN, and its use as a key scaling parameter

The model presented above shows the importance of the initial acceleration, a_0, in the estimation of the impact velocity and impact time (Equations 4.9, 4.11, and 4.13). We observe that the magnitude of the normalized acceleration, a_0/g, depends strongly upon the three principal mass and drag parameters, m, A, and C_d, through the relationship

$$\frac{a_0}{g} = \frac{Drag}{mg} = \frac{1}{2mg}\rho_{air}(V_\infty)^2 C_d A = \left[\frac{1}{(mg/C_d A)}\right][\tfrac{1}{2}\rho_{air}(V_\infty)^2] \qquad (4.21)$$

Three parameters are collected together as a group in the bracketed denominator of Equation 4.21, and this group is defined to be the ballistic number

$$BN \equiv \frac{mg}{C_d A} = \frac{\text{weight}}{\text{(effective drag area)}} \qquad (4.22)$$

The ballistic number is not dimensionless; it has units of force/area. If we use the free-stream dynamic pressure $q_\infty = [\tfrac{1}{2}\rho_{air}(V_\infty)^2]$, then a simple expression for the normalized acceleration of a moving object is given by

$$\frac{a_0}{g} = \frac{q_\infty}{BN} \qquad (4.23)$$

Thus the normalized acceleration is equal to the ratio of the dynamic pressure q_∞ to the ballistic number BN. Thus BN is a measure of the relative sensitivity of an object's acceleration to the aerodynamic drag force applied to a given mass. A very high value of BN (relative to q_∞) produces low accelerations, and a lower BN yields high accelerations. For our foam acceleration problem, we estimated $a_0/g = 219.7$, which is very high. For a Shuttle velocity of 708 m/s at an altitude of about 20 km, the free-stream dynamic pressure $q_\infty = 2.3 \times 10^4$ Pa. Hence the magnitude of BN for our foam problem is

$$BN = \frac{q_\infty}{a_0/g} = \frac{2.3 \times 10^4}{219.7} \approx 105 \text{ Pa} \qquad (4.24)$$

In trajectory analyses, a BN of about 100 Pa is considered to be a low value. BN values for manned vehicles reentering the earth's atmosphere are considerably higher (by several orders of magnitude) because, to protect the vehicle and crew from high reentry deceleration forces, they are designed to produce much smaller accelerations.

4.3.3.8 Uncertainties in estimating velocity and impact time related to uncertainty in BN

As we have seen, the accuracy of our BotE calculations for the impact velocity v_c and impact time t_c (Equations 4.17 and 4.19) depends significantly on how well we can estimate m, A and C_d which, when combined, form the ballistic number, BN (Equation 4.22). As a result of this convenient mathematical grouping, the size of our relative uncertainty in the calculated magnitude of v_c (let's define it as $\delta v_c/v_c$) can be specifically related to our uncertainty in the ballistic parameter $[\delta(BN)/BN]$.

The dependence of v_c on BN is obtained by replacing the acceleration a_0 in Equation 4.17 with the equivalent expression given in Equation 4.23, from which we obtain the specific dependence of v_c on BN at a particular distance $x = D$

$$v_c = \sqrt{2a_0 x} = \sqrt{\frac{2g \times q_\infty}{BN}} = V_\infty \sqrt{\frac{\rho_\infty g D}{BN}} \qquad (4.25)$$

In Figure 4.4 we plot v_c as a function of BN using the values for D and ρ_∞ in Table 4.1.

After taking the differential of both sides of Equation 4.25, the relative error (or perturbation) in the overall impact velocity, v_c, is found to be simply proportional to the negative of the relative error (or perturbation) in BN

$$\frac{\delta v_c}{v_c} = -\frac{1}{2}\frac{\delta(BN)}{BN} \qquad (4.26)$$

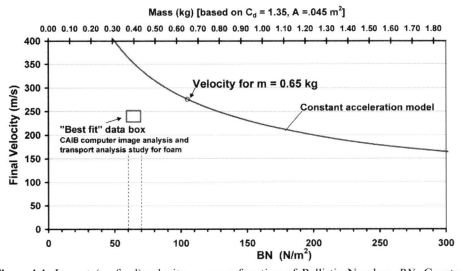

Figure 4.4. Impact (or final) velocity, v_c, as a function of Ballistic Number, BN. Constant acceleration model (Equation 4.25) compared to the CAIB "best fit" data uncertainty box [1, 7]. Note that our initial estimate for v_c (Equation 4.17 evaluated for a mass $m = 0.65$ kg) is plotted as a small circle centered at coordinates $v_c = 276$ m/s and $BN = 105$ N/m^2.

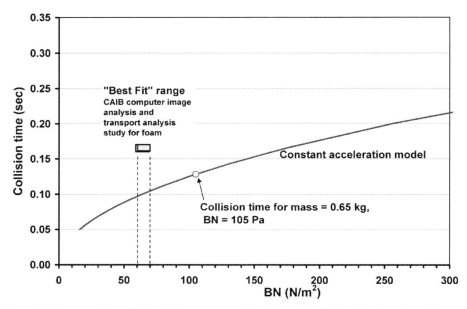

Figure 4.5. Impact (or collision) time, t_c, as a function of Ballistic Number, BN. Constant acceleration model (Equation 4.17) compared to "best fit" data uncertainty range [1, 7]. Our initial estimate for t_c with mass $m = 0.65$ kg (Equations 4.17 and 4.19) is plotted as the "circle" at collision time of 0.13 s corresponding to $BN = 105$ Pa for the values of C_d and A given earlier.

As an example of applying Equation 4.26, we observe that a +18% bias, or a positive shift in the estimated BN (which is the estimated BN uncertainty deduced by CAIB analysts for our foam problem) produces a −9% bias, or a negative shift, in our estimate of impact velocity v_c. This is confirmed in Figure 4.4, where the CAIB "best fit" error box is shown as a rectangle with "height" of 10% $[\delta v_c/v_c]$ for a "base" dimension of 18% $[\delta(BN)/BN]$.

The corresponding expression for the dependence of impact time t_c on BN is readily derived from Equations 4.13 and 4.25 as

$$t_c = \frac{2D}{v_c} = \frac{2}{V_\infty}\sqrt{\frac{(BN)D}{\rho_\infty g}} \qquad (4.27)$$

Figure 4.5 (the counterpart of Figure 4.4) plots t_c as a function of BN using the values for g, D, and ρ_∞ in Table 4.1.

The relative perturbation relationship for t_c is similar to Equation 4.26 but with a positive sign

$$\frac{\delta t_c}{t_c} = +\frac{1}{2}\frac{\delta(BN)}{BN} \qquad (4.28)$$

where a positive perturbation in BN relative to a reference value will produce a positive perturbation in t_c, but with half the relative change.

4.3.4 Summary of results for constant acceleration model compared to data

When we allow for the uncertainties in our estimates of the parameters which contribute to the ballistic number, we see that our constant acceleration model calculations for impact velocity v_c as a function of BN are about 50% *above* the best fit data range plotted in Figure 4.4. Our constant acceleration model results for collision time vs BN (Figure 4.5) lie *below* the corresponding best fit data for t_c by about 50% because t_c is inversely proportional to v_c. While our simplest BotE solution is "in the ball park" and exhibits the basic trends in the data, we are led to ask whether we can obtain better agreement with the published results of the CAIB by simply changing one or more of our assumptions or parameters?

We will show in the next section that a significant improvement in our results can be made by *dropping* the constant acceleration assumption. Recall that we made this assumption for mathematical simplicity and because it would help us clearly understand the importance and dependence of our results on the key input parameters. By finding the "closed-form" solution to Equation 4.6 where the drag force is no longer presumed to be constant, we will be able to quantify the level of improvement. We will be guided by an analytical integral expression that can be readily derived by first-year calculus students.

4.3.5 The non-constant acceleration solution

If we solve Equation 4.6 for the differential distance, ds, and then integrate both sides of the resulting equation from the foam breakaway point at $s = 0$, $v = 0$ to a value of $s = x$ and $v = v(x)$ we obtain the following expression

$$\left(\frac{\rho_{\text{air}} C_d A}{2m}\right) \int_0^x ds = \int_0^{v(x)} \frac{v}{(V_\infty - v)^2} dv \qquad (4.29)$$

We can rewrite this in a convenient non-dimensional form to facilitate finding an exact integral of the right-hand side by defining the following non-dimensional quantities

$$\tilde{v} \equiv \frac{v}{V_\infty} \qquad (4.30)$$

$$\tilde{x} \equiv \left(\frac{\rho_{\text{air}} C_d A}{2m}\right) x = \frac{x}{[2(BN)/(\rho_{\text{air}} g)]} = \frac{x}{X_s} \qquad (4.31)$$

where the effective distance scale X_s is defined by $X_s = [2(BN)/(\rho_{\text{air}} g)] = V_\infty^2/a_0$. With these definitions, the non-dimensional scaled version of Equation 4.29 is

$$\tilde{x} = \int_0^{\tilde{v}(\tilde{x})} \frac{\tilde{v}}{(1 - \tilde{v})^2} d\tilde{v} \qquad (4.32)$$

An exact relationship between $\tilde{x}$ and $\tilde{v}$ is obtained from Equation 4.32 by the use of a closed-form expression for the integral on the right side of the equation. The inverse result, $\tilde{v}$ as a function of $\tilde{x}$, can be obtained by solving the resulting transcendental equation for $\tilde{v}$ in which $\tilde{x}$ is treated as the independent variable.

Dwight's table of integrals [9] gives the indefinite integral on the right of Equation 4.32 as

$$I(z) = \int \frac{z}{(1-z)^2} dz = \ln(1-z) + \frac{1}{(1-z)}$$

Applying the upper and lower limits shown in Equation 4.32 to the evaluation of the definite integral form of $I(z)$, we arrive at the following equation for $\tilde{x}$ as a function of $\tilde{v}$

$$\tilde{x} = \left[\ln(1-z) + \frac{1}{(1-z)}\right]_0^{\tilde{v}} = \ln(1-\tilde{v}) + \frac{\tilde{v}}{(1-\tilde{v})} \quad (4.33)$$

For very small velocities compared to 1, we can expand the right-hand side of Equation 4.33 in a Taylor's series to obtain the following small $\tilde{v}$ approximation for $\tilde{x}(\tilde{v})$

$$\tilde{x} = -\left[\tilde{v} + \frac{\tilde{v}^2}{2} + \frac{\tilde{v}^3}{3} + \cdots\right] + \tilde{v}[1 + \tilde{v} + \tilde{v}^2 + \cdots] = \frac{\tilde{v}^2}{2} + \frac{2\tilde{v}^3}{3} + \cdots \quad (4.34)$$

Note that the leading quadratic term on the far right, valid when $v/V_\infty \ll 1$, is the non-dimensional equivalent to our constant acceleration solution $x = \frac{v^2}{2a_0}$.

Figure 4.6, a plot of $\tilde{v}$ as a function of $\tilde{x}$, using Equation 4.33 with $\tilde{x}$ as the independent variable, shows us, that the rate of change of velocity with $\tilde{x}$ is noticeably lower than that of the constant acceleration solution, $\tilde{v} = \sqrt{2\tilde{x}}$, for values of $\tilde{x}$ greater than about 0.01.

One might ask why the non-constant acceleration solution produces smaller velocities at a given displacement distance than were obtained from our approximate constant acceleration solution? The answer comes from the physics and mathematical model for the drag force (Equations 4.2 and 4.3).

Shortly after being set "free", the foam particle accelerates (closely in accord with Equation 4.11) in response to a nearly constant drag force that has a strength proportional to $(V_\infty)^2$. However, as the particle velocity increases, the drag force on the "particle" begins to diminish, thereby slowing the acceleration of the foam. This continuous reduction in drag arises because as the particle accelerates, its velocity relative to the free stream velocity, $V_{\text{rel}} = [V_\infty - v(x)]$, diminishes. Because the drag force is proportional to the square of the relative velocity, $F_d \sim (V_{\text{rel}})^2$, it is evident that for particle velocities that have accelerated to only 10% of V_∞ (about 70 m/s) the "exact" drag force is reduced to 81% of its initial value. For velocities of order 20% of V_∞ the drag force drops to 64% of its initial value. The reduction in drag force with distance accounts for the lower rate of acceleration shown in Figure 4.6 for our non-constant acceleration solution (Equation 4.33). Observe the widening divergence in velocity compared to the constant acceleration solution (Equation 4.11).

For the length and mass dimensions we selected for our numerical example, the difference in velocity is almost 60 m/s at a dimensional distance $D = 17.7$ m for a ballistic number of 105 Pa. Note that if $D = 17.7$ m is normalized according to

Sec. 4.3] Modeling the impact velocity of a piece of foam debris 191

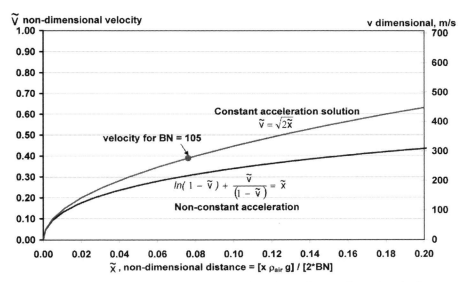

Figure 4.6. Non-dimensional velocity $\tilde{v}$ as a function of non-dimensional distance $\tilde{x}$. Comparison of exact solution to approximate constant-acceleration solution for velocity as a function of distance.

Equation 4.31, it corresponds to the non-dimensional distance $\tilde{x} = 0.076$. If we lower BN to about 60 Pa (the approximate average value proposed by the CAIB, as shown in Figure 4.4), the difference in velocity increases to about 98 m/s. For this value of BN our "exact" solution yields an impact velocity of 267 m/s (i.e. for $\tilde{x} = 0.133$ in Figure 4.6). This "exact" velocity is within 13% of the calculated best estimated impact velocity of 236 m/s published by the CAIB (Equation 4.18) for an estimated $BN = 60$.

Another way of showing the improvement in our extended BotE results for the impact velocity is presented in Figure 4.7. This figure plots impact (or final) velocity as a function of BN with distance $x = D$ fixed, and is similar to Figure 4.4 but has been amended to include our "exact" non-constant acceleration solution. Note that with the left-hand side of Equation 4.33 written as a ratio of *dimensional* terms with BN in the denominator, the direct dependence of impact velocity ratio v/V_∞ on BN is clearly seen in the following expression

$$\left. \begin{array}{c} \ln(1-\tilde{v}) + \dfrac{\tilde{v}}{(1-\tilde{v})} = \tilde{x} = \dfrac{D}{[2(BN)/(\rho_{\text{air}} g)]} \\ \text{where} \quad \tilde{v} = \dfrac{v}{V_\infty} \end{array} \right\} \quad (4.35)$$

We observe from Figure 4.7 that the "exact" non-constant acceleration curve $v_c(BN)$ lies below the constant acceleration approximation, and is closer to the CAIB "best fit" data box range than is our initial constant acceleration result.

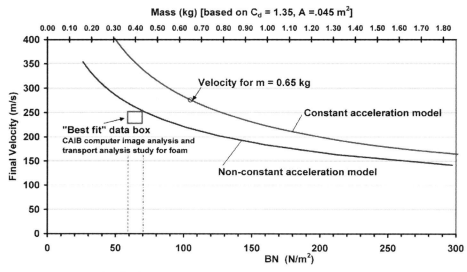

Figure 4.7. Impact (or final) velocity, v_c, as a function of Ballistic Number, BN. The non-constant acceleration solution (Equation 4.35) and the constant acceleration solution (Equation 4.25) are compared to the CAIB "best fit" data uncertainty box [1]. Note that our initial constant acceleration estimate for v_c with mass $m = 0.65$ kg (Equation 4.17) is plotted as the circle with axial coordinate $BN = 105$ N/m^2.

The non-constant acceleration solution for the dependence of impact time t_c on ballistic number BN, can readily be derived from the non-dimensional form of Newton's second law (Equation 4.4) and from the time rate of change or particle displacement, $\dfrac{d\tilde{x}}{d\tilde{t}} = \tilde{v}$; this is left as an exercise for the reader.

The resulting equation for non-dimensional time $\tilde{t}$ as a function of $\tilde{x}$ (which is similar to Equation 4.33, relating non-dimensional velocity $\tilde{v}$ to $\tilde{x}$) is given by

$$\left. \begin{array}{c} \tilde{t} - \ln(1+\tilde{t}) = \tilde{x} = \dfrac{D}{[2(BN)/(\rho_{\text{air}}g)]} \\ \text{where} \quad \tilde{t} = \dfrac{t}{T_s} = \dfrac{t}{X_s/V_\infty} \end{array} \right\} \quad (4.36)$$

and the effective time scale T_s is defined by $T_s = [2(BN)/(\rho_{\text{air}}gV_\infty)] = V_\infty/a_0$.

In Figure 4.8, the dimensional impact time, t_c, is plotted as a function of BN using the values for g, D, and ρ_∞ in Table 4.1. Both the non-constant acceleration model and the constant acceleration model are shown. It is the companion plot to Figure 4.7. For the *non-constant* acceleration model, the impact (or collision) times exceed those given by the *constant* acceleration results by about 0.02 s. This increase in modeled impact time is as expected because the non-constant acceleration solution

Sec. 4.3] Modeling the impact velocity of a piece of foam debris 193

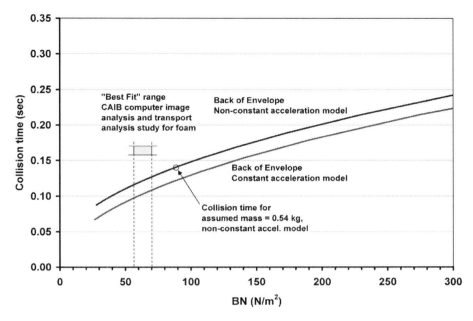

Figure 4.8. Impact time, t_c, as a function of Ballistic Number, BN. Comparison of non-constant acceleration solution (Equation 4.36) with constant acceleration solution (Equation 4.27) and with the "best fit" range [7].

yields smaller impact velocities as a function of distance in comparison to the constant acceleration model. The more "exact" results follow quite closely the simple reference trends with BN, Equation 4.27, i.e. t_c increases nearly proportional to $\sqrt{BN}$ or $\sqrt{\text{mass}}$ for a wide range of foam mass and drag parameters. In comparison to the "best fit" range for BN shown in Figure 4.8, the non-constant acceleration solution for t_c underestimates that of the CAIB image analysts by about 25% to 30%. This difference, or bias, may possibly be related to the fact that the foam did not actually follow a straight-line path parallel to the Shuttle centerline. Its initial trajectory, particularly near the breakaway point on the left bipod ramp of the external tank (see Figure 4.1) was rather more three-dimensional in character.

Nevertheless, our BotE results for the time to impact, based on a simplified one-dimensional trajectory, are certainly within the "ball park" of the published CAIB data; 0.12 to 0.13 seconds from our simple physics model compared to 0.16 to 0.17 seconds from the CAIB image analysis team.

In the next section we will use the CAIB estimate of approximately 0.16 s for the time to impact as a *known* quantity to input into our simple constant acceleration model in order to deduce the impact velocity and estimate the mass of the foam. This is the so-called "inverse" or Step 2 problem.

4.3.6 An estimate of impact velocity and particle mass, taking the time to impact as given (the "inverse" problem)

David Osheroff, Nobel Laureate, while serving as a member of the CAIB panel, produced a Back-of-the-Envelope calculation for the foam "strike" velocity which was an order of magnitude greater than the 50 miles/hour (22.4 m/s) that NASA administrator Sean O'Keefe suggested at a 2003 congressional hearing on the Columbia accident. According to a review article about Osheroff's efforts in support of the CAIB [3, Stanford Scientific Review, 2006], O'Keefe, in an effort to downplay the foam strike as a cause of the accident, testified that: "The piece (of foam) came off (and) dropped roughly 40 feet at a rate of something like 50 mph, so it's the functional equivalent, as one astronaut described to me, of a Styrofoam cooler blowing off of a pickup truck ahead of you on a highway!"

Osheroff's simple BotE pre-calculus physics calculation (featured in an original draft of the Stanford article) was an example of an "inverse" problem. As will be shown below, Osheroff came up with a significantly higher impact velocity estimate, one that might have lead NASA to quite a different conclusion as to the possible effect of the foam strike.

Osheroff used NASA's post-launch photographic analysis results to generate the given information for the inverse velocity estimation problem. We outline his solution here using the constants that he selected.

We are given

- Time from foam breakaway to foam impact: $t_c \approx 0.16$ s
- Distance from the bipod ramp to the leading edge of the wing, $D \approx 19$ m

The BotE calculation for v_c is

$$v_c = \frac{2D}{t_c} = \frac{2(19 \text{ m})}{0.16 \text{ s}} \approx 238 \text{ m/s} \tag{4.37}$$

Osheroff estimated the "strike velocity" (the same as our v_c) using a constant acceleration solution that is equivalent to our Equation 4.14. This was apparently derived using a constant drag-induced acceleration model in a moving Shuttle-based coordinate system. In fact, Osheroff calculated the constant acceleration magnitude a_0 first (our Equation 4.15), as noted in the Stanford Scientific Review article, using

$$a_0 \approx \frac{2D}{(t_c)^2} = \frac{2(19)}{(0.16)^2} = 1{,}484 \text{ m/s}^2 \quad \text{(or } 151g\text{)} \tag{4.38}$$

then evaluated v_c from

$$v_c = a_0 t_c = 1{,}484(0.16) \approx 238 \text{ m/s} = 532 \text{ mph} \tag{4.39}$$

The results, (Equations 4.37 and 4.39), are both algebraically and numerically identical, as is quite evident.

According to Robert Kennedy, P.E., who also wrote about Osheroff's BotE efforts [4], his calculations also included an estimate for the mass of the foam

Sec. 4.3] Modeling the impact velocity of a piece of foam debris 195

block, m. While the exact details are not available, we can make an educated guess as to how the foam mass might have been calculated by substituting into our Equation 4.16 Osheroff's initial estimates for air density, free stream velocity, drag coefficient, and effective drag area.

To calculate m in the "inverse" BotE problem using Equation 4.16

$$m = \frac{1}{2a_0} \rho_{\text{air}} (V_\infty)^2 C_d A = \frac{(t_c)^2}{4D} \rho_{\text{air}} (V_\infty)^2 C_d A$$

we need four additional parameters to be specified. Based on notes attributed to Osheroff [4], these are:

- Air density at altitude of foam shedding, $\rho_{\text{air}} \approx 0.1 \text{ kg/m}^3$
- Orbiter speed at time of foam shedding, $V_\infty = 700 \text{ m/s}$
- Drag coefficient, $C_d \approx 0.9$ [our guess based on Osheroff's result for mass]
- Foam effective frontal area, $A \approx 0.05 \text{ m}^2$ [i.e. for a rectangular foam frontal area of approximately $0.1 \times 0.5 \text{ m}$]

Using these parameters and Equation 4.38 for the magnitude of a_0, Osheroff's numerical estimate for mass is

$$m = \frac{1}{2a_0} \rho_{\text{air}} (V_\infty)^2 C_d A = \frac{1}{2(1,484)} [(0.1)(700)^2 (0.9)(.05)] = 0.74 \text{ kg} \quad (4.40)$$

The corresponding ballistic number is

$$BN = \frac{mg}{C_d A} = \frac{(0.74)(9.81)}{(0.9)(0.05)} = 162 \text{ N/m}^2 = 162 \text{ Pa} \quad (4.41)$$

An alternate expression for BN and mass showing the direct dependence on t_c (obtained from either Equation 4.27 or Equation 4.16) which also could be used in the calculations is

$$BN = \frac{(t_c)^2}{4D} \rho_{\text{air}} (V_\infty)^2 g \quad (4.42)$$

$$m = \frac{(t_c)^2}{4D} \rho_{\text{air}} (V_\infty)^2 C_d A \quad (4.43)$$

4.3.7 Comparing Osheroff's "inverse" calculations to our "direct" estimate results

In principle, for the constant acceleration approximation the derived mathematical relationships between the key variables t_c, v_c, and m, are guaranteed to yield the same results regardless of whether one adopts a "direct" or "inverse" calculation method, provided that all the "given" parameters are in effect the same. But it is also evident that the real-world input information often available to us in a given situation is almost always *not* the same; it depends not only on the type and accuracy of the data at hand, but also on the nature of the assumptions and approximations made when working either the direct or inverse problems.

Table 4.2. A comparison of the "direct" (given the foam mass) and "inverse" (given the time to impact; Osheroff [3, 4] BotE input parameters, together with a comparison of the magnitudes of the computed foam trajectory time, velocity, and acceleration.

	Direct (Equations 4.9–4.22)		Inverse/Osheroff: Eqs. 4.14 and 4.16	
	Given	Calculated	Given	Calculated
Drag coeff: C_d	1.35		0.9	
Drag area: A (m^2)	0.045		0.05	
Distance to wing: D (m)	17.7		19.0	
Air Density: ρ_{air} (kg/m^3)	0.092		0.100	
Orbiter velocity: V_∞ (m/s)	708.0		700.0	
Mass: m (kg)	0.65			0.74
Time to impact: t_c (sec)		0.13	0.16	
Velocity at impact: v_c (m/s)		276		238
Acceleration: a_0 (m/s^2)		2,160		1,484
Ballistic number: BN (Pa)		105		162

To illustrate these differences, let's examine Table 4.2 which shows the inputs and outputs based on the different BotE foam trajectory calculations that have been presented for the direct and inverse methods. In comparisons of various BotE estimations, the mathematical models employed might also differ, such as when one compares a baseline solution to a non-constant acceleration model or to a three-dimensional least-squares trajectory analysis. However, for this basic comparison we will ignore this added complication.

The biggest difference in the inputs between the two sets of calculations is in the uncertainty of the estimate for the drag coefficient. Both are of order unity, but the C_d for our direct input is 50% greater than the value inferred from Osheroff's inverse calculation. This directly affects the calculation of the initial acceleration a_0 and ballistic number. Note that BN is inversely proportional to C_d in the direct method (m given) and is totally independent of C_d in the inverse method (t_c given, as in Equation 4.42). Because a smaller C_d in Equation 4.43 yields a smaller particle mass, the effect of the lower assumed C_d on the calculation of m by the inverse method is mostly offset in the Osheroff case by the fact that the assumed impact time (0.16 s) for the inverse method is 23% *larger* than the value of t_c calculated using the direct method. That is, the calculated mass is proportional to $C_d(t_c)^2$ in Equation 4.43 for the inverse method. As a result of these two competing effects, the calculated masses for the two methods are quite close together: 0.65 kg vs 0.74 kg. The calculated impact velocities are also not far apart: 276 m/s vs 238 m/s, only a 16% difference.

4.3.8 Concluding thoughts on the impact velocity estimate

We now return to the question posed at the start of this chapter:

- *How could a lightweight piece of foam travel so fast and hit the wing of the Orbiter at over 500 miles/hour ($v = 223$ m/s) in under 0.2 seconds?*

The answer, as we have seen by our BotE analysis, is that a foam mass of about one kilogram and with less than $0.1\,\text{m}^2$ in cross-sectional area, was acted on by such a strong aerodynamic drag force (at a time of peak dynamic pressure in the Shuttle ascent phase) that it was ripped away from the external tank and then accelerated to over 200 m/s in less than 0.2 seconds.

We have also seen that differing BotE approaches, based on two similar physical models but somewhat dissimilar input information sets, can still produce estimates of relevant engineering quantities, like velocity and time, that are quantitatively and qualitatively in the same engineering ball park—in this example, our "direct" calculated value for impact time differed by approximately 25% from the *a-priori* time provided as a given for the "inverse" problem.

One key lesson demonstrated here is that when engineers are creating an engineering or physical model, they must: (a) always make sound judgments as to the physics of the problem, and (b) select "reliable" sources (data, tabulated reference dimensions, etc) as input to their calculations. Considerable practice and experience in working with Back-of-the-Envelope techniques is required to develop the confidence and understanding necessary to estimate and interpret the behavior of real-world engineering systems, devices, and processes.

In the next section we use our basic BotE methods and principles to estimate the load on the leading edge of the Orbiter's wing due to the impact of the foam. In Section 4.5 we then use our estimate of the magnitude of the foam/wing impact load, together with the known material properties of the leading-edge, to determine whether damage to (or failure of) the leading-edge is possible.

4.4 MODELING THE IMPACT PRESSURE AND LOAD CAUSED BY IMPACT OF FOAM DEBRIS WITH AN RCC WING PANEL

4.4.1 The impact load

The goal of this section is to create a simple physics-based framework for calculating the impact force, or load P_L, produced by a high velocity collision of a low density material (foam) with the high-density fairly-rigid wing material (RCC) on the wing of the Orbiter. In the subsequent section, we develop a simple BotE model for the resulting stresses in the RCC wing panel that indicates the possibility of structural failure of the wing, under certain conditions.

4.4.2 Impact overview

From our own experience and judgment we can expect that a piece of foam, on impact with a strong surface, will be easily crushed if the impact velocity normal to the wing surface, v_n, is greater than about 100 miles/hour (45 m/s). We've all seen the serious damage inflicted on automobiles and their occupants in a high-speed collision. But it is not so obvious that the foam-on-wing impact force, P_L, (which calculations and experiments indicate to be in the range of 6,000 to 8,000 lbf, or 27 kN to 36 kN) will be of sufficient magnitude to seriously damage a strong, stiff high-density material such as the reinforced carbon-carbon leading edge of the Shuttle Orbiter's wing. In fact, foam/wing impact experiments undertaken post-flight by NASA and the Columbia Accident Investigation Board (CAIB) revealed that leading edge RCC structural damage and indeed failure can occur, but only for specific foam/wing incidence angles and impact speeds [2]. The exact structural mechanism for such failure appeared to be fairly complex. Localized bending, buckling, and deformation stresses for the RCC panel (calculated in [2] from large-scale 3D finite-grid solid mechanics simulations of the foam impact event) were found to be so high that they exceeded published criteria for RCC material failure.

Although the duration of the impact of the foam on the panel is brief (only a few milliseconds, as we will show), the compressive loading pressure, or stress σ (N/m^2 or Pa) created at the interface between the two colliding materials, can be quite high. We first model σ for elastic collisions, and show that the magnitude of the loading pressure is directly proportional to the *perpendicular* component of the impact velocity, v_n. We then extend this simple stress model for foam compression to the elastic-plastic regime, and demonstrate a much-reduced dependence of σ on v_n in the case of impact stresses which exceed the crush (or yield) stress of foam. This significantly improves the accuracy of our impact load calculation. And finally in Section 4.5, a well-known theoretical solution for the "static" bending stress in a thin, flat plate produced by a uniformly distributed external load is used to model the impact stresses within the thin RCC wing panel. This solution is used to determine the range of impact speeds likely to produce failure-level stresses in the RCC material. We will show that the calculated stresses in our model wing "plate" or panel (with a thickness equal to that of the wing leading edge) may well exceed the measured failure stress levels published for RCC. We present a simple failure criterion for the leading edge of the wing based on two primary variables: (1) the magnitude of the impact velocity, and (2) the angle of incidence of the impact relative to the wing. For low velocities and low angles of incidence there is no failure, but we will see that failure is possible for high velocities and high incidence angles above the prescribed failure threshold.

We will compare the velocities likely to cause RCC wing failure to the foam impact velocities previously calculated for the foam drag problem, and it will be shown that these calculated impact velocities are sufficient to cause failure of the

Sec. 4.4]   Modeling the impact pressure and load caused by impact of foam   199

wing RCC material, but only for those incidence angles that lie above the stress-failure curve that we will develop from our models. Our failure estimates will be shown to be in reasonable agreement with the conclusions drawn by the CAIB and its investigation team for the Columbia accident event.

4.4.3 Impact load mathematical modeling

The product of the compressive impact stress or pressure, σ, with the effective loading area A_L determines the short-duration load P_L (newtons) delivered to the wing leading edge panel and its supporting rib structure

$$P_L = \sigma A_L \tag{4.44}$$

In Equation 4.44 we define A_L to be the area on the wing leading edge over which the foam/wing collision occurs; i.e. of order of $0.1\,\text{m}^2$.

To calculate the impact stress σ we proceed in two steps.

First, we model the dynamical impact process using one-dimensional elastic stress-wave mathematics, and show that the peak interfacial stress, or pressure, generated by the foam-RCC impact for the range of normal impact speeds of interest (v_n of order 40 to 100 m/s) typically exceeds the yield strength (i.e. the crush strength) for foam, Y or σ_{YFoam}. For high speed foam impacts against a nearly solid wall (such as a wing panel), a fast elastic compression wave is first generated in the foam object and is typically followed, milliseconds later (in certain materials), by a slower non-elastic stress wave or shock. This *two wave* compression process is a consequence of the bilinear character of the measured stress-strain curve for foam (Figure 4.9). Note the high stress-strain slope in the elastic region of this figure and the much shallower slope in the plastic region above the yield or "crush" stress, σ_{YFoam}. Since stress wave

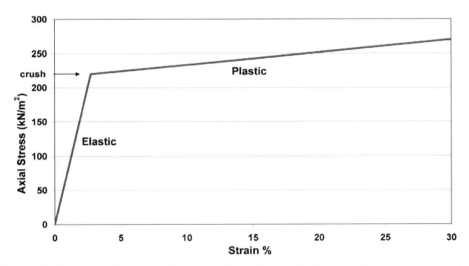

Figure 4.9. Stress as a function of strain for foam. The plot is based on the foam compression test data given in [2]; modeled here by a bilinear Hooke's law.

propagation speeds, C_0, are proportional to the square root of the stress-strain slope, $d\sigma/d\varepsilon$, the elastic precursor wave will propagate into the foam considerably faster than its non-elastic (plastic) wave companion. For the stronger RCC material, we will presume that the stresses generated within the material are below the RCC crush limit and therefore are elastic—at least in the early stages of the impact process, e.g. for post-impact times less than one millisecond.

It turns out that the equation we derive for the resulting interface stress σ, in the presence of a modeled plastic wave with an elastic precursor, is surprisingly similar to the expression for an elastic impact alone. We simply add a model for the jump in stress across the plastic wave to the magnitude of the yield stress for foam in order to arrive at our estimate for the total impact stress.

4.4.4 Elastic model for the impact stress

We start by constructing a basic elastic wave propagation model for the impact of a semi-infinite one-dimensional foam rod with a solid wall, which gives the rather simple result that the loading stress is proportional to the normal component of impact velocity v_n

$$\sigma_{\text{elastic}} = \rho_{\text{foam}} C_{\text{foam}} v_n \qquad (4.45)$$

We obtain Equation 4.45 by analyzing the problem of the impact of a purely elastic rod (initially unstressed and moving at velocity v_0) with a stationary rigid wall—and we will later argue that a rigid wall approximates the response of the RCC panel. The simple rigid wall impact is depicted in Figure 4.10 (we follow the schematic and derivation of Zukas [10]).

Note that the rigid wall (at $x = 0$) imposes the condition that the velocity of the material particles in the impacting rod that are closest to the wall must vanish (i.e.

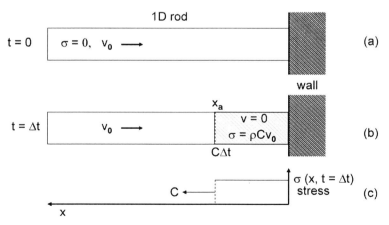

Figure 4.10. Schematic of unstressed one-dimensional rod at velocity v_0 impacting a rigid wall, (a) rod just prior to impact, $t = 0$, (b) rod at time Δt after impact; elastic wave has propagated a distance $x_a = C \Delta t$, (c) stress (x) at $t = \Delta t$.

$v = 0$ at $x = 0$). We know that to instantaneously drop the velocity of the nearest material particles in the rod to zero a compressive force F (or a corresponding impulse $\int F\,dt$) is required. Recall that one of the forms of Newton's second law of motion is that the change in momentum of a body (Δmv) is equal to the impulse I (the integral of the force on the body over time), i.e.

$$\Delta(mv) = I = \int F\,dt \qquad (4.46)$$

Let's think of the rod as being made up of a number of thin layers of material parallel to the end of the impacting rod. For $t > 0$, the deceleration to zero velocity of the first layer of particles must be balanced by a compressive force at the end of the rod which equals the product of the induced stress, or pressure σ, and the cross-sectional area of the rod, A. The force generated in the first layer leads to the deceleration of the second layer, extending the compression zone to the left. The second layer then decelerates the third layer, and so on. When the stresses in the first and second layers become equal, the motion between them ceases. This cessation of motion then moves to the third layer, and so on. In this way, a stress or velocity wave (that might be considered to be a propagating stress/velocity "shock" jump) develops and propagates along the axis of the rod to the left. This wave propagates at a finite speed, C (or C_e), because for an elastic material the layer-to-layer compression and deceleration process does not occur instantaneously. So C is defined to be the elastic wave velocity. Using Newton's second law and a simple linear Hooke's stress–strain relationship ($\sigma = S\varepsilon$), it can be shown that $C^2 = S/\rho$ where S is the slope of the stress–strain curve, and ρ is the density of the rod material.

In Figure 4.10 we use Newton's algebraic impulse relationship (Equation 4.46) in order to derive the simple dependence of the stress jump across the elastic wave to the change in the velocity across the wave produced by the moving rod encountering a solid stationary wall. This approximation is in keeping with our Back-of-the-Envelope approach. Using Figures 4.10b and 4.10c in a period of time Δt after impact, the stress wave generated in the impacting rod will have traveled a distance $C\Delta t$ to the plane x_a. The mass, m, of that portion of the rod, decelerated to zero velocity, is then $m = \rho AC\Delta t$. The momentum change for this portion of the rod is

$$\Delta(mv) = m\,\Delta v = (\rho AC\,\Delta t)(v_0 - 0) = \rho AC\,\Delta tv_0 \qquad (4.47)$$

The corresponding impulse required to produce this change in momentum is

$$I = F\,\Delta t = \sigma A\,\Delta t \qquad (4.48)$$

Equating 4.47 and 4.48 according to the impulse form of Newton's second law (Equation 4.46), we obtain the equation for the stress σ in the region behind the advancing stress wave

$$\sigma = \rho C v_0 \qquad (4.49)$$

Following Zukas [10], if the initial stress in the rod prior to the strike was other than zero, and if the particle velocity behind the wave was non-zero, then σ and v_0 in Equation 4.49 can be replaced by their changes from their pre-impact levels, producing the more general equation

$$\left. \begin{array}{c} \Delta\sigma = \rho C \Delta u \\ \text{or} \quad (\sigma_- - \sigma_+) = \rho C(u_+ - u_-) \end{array} \right\} \quad (4.50)$$

where u_+ is the particle velocity ahead of the wave (particle velocity is defined as positive when opposite the direction of motion of the wave) and u_- is the particle velocity behind the wave. In the same manner σ_+ is the material stress ahead of the wave and σ_- is the material stress behind the wave.

Following Zukas [10], the model for a single elastic rod can be extended to treat the case of the elastic impact of *two* separate rods made of different materials; for example a moving foam rod impacting a stationary RCC rod. Using this analysis, it can readily be shown that for a normal foam impact velocity $v_0 \approx 50\,\text{m/s}$, the velocity induced in the RCC rod (equivalent to u_-) is of order

$$(\rho C v_0)_{\text{foam}} / (\rho C)_{\text{RCC}} \approx 14\,\text{cm/s}$$

which is quite a low velocity mainly because the ratio of foam density to RCC density is so small. Since $14\,\text{cm/s} \ll 50\,\text{m/s}$, it is apparent that the *rigid-wall* approximation (Equation 4.49) yields a reasonable first-order estimate of the *elastic* impact stress for the foam/RCC impact. A rigid-wall assumption not only considerably simplifies our calculations of impact pressure but also the extended analysis where we consider non-elastic impact effects.

The magnitude of the elastic stress produced by a foam rod with density $\rho = 38.5\,\text{kg/m}^3$ and elastic wave speed $C = 450\,\text{m/s}$ impacting a rigid wall at $50\,\text{m/s}$ is calculated directly from our simple solution (Equation 4.49) to be

$$\left. \begin{array}{l} \sigma = \rho C v_0 = (38.5\,\text{kg/m}^3)(450\,\text{m/s})(50\,\text{m/s}) = 8.7 \times 10^5\,\text{N/m}^2 \\ \text{or} \quad \sigma = 126\,\text{lb/in}^2 = 8.6\,\text{atmosphere} \end{array} \right\} \quad (4.51)$$

We note that this impact stress, or equivalent pressure on our modeled wing surface, is quite large (more than 100 psi). This elastic estimate for σ (8.7×10^5 N/m^2) exceeds by about a factor of four the measured crush stress (or yield stress, Y) for foam $\approx 2.2 \times 10^5$ N/m^2. Y is indicated in Figure 4.9 by the arrow annotated with the caption "crush". It is evident that our elastic solution considerably over-estimates the impact stress or interface pressure, since we have not considered the stress relief (for a given impact velocity or induced strain) that occurs when foam is compressed beyond the yield point into the plastic regime. A somewhat more complex stress model is therefore needed to better estimate the impact stress beyond the crush limit for foam. Using the bilinear properties of the stress–strain curve for foam (shown in Figure 4.9), a simplified plastic or bilinear elastic model will be developed in the next sub-section to better estimate σ for impact stresses above the foam crush limit. We again focus our attention on stress waves solely generated on the foam side of the impact problem, utilizing the rigid-wall approximation that we employed above for the case of a purely elastic collision. This revised elastic/plastic model for the impact stress, together with an estimate of the impact area, provides us with an improved estimate for the maximum foam/RCC impact load.

4.4.5 Elastic–plastic impact of a one-dimensional rod against a rigid wall

The bilinear stress–strain relationship for an unstressed one-dimensional semi-infinite foam rod, moving at velocity v_0, impacting a stationary rigid wall is depicted in Figure 4.9. A schematic of the impact and the resulting compressive waves is shown in Figure 4.11. Similar modeling of high-speed impact problems is presented in [11–13]. In terms of the foam schematic of Figure 4.11, a high velocity impact will propagate two waves to the left. The fast elastic precursor wave, moving at speed C_e (about 450 m/s), produces a compressive stress jump to the yield stress for foam, σ_{YF}, of about 220 kPa (2.2×10^5 N/m^2). This stress jump to σ_{YF} creates an instantaneous reduction in the foam particle velocity behind the elastic wave, designated as δ in the bottom plot of Figure 4.11.

From Equation 4.50 the velocity jump Δu across the elastic wave is related to the stress jump $\Delta \sigma$ by the expression

$$\Delta u = \Delta \sigma / \rho C$$

and the velocity reduction δ shown in Figure 4.11 is given by

$$\delta = \sigma_{YF} / \rho C_e \qquad (4.52)$$

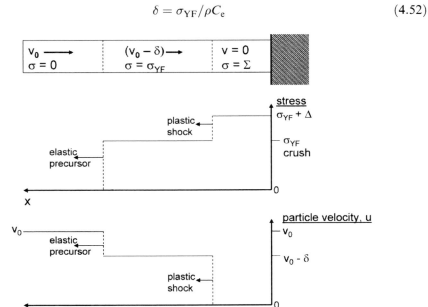

Figure 4.11. Schematic of elastic–plastic rod, moving at velocity v_0 after impact with a rigid wall ($v = 0$). An elastic precursor wave compresses the rod to the yield stress and reduces the particle velocity by δ. The trailing slower plastic "wave" further increases the stress by an amount Δ. The full impact stress is therefore $\sigma_{YF} + \Delta$.

As indicated by the reduction in slope of the stress–strain curve in Figure 4.9 above the yield point, a slower plastic "wave", moving at approximately the bulk wave speed for foam, $C_B \approx 50\,\text{m/s}$, follows the elastic precursor. This plastic wave produces a second compressive jump or increase in the foam stress, Δ. Using our rigid wall approximation, the particle velocity behind the plastic wave drops to the velocity set by the rigid wall boundary (namely zero).

4.4.6 The elastic–plastic model

The methodology employed here to calculate the total impact stress σ follows, in spirit, the development of the analytical model presented by James Walker in the report prepared for the Columbia Accident investigation of 2003 [2, CAIB report volume II, app D12, Impact Modeling].

We again simplify our solution by employing a rigid-wall approximation for the impact of foam and RCC materials.

We assume that the stress jump Δ, across the slow plastic wave is the same as for any equivalent elastic wave but with a wave speed C_B considerably less than the elastic wave speed C_E

$$\Delta = (\sigma_- - \sigma_+) = \rho C_B (u_+ - u_-) \qquad (4.53)$$

In Equation 4.53, u_+ is the foam particle velocity in front of the plastic wave, of necessity equal to the velocity behind the elastic foam precursor, and u_- is the velocity behind the plastic wave, which has been forced to be zero by the rigid-wall boundary. A more correct representation for the plastic "wave" jump would necessitate the modeling of a complex plastic *shock* wave process within the foam material. Such a shock jump relationship for Δ would require additional shock physics modeling and knowledge of the foam material properties for a solid undergoing a large deformation [11–13]. For our BotE approach we utilize the simpler "equivalent" elastic wave model given by Equation 4.53. From Figure 4.11, the velocity in the foam rod at negative infinity is shown to be equal to v_0. Thus the velocity u_+ in front of our modeled plastic wave can be found by subtracting the velocity drop, δ, from v_0. As previously noted in the discussion leading to Equation 4.52 for δ, the velocity drop δ results from the elastic-precursor wave compressing the foam to its yield point, i.e. $\delta = \sigma_{YF}/(\rho C_e)$.

Thus the expression for u_+ to be applied in Equation 4.53 is

$$u_+ = v_0 - \sigma_{YF}/(\rho C_e) \qquad (4.54)$$

We also note that behind the plastic wave (at the wall) the rigid boundary condition forces the velocity to vanish. Therefore we set

$$u_- = 0 \qquad (4.55)$$

The total impact stress σ (for high impact speeds and compressions beyond the elastic limit) can then be written as the sum of the yield stress for foam, σ_{YF}, and the

Sec. 4.4]   Modeling the impact pressure and load caused by impact of foam   205

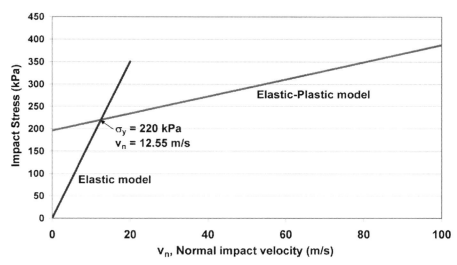

Figure 4.12. Foam impact stress (or pressure) as a function of normal velocity v_n. Model solutions for elastic (Equation 4.49 with impact velocity $= v_n$) and elastic–plastic stresses (Equation 4.56 with impact velocity $= v_n$).

plastic wave jump in stress Δ (Equations 4.53 to 4.55)

$$\sigma = \sigma_{YF} + \rho[C_B][v_0 - \sigma_{YF}/(\rho C_e)] \tag{4.56}$$

We take ρ to be the unperturbed density of the foam, whose tabulated value is $\rho = \rho_0 = 38.45$ kg/m^3 according to Walker [2]. Our final expression for the total impact stress σ (Equation 4.56), is quite close in form to that given by Walker in his treatment of this problem [2, 14].

Numerical estimates for the total impact stress produced by the combined elastic–plastic model (Equation 4.56), are presented in the following section and are plotted in Figure 4.12.

4.4.7 Numerical results and plotted trends

Our goal, in this section, is to determine a first-order numerical estimate for the magnitude of the impact load $P_L = \sigma \cdot A_L$ for the foam/RCC collision expressed in units of pounds-force or newtons. We numerically evaluate the loading pressure or impact stress σ, using Equation 4.56 for a range of velocities and impact angles. We then make an educated guess as to the size of the loading area A_L, based on the reported video observations of the foam dimensions prior to impact, in order to finally arrive at an estimate for the impact load.

To begin this numerical effort, we calculate the loading pressure or impact stress σ as a function of the normal component of the impact velocity v_n for a foam rod impacting a solid wall over a range of *normal* impact speeds from 0 to 100 m/s. Table 4.3 lists values for Shuttle-insulation foam density, yield stress, and foam elastic and

Table 4.3. Foam empirical data [2, 15].

Foam impact parameter	Experimental values 1999 SwRI tests
Density (reference, unstressed)	38.45 kg/m^3
Yield or crush stress, Y	$2.2 \times 10^5 \text{ N/m}^2$ (31.9 psi)
Elastic wave speed	456 m/s
Plastic wave speed (see Figure 4.2)	49.8 m/s

plastic wave speeds. These values were determined in a series of foam crush tests undertaken by the Southwest Research Institute and issued in 1999 [15].

Figure 4.12 illustrates the dependence of foam impact stress on the normal component of the impact velocity vector to the impacted surface, and shows the steep-slope elastic solution (Equation 4.49) which is valid below the yield or crush stress at $v_n = 12.55$ m/s, and the smaller-slope elastic–plastic solution (Equation 4.56), used to calculate σ above $\sigma_{YF} = 220 \times 10^3$ N/m^2 (or 220 kPa).

We now extend our model to deal with two-dimensional impacts, and typical impact geometries in which the foam hits the wing at an oblique angle. For such impacts at incidence angles, α_i, less than 90° (see sketch below) we resolve the impact or collision velocity vector into components normal and tangential to the impacted surface (e.g. the wing leading edge).

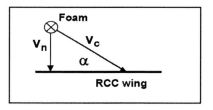

We make the assumption that only the normal component of the foam velocity vector, with magnitude

$$v_n = v_c \sin \alpha_i \qquad (4.57)$$

generates a stress wave in the foam. This wave is the equivalent of a one-dimensional compression wave generated on impact. This concept was first formulated by Walker [2]. As the tangential velocity component, $v_c \cos \alpha_i$, simply translates the foam along the surface without crushing it, we ignore the tangential component's contribution to the calculation of the total impact load. Calculated values of impact stress as a function of foam impact speed v_c are shown parametrically in Figure 4.13 for angles of incidence in the range 10° to 30°; this range being based on the assumption that the breakaway foam would follow a trajectory that is nearly parallel to the external tank before striking the Orbiter's swept-back left wing (Figure 4.1). Figure 4.13 is really another way of plotting Figure 4.12, but with v_c on the abscissa as the independent variable and α_i a parameter for each curve.

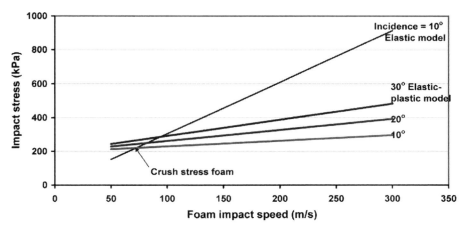

Figure 4.13. Impact stress dependence on foam impact speed, v_c, for several incidence angles. Comparison of model solutions for elastic (Equation 4.49, with $v_n = v_c \sin \alpha_i$) and elastic–plastic stress (Equation 4.56, with $v_n = v_c \sin \alpha_i$).

The elastic stress line at 10° incidence is the steepest of the lines in Figure 4.13, and applies only below the yield stress for foam. The elastic–plastic lines for 10°, 20°, and 30° show a weak dependence on foam impact speed, and a somewhat stronger dependence on impact angle. Notice that for impact speeds v_c in the expected range of 200 to 250 m/s and for impact angles α_i on the order of 20°, the impact stress is about 360 kPa, or about 1.6 times the yield stress of foam.

We now proceed to calculate the corresponding impact load based on the following estimation of the impact area (or loading footprint).

4.4.8 Impact area estimate

Let's assume, as indicated by the video evidence reported by the CAIB [1, 7], that a tumbling rectangular piece of foam, approximately 12 × 21 inches in size, strikes a glancing blow on the swept wing of the Orbiter. By this we mean that the foam undergoes the following hypothesized series of dynamic events: it is crushed by the initial impact with the wing and the residual material (perhaps a group of foam pieces) then moves past the wing at a tangential velocity close to its pre-impact value $\approx v_c \cos \alpha_i$. Figure 4.14 shows impact footprints created by the oblique impact of a numerical foam block against two numerically-generated finite-grid RCC panels. [These footprints were extracted from the large-scale numerical simulations conducted by the Southwest Research Institute for the CAIB report [2], Volume II, App D12, "Impact modeling"].

If the impact footprint is triangular in shape, as in the left-hand graphic in Figure 4.14, with a base of 12 inches and height of 24 inches, the load impact area is

$$A_L(\text{triangular}) = \tfrac{1}{2}(12 \times 24 \text{ in}) = 144 \text{ in}^2 = 0.093 \text{ m}^2$$

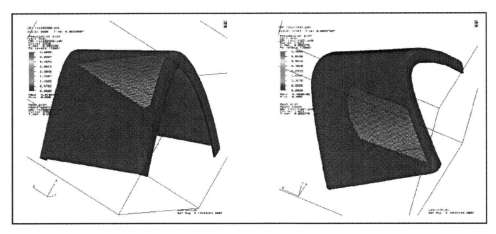

Figure 4.14. Footprints of impacts against RCC panels #6 (left) and #8 (right) from the CAIB report [2, App. D12, p. 78]. RCC panel thickness $\approx \frac{1}{4}$ "$= 0.6\,\text{cm}$".

For a slanted quadrilateral footprint, as in the right-hand graphic in Figure 4.14, with a base of 12 inches and a height of about 20 inches, the load impact area is

$$A_\text{L}(\text{quadrilateral}) \approx (12'' \times 20'') = 240\,\text{in}^2 = 0.155\,\text{m}^2$$

As we really don't know the actual shape or size of the footprint, we will just guess the magnitude of A_L based on these two values. Thus, for the load calculations to follow we simply assume the load area to be of order

$$A_\text{L} \approx 0.1\,\text{m}^2 \tag{4.58}$$

4.4.9 Load estimate

Using the above estimate for the load area, we calculate the foam/RCC impact load, P_L, from Equation 4.44, using the speed/incidence-angle-dependent stress, σ, from Equation 4.56 and 4.57 (Figure 4.13), and A_L from Equation 4.58

$$P_\text{L} = \sigma(v_\text{c}, \alpha_\text{i}) \cdot A_\text{L} \tag{4.59}$$

Impact load as a function of foam impact speed v_c is plotted parametrically in Figure 4.15 for incidence angles ranging from 10° to 30° using the elastic–plastic model for loads above the crush load for a $0.1\,\text{m}^2$ block of foam (i.e. $P_\text{L} > 4{,}946\,\text{lbf}$). For comparison with the load measurements and theory presented in the CAIB report [2], P_L is presented in Figure 4.15 in units of pounds force (lbf, where $1\,\text{lbf} = 4.448\,\text{N}$).

The loads depicted in Figure 4.15 represent a best estimate of the *maximum* load delivered by the foam impact on the RCC panel. The load is initially zero just prior to impact. It then builds up to a maximum level P_L, and subsequently drops rapidly back to zero. This trend with time is primarily geometrical in nature (for a nearly

Sec. 4.4] Modeling the impact pressure and load caused by impact of foam 209

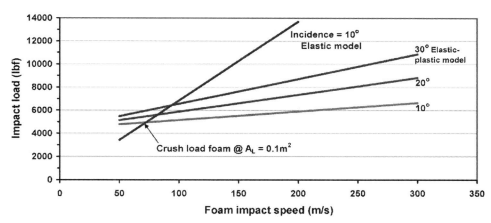

Figure 4.15. Impact load dependence on foam impact speed, v_c, for several incidence angles. Comparison of load for elastic and elastic–plastic loads based on impact load area $A_L = 0.1 \, \text{m}^2$.

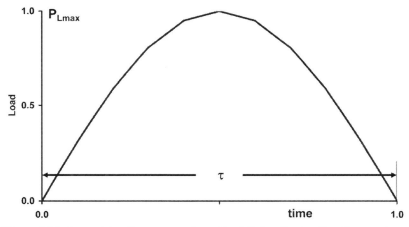

Figure 4.16. Impact loading curve schematic defining the loading time scale τ.

constant foam pressure) because in a glancing blow, the impact area first increases to a maximum, then decreases as the foam pieces move past the wing surface. We define τ as the characteristic time that it takes for the impact load to rise to its maximum and fall to zero as sketched in Figure 4.16.

We can estimate the magnitude of τ using a BotE approach based on the arguments presented by Walker for the impact process [2].

4.4.10 Impact loading time scale (BotE)

Consider the time required for a block of foam to tangentially traverse a distance Δx equal to an estimated impact length of about 2 feet. We therefore set $\Delta x \approx 24$ inches

210   Columbia Shuttle accident analysis with Back-of-the-Envelope methods   [Ch. 4

(0.61 m). For a given incidence angle, say 20°, the tangential velocity v_t at impact is

$$v_t = v_c \cos \alpha_i = v_c \cos 20°$$

as shown in the vector sketch above (immediately preceding Equation 4.57). We further assume that impact speeds v_c are in the expected range of 200–250 m/s and select the median speed for v_c as equal to 225 m/s.

Using the above values for distance and velocity, the loading and unloading time scale τ is then estimated to be

$$\tau = \Delta x/(v_c \cos \alpha_i) = (0.61 \text{ m}) \Big/ \left(211.4 \frac{\text{m}}{\text{s}}\right) = 2.9 \times 10^{-3} \text{ s}$$

Thus the loading time scale is of order 3 milliseconds. We will now compare the strength of the predicted maximum load (Figure 4.15) of $P_L \approx 8,000$ lbf, and the time scale $\tau = 3$ ms with the results of numerical simulations carried out by the CAIB impact study team.

4.4.11 Loading time histories, numerical simulations

Figure 4.17 shows the foam/wing panel loading time history calculated from the large scale finite-element computer simulations undertaken for the CAIB by the Southwest Research Institute [2].

Note that in Figure 4.17 the calculated impact load on the panel face reaches its peak of 9,150 lbf (or 407×10^2 newtons, as shown by the left hand scale) at about 2.2

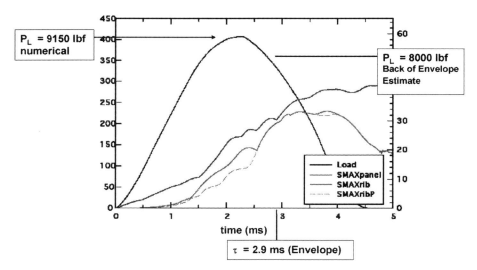

Figure 4.17. Impact loading curve for panel #8 from a large-scale explicit finite-element numerical simulation developed at the Southwest Research Institute [2]. The numerically calculated maximum load, the peak of the parabolic-like curve, and the overall time scale (x-axis) are compared with the nominal Back-of-the-Envelope estimates of 8,000 lbf. The curves labeled SMAXpanel, SMAXrib refer to the calculated maximum stresses in the wing RCC panel face and supporting side-panel rib [2].

milliseconds. The overall loading time, evaluated at the point where the load drops back to zero, is approximately 4.5 ms. Our corresponding BotE estimate for P_L is 8,000 lbf (or 356×10^2 Newtons) with $\tau = 2.9$ ms. Clearly, these BotE values are in the same ballpark as the measurements. The blue and red curves curves in Figure 4.17 are the calculated maximum principal stresses for the RCC panel face and for the supporting RCC side-panel rib (see Figure 4.14). These curves represent the output of the large-scale 3D elastic simulation made by the Southwest Research Institute, whose numerical stress dynamics code did not include any failure criteria due to cracks or large-scale buckling, but did calculate expected maximum stresses in the RCC panel just prior to catastrophic failure using an elastic isotropic material model.

The largest calculated RCC panel stress (the blue curve at 5 milliseconds for panel #8) is about 43×10^3 psi (right-hand scale) or 296 MPa (left-hand scale). According to Walker, it is possible that a *crack generated failure* actually occurs in the RCC panel face near the time of the peak load (≈ 2.7 milliseconds), when the maximum principal stress in the panel face is about 30×10^3 psi (30 ksi, or 207 MPa). Note that 1 ksi = 1,000 lbf/in^2.

In the next section we pose the final BotE challenge related to the foam/RCC impact problem: Determine a criterion for possible wing damage or failure.

4.5 DEVELOP A BACK-OF-THE-ENVELOPE ENGINEERING STRESS EQUATION FOR THE MAXIMUM STRESS IN THE RCC PANEL FACE FOR A GIVEN PANEL LOAD

We will now develop a simple engineering model for the maximum stress in the RCC panel face as a function of impact load, for a given set of impact variables and system parameters (e.g. impact velocity and incidence angle). We will also invert the resulting equation for the panel face stress and construct a parametric condition for wing failure based on the criterion that the maximum stress in the RCC reaches or exceeds a certain prescribed high threshold. This maximum stress threshold will then be set, empirically, using results obtained from basic laboratory tests for RCC material failure (e.g. set $\sigma_{\text{fail}} \approx$ order 30,000 lbf/in^2 or 30 ksi). This simple failure curve provides a unique relationship between the incidence angle α_i for failure and the given impact speed v_c; or, alternatively, a relationship between the impact speed v_c for failure and a given incidence angle. That is, failure of the RCC panel can be expected for a coordinate pair or point $[\alpha_i, v_c]$ that lies above the curve, but no failure for a point below the curve. This therefore serves as a reasonable parametric failure criterion for the Columbia impact problem [2].

The estimated RCC panel stress model and the derivative RCC failure model are presented in the following subsections.

4.5.1 BotE panel stress model

To calculate our BotE estimate of the impact-induced stresses within the RCC panel (generated by the impact loads depicted in Figure 4.15), we first adopt a simple thin flat plate idealization for the panel. To construct this idealization, we first consider an experimental analog of the foam/wing impact. In this idealization the curved wing panel of Figure 4.14 is replaced by a stationary flat-plate made of RCC, with a thickness of about 1/4 inch. The foam is treated as a moving cylinder that has a diameter comparable to the flat plate, as depicted in Figure 4.18a. This impact "experiment" was actually conducted by NASA engineers in a dynamics materials laboratory using actual foam and RCC samples. A parallel numerical investigation was conducted using a large-scale finite-element dynamic stress/displacement code called LS-DYNA [17] that allowed for a generalized input of material properties over a full range of stresses and strains. Possible RCC plate failure modes (cracks and holes) were observed and measured for a range of impact speeds, both in the physical tests and in the numerical simulations.

Solving the cylinder/flat plate dynamic (or vibrational) stress impact problem of Figure 4.18a represents a considerably more difficult undertaking than we are able to pursue with our BotE analysis. However, we can considerably simplify our stress analysis effort somewhat by solving a "static" and elastic version of this short-lived loading problem; namely the static uniform loading of a square flat plate as sketched in Figure 4.18b.

Of course we will be significantly increasing the uncertainty of our estimate by adopting this idealized representation for the RCC stress impact problem. But as we will show, the solution of the static thin-flat-plate bending approximation used here provides an order-of-magnitude estimate for the maximum bending stresses in the RCC leading edge panel that is consistent with NASA's numerical simulations and laboratory measurements for a fully three-dimensional leading-edge panel of the wing (Figures 4.14 and 4.17). In our simplified model problem we will set the static load applied to the thin, flat plate to be equal to our previously calculated dynamic foam impact load (of order of 8,000 lbf). A sketch of the corresponding

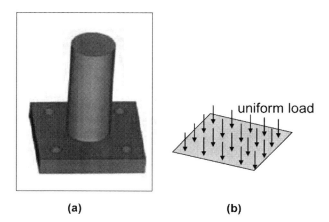

Figure 4.18. (a) LS-DYNA finite difference code impact of foam cylinder on an RCC flat plate target [16], (b) Bending of an idealized thin plate subject to uniform pressure loading (clamped edges).

static problem for a flat plate loaded uniformly over its entire area is depicted in Figure 4.18b.

We assume that a distributed lateral external load (i.e. a load normal to the surface) produces a bending (or bowing) of the plate that can be calculated by the classical theory for the vertical displacement of thin elastic plates. This was first attempted by Galileo in the 17th century but the correct theory had to await Navier in 1820, who derived a two-dimensional fourth-order partial differential equation for the vertical displacement $z(x, y)$ of a plate in the x–y plane subject to a given distributed loading [18]. The amount of in-plane stretching of material plate fibers (in a particular direction) due to the bending of the plate is a measure of the local strain of the material, which for an elastic plate is proportional to the in-plane stress in that direction. This in-plane elastic stress can be calculated using a simple formula, readily available in the engineering solid mechanics literature, that describes the maximum bending stress in a flat plate under a uniform load or pressure.

For a square plate, William Griffel [18, "Plate Formulas"] presents the following generic formula which stipulates that the *maximum* in-plane bending stress σ_{max} in the plate (located at the plate edges) is simply proportional to the external load P divided by the square of the plate thickness t

$$\sigma_{max} = K(P_L/t^2) \tag{4.60}$$

where K is the non-dimensional proportionality constant.

For a square plate *with all edges clamped*, Griffel lists the following value for the constant K in Equation 4.60.

$$K = 0.3 \tag{4.61}$$

This value for K is based on published solutions of Navier's equation for a thin elastic plate under a uniform pressure loading (i.e. similar to that depicted in Figure 4.18b). The numerical results for calculating stresses and deflections in clamped rectangular plates (obtained by solving Navier's equation) is presented in Timoshenko's classic 1959 text entitled "Theory of Plates and Shells" [19]. The value of K derived in Chapter 6 of Timoshenko's book is $K = 0.3078$.

Hence all we need to do to solve for σ_{max} (the maximum plate bending stress) is to input into Equation 4.60 approximate values for the maximum load P and the thickness t and apply the constant $K = 0.3$.

On the other hand, we can solve for the magnitude of the impact load P that is required to produce the prescribed *maximum* bending stress in the RCC plate. We will set this maximum stress equal to the measured yield or failure stress for the RCC material, $\sigma_{Y_{RCC}}$ or $\sigma_{failure}$ obtained from laboratory or numerical experiments.

4.5.2 Estimates for the allowable maximum stress or critical load parameters for failure

4.5.2.1 Maximum stress in equivalent RCC plate for a given load P_L

Let's first calculate the magnitude of the maximum stress in a model RCC square plate ($K = 0.3$) using Equation 4.60 with a load P_L of 8,000 lbf (see Figure 4.15) and a thickness t of 1/4 inch (typical for the geometry of an Orbiter leading edge wing panel of RCC [2]). The estimated maximum stress in the RCC wing (i.e. at the edges of our model plate) is

$$\sigma_{\max} = K(P_L/t^2) = 0.3\left(\frac{8{,}000}{(0.25)^2}\right) = 38{,}400\,\frac{\text{lbf}}{\text{in}^2} = 38.4\,\text{ksi} = 264\,\text{Megapascals} = 264\,\text{MPa} \tag{4.62}$$

4.5.2.2 Threshold load P_L at failure onset

Numerical and experimental studies by CAIB engineers (performed by Walker et al. at the Southwest Research Institute [2]) indicate that the peak stress within the RCC panel face when a crack first forms, or some localized material failure occurs, is in the vicinity of

$$\sigma_{\text{failure}} = 27.0\,\text{ksi} \quad \text{or}\quad 185\,\text{MPa}$$

This stress produces slightly under a 1% maximum strain in the RCC panel. If we adopt this peak stress as the upper stress limit denoting the *onset of failure* of the RCC panel face, sometimes also designated as the RCC yield stress σ_Y, then we can solve Equation 4.60 and determine the following expression for and magnitude of the threshold load P_t

$$P_t = \frac{\sigma_Y t^2}{K} = \frac{\left(27{,}000\,\frac{\text{lbf}}{\text{in}^2}\right)(0.25^2\,\text{in}^2)}{0.3} = 5{,}625\,\text{lbf} = 25.0\,\text{kN} \tag{4.63}$$

If we now fix this level of P_L as the load for "failure onset", we can reverse the dependent and independent variables in Figure 4.15 and solve for the magnitude of impact speed, v_{fail}, that would produce failure of an RCC wing panel face for a prescribed incidence angle, α_i. For example, for an impact angle $\alpha_i = 20°$ and a maximum load $P_L = 5{,}625$ lbf (as given by Equation 4.63), Figure 4.15 gives as the approximate foam impact speed for the onset of failure $v_{\text{fail}} \approx 90$ m/s. Next, the equivalent *normal* impact speed perpendicular to the surface of the wing or plate for a 20° impact angle is $v_n = v_{\text{fail}}(\sin \alpha_i) = 30.8$ m/s.

Independently of the above considerations we have shown that for an impact speed $v_c = 230$ m/s and a shallow incidence angle of 20°, the estimated impact load is about 7,800 lbf (see Figure 4.15).

We therefore conclude that failure of the RCC wing panel is "very" likely at an impact speed of $v_c = 230$ m/s and an incidence angle of 20° because the estimated panel stresses would exceed the 27 ksi yield stress for a thin RCC plate by over 50% and hence $v_c > v_{\text{fail}}$.

4.5.2.3 Parametric failure curve

We can now produce an imminent failure curve of maximum speed at failure vs incidence angle for a range of angles and speeds. This parametric curve is solely based on the normal velocity v_n at which the predicted impact load is equal to the imminent RCC failure load, $P_t = 5{,}625$ lbf or 25.0 kN (as given by Equation 4.63).

Sec. 4.5]   Develop a Back-of-the-Envelope engineering stress equation   215

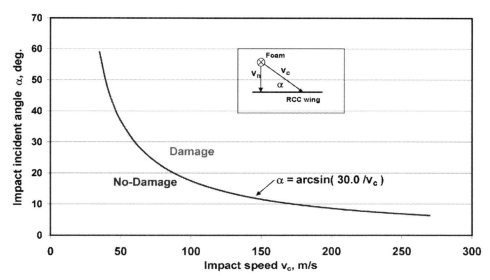

Figure 4.19. Damage/No-damage model transition curve for Shuttle wing RCC failure due to foam impact at incidence angle α_i and speed v_c (after Walker [2]).

This failure point occurs when the predicted bending stress of a 1/4 inch loaded flat RCC plate exceeds 27 ksi. The maximum normal velocity producing the 25.0 kN force, as noted above, is approximately $v_n \approx 30$ m/s, according to our elastic–plastic impact model (see Section 4.5.2.2). For this normal velocity our parametric expression simply becomes

$$v_c \sin \alpha_i = v_{n_{fail}} = 30 \text{ m/s} \tag{4.64}$$

where v_c is the impact speed of the foam as given in Figure 4.19.

We can solve Equation 4.64 for the onset incidence angle α_i that leads to wing failure following the approach of Walker employed in his Shuttle tile damage study [2], i.e.

$$\alpha_i = \arcsin\left(\frac{v_{n_{fail}}}{v_c}\right) = \arcsin\left(\frac{30 \text{ m/s}}{v_c}\right) \tag{4.65}$$

Figure 4.19 shows a plot of incidence angle α_i as a function of impact speed v_c from Equation 4.65. This *failure* transition curve can be used to determine whether certain combinations of impact angle and speed suggest the possibility of RCC structural damage, since damage is possible for velocity and angle combinations that lie above the curve. Alternatively, no damage is likely for those velocity and angle combinations below the curve.

4.5.3 Final comments on the prediction of possible wing damage or failure

Our simplified BotE damage model (Figure 4.19) indicates that failure is quite sensitive to the angle of incidence of the impact, particularly if the impact speed falls below 60 to 70 m/s. Note also that if the incidence angle is approximately 10°, damage is possible only for impact speeds exceeding 170 m/s. These trends support the CAIB findings that while serious wing damage is possible for certain foam breakaway conditions, it is only a particular range of foam incidence angles and speeds which are likely to create the impact forces and stresses that result in damage or failure of the leading edge of the Orbiter's wing. One such possible combination is $v_c = 200$ m/s and all incidence angles exceeding 10°. Even if the probability of these "failure" modes is small, the implications for the Shuttle can, for missions like Columbia flight STS-107, be potentially catastrophic.

While our BotE analysis is quite approximate, this kind of preliminary result may lead a practicing engineer to bring his safety concerns to the attention of management in support of a recommendation for more in-depth numerical and experimental studies designed to try to further quantify the possibility of failure.

Further analyses of this kind (using Back-of-the-Envelope and other more advanced techniques) can be used to develop mitigation schemes aimed at reducing the possibility of catastrophic system failures for future launch systems similar to the Space Shuttle.

4.6 SUMMARY OF RESULTS FOR SECTIONS 4.2, 4.3, AND 4.4

We conclude this chapter by briefly summarizing all of our BotE results in tabular form. Table 4.4 gives the key results for the direct foam impact trajectory problem. Table 4.5 summarizes the results for the foam impact load. Table 4.6 has the results for the impact induced stress in the wing leading edge RCC material.

Table 4.4. BotE input parameters and brief summary of computed foam trajectory velocity, time, and acceleration scales. Dependencies on BN.

		Direct trajectory results (Equations 4.9–4.27)	
	Given	Calculation	
Orbiter velocity: V_∞ (m/s)	708.0		
Air Density: ρ_{air} (kg/m^3)	0.092		
Distance to wing: D (m)	17.7		
Drag coeff: C_d	1.35		
Drag area: A (m^2)	0.045		
Mass: m (kg)	0.65		
Acceleration: a_0 (m/s^2)		$a_0 = \dfrac{1}{2m}\rho_{\text{air}}(V_\infty)^2 C_d A = 2{,}160 \text{ m/s}^2$	(4.9)
Velocity at impact: v_c (m/s)		$v_c = \sqrt{2a_0 D} = 276 \text{ m/s}$	(4.17)
Time to impact: t_c (s)		$t_c = \sqrt{\dfrac{2D}{a_0}} = \dfrac{2D}{v_c} = 0.13 \text{ s}$	(4.19)
Ballistic number: BN (Pa)		$BN \equiv \dfrac{mg}{C_d A} = 105 \text{ Pa}$	(4.22)
v_c as a function of BN		$v_c = V_\infty \sqrt{\dfrac{\rho_\infty g D}{BN}}$	(4.25)
t_c as a function of BN		$t_c = \dfrac{2}{V_\infty}\sqrt{\dfrac{(BN)D}{\rho_\infty g}}$	(4.27)

Table 4.5. BotE input parameters and brief summary of computed foam-wing collision impact stress, impact load, and loading time scale.

		Impact stress and load results (Eqns. 4.44–4.57)	
	Given	Calculation	
Foam Density: ρ_0 (kg/m³)	38.45		
Elastic wave speed in foam: C_e (m/s)	456		
Incidence angle: α_i (deg)	10		
Impact speed: v_c (m/s)	276		
Normal velocity for a 10° incidence angle		$v_n = v_c \sin \alpha_i = 47.9$ m/s	(4.57)
Impact load area: A_L (m²)	0.10		
Impact stress: σ (N/m²) Elastic model		$\sigma = \rho_0 C_e v_n = 8.4 \times 10^5$ N/m²	(4.49), (4.51)
Impact load: P_L (N) Elastic model		$P_L = \sigma \cdot A = 8.4 \times 10^4$ N $= 18.9 \times 10^3$ lbf	(4.44)
Yield stress for foam: Y (N/m²)		$Y = 2.2 \times 10^5$ N/m²	(Table 4.3)
Compare σ to Y: Is $\sigma > Y$? If yes, use Elastic–Plastic model		Yes	
Plastic wave velocity = foam bulk wave speed: C_B, where $\sigma > Y$, (m/s)	49.8		
Impact stress: σ_{EP} Elastic–Plastic model (N/m²)		$\sigma_{EP} = Y + \rho_0 C_B \left[v_n - \dfrac{Y}{\rho_0 C_e} \right]$ $= 2.87 \times 10^5$ N/m²	(4.56)
Impact load: P_L (N) Elastic–Plastic model		$P_L = \sigma_{EP} \cdot A = 2.87 \times 10^4$ N $= 6.46 \times 10^3$ lbf	
Estimated impact length: Δx (m)	0.61 (2 ft)		
Estimated impact loading time scale: τ (s)		$\tau = \Delta x / (v_c \cos \alpha_i) = 2.24 \times 10^{-3}$ s	

Sec. 4.6]   **Summary of results for Sections 4.2, 4.3, and 4.4**

Table 4.6. BotE results for the maximum impact stress in the wing leading edge RCC material. Evaluation of failure criterion.

	Max RCC wing stress results (Eqns. 4.56–4.63)	
	Given	*Calculation*
Impact load: P_L (lbf)	6.46×10^3	see Table 4.5
Impact load: P_L (N)	2.87×10^4	see Table 4.5
RCC leading edge thickness (inches) [1, 2]	0.25 inch or 6.4×10^{-3} m	
Max bending stress scaling parameter: K	0.3 square plate	
Maximum RCC bending stress for square plate: σ_{max} (lbf/in^2)		$\sigma_{max} = K(P_L/t^2)$ $= 0.3 \left(\dfrac{6,460}{(0.25)^2} \right) = 31,000 \dfrac{\text{lbf}}{\text{in}^2}$ $= 214 \text{ Megapascal} = 214 \text{ MPa}$ Equation 4.60
Failure stress for RCC σ_{fail} (lbf/in^2)	27,000 186 MPa	
Is estimated RCC stress $\sigma_{max} > \sigma_{fail}$		Yes
Maximum impact load leading to onset of RCC stress failure (lbf)		$P_t = \dfrac{\sigma_{fail} t^2}{K} = \dfrac{27,000(0.25^2)}{0.3} = 5,625 \text{ lbf}$ $= 25.0 \text{ kN}$ Equation 4.63
Maximum normal velocity leading to failure (m/s)		$v_n = \dfrac{(\sigma_{EP} - Y)}{\rho_0 C_B} + \dfrac{Y}{\rho_0 C_e}$ Using (4.56) $= 28.2 \text{ m/s}$ where $\sigma_{EP} = P_t/A = 2.50 \times 10^5 \text{ N/m}^2$
Maximum impact speed leading to failure for 10° incidence angle		$v_{fail} = \dfrac{v_n}{\sin \alpha_i} = \dfrac{28.2}{\sin 10°} = 162 \text{ m/s}$
Does our estimated calculation for impact speed (v_c in Table 4.4) exceed this maximum speed for failure?		Yes., damage may be possible since $v_c > v_{fail}$ for $\alpha_i = 10°$ i.e. 276 m/s > 162 m/s

4.7 REFERENCES

[1] Gehman, H.W. et al., "Columbia Accident Investigation Board Report," Volume 1, U.S. Government Printing Office, Washington DC, August, 2003.
[2] Gehman, H.W. et al., "Columbia Accident Investigation Board Report," Volume 2, Appendix D12, Impact Modeling, U.S. Government Printing Office, Washington DC, August, 2003.
[3] Yu, Kristine, "Tracing the loss of Columbia," Volume II, Stanford Scientific Review, pp. 12–17, 2006.
[4] Kennedy, R, "It ain't rocket science, folks." [Online. First appeared, Feb. 2003, revised July 2005]. www.ultimax.com
[5] *Top Chef*, Television show, Bravo Network, Season 1, March 2006.
[6] Prandtl, L. and Tietjens, O.G., *Applied Hydro-and Aero-mechanics*, reissued by Dover, New York, 1934.
[7] Gehman, H.W. *et al.*, "Columbia Accident Investigation Board Report," Volume 2, Appendix D08, Debris Transport Analysis, U.S. Government Printing Office, Washington DC, August, 2003.
[8] Hoerner, S.F., *Fluid Dynamic Drag*, Published by the author, Midland Park, New Jersey, 1958.
[9] Dwight, H. B., *Tables of Integrals and Other Mathematical Data*, The Macmillan Co, New York, 1961.
[10] Zukas, J. A. (editor), *High Velocity Impact Dynamics*, Wiley & Sons, New York, 1990.
[11] Asay, J.R. and Shahinpoor, M. (editors), *High-Pressure Shock Compression of Solids*, Springer-Verlag, New York, 1992.
[12] Drumheller, D.S., *Introduction to Wave Propagation in Nonlinear Fluids and Solids*, Cambridge University Press, 1998.
[13] Molinari, A. and Ravichandran, G., "Fundamental structure of steady plastic shock waves in metals," *Journal of Applied Physics*, Volume 95, No. 4, 2004.
[14] Walker, J.D., "Errata to CAIB Report Vol. II, Appendix D.12, Impact Modeling", 2003. http://www.nasa.gov/columbia/caib/html/VOL2.html
[15] Goodlin, D.L., "Orbiter Tile Impact Testing," Southwest Research Institute Final Report #18-7503-005 prepared for NASA JSC, San Antonio, Texas, March 5, 1999.
[16] Melis, M.E., Brand, J., Pereira, M., Revilock, D.M., "Reinforced Carbon-Carbon Subcomponent Flat Plate Impact Testing for Space Shuttle Orbiter Return to Flight," NASA TM 2007-214384, January 2007.
[17] Anon., "LS-DYNA Keyword User's Manual Volume I and II- Version 960," Livermore Software Technology Company, Livermore, CA, March 2001.
[18] Griffel, W. *Plate Formulas*, F. Ungar Publishing Company, 1968.
[19] Timoshenko, S., Woinowsky-Krieger, S., *Theory of Plates and Shells*, McGraw-Hill, New York, 1959.

5

Estimating the Orbiter reentry trajectory and the associated peak heating rates

5.1 INTRODUCTION

The Space Shuttle was specifically designed to provide a multi-mission reusable orbital payload delivery capability. Overall, the system was required to have the capability to launch into orbit a number of large payloads and carry out a variety of complex orbital operations. After completing the space portion of its mission, the Orbiter was designed to be able to reenter the earth's atmosphere at a very high speed, slow down significantly, and land intact. The Orbiter was then to be quickly re-integrated with various Shuttle launch components in order to fly a subsequent mission.

Reentry is one of the most stressful portions of the mission. The fatal system failure of Columbia (STS-107) is a stark reminder of the dangerous thermal and aerodynamic environments encountered during reentry. The Orbiter reenters the atmosphere at an altitude of about 120 km (or 400,000 feet) at only slightly less than the orbital speed of 8,000 m/s or 29,000 km/hr. A robust thermal protection system (TPS) and a versatile aerodynamic control capability were developed to protect the structure from the very high levels of aerodynamic heating. Indeed, as stated by Williams and Curry [1], "The key to development of a low-cost, weight-effective Orbiter thermal protection system design was the selection of unique reusable materials which could withstand the high temperature reentry environment and also provide adequate insulation for the structure and internal systems of the Orbiter." They also spoke of the importance of making accurate predictions of reentry heating when drawing up the stressing requirements: "The ability to predict the Orbiter thermal environment was of the utmost importance in the TPS design."

I. E. Alber, *Aerospace Engineering on the Back of an Envelope*.
© Springer-Verlag Berlin Heidelberg 2012.

5.2 THE DEORBIT AND REENTRY SEQUENCE

During the deorbit sequence, the Orbiter is initially rotated to place its tail in the direction of its orbital velocity vector, using its low-thrust reaction control system engines (see Figure 5.1). In this orientation, the OMS engines are fired for about 3 minutes to decrease the vehicle's velocity below the mission's orbital speed. The spacecraft then rotates to point its nose forward.

The deorbit process is typically begun about half-way around the globe from the intended landing site, and at an orbital altitude of about 200 to 300 km. The spacecraft's entry angle is set to a value between $1.1°$ and $1.5°$. The altitude begins to drop at a rate of about 10 km per minute. Effectively, the Orbiter first enters the upper atmosphere at an altitude of about 120 km at a speed just below orbital velocity, oriented nose-forward with an angle of attack of $40°$. As the air density begins to increase, the aerodynamic pressures on the vehicle build up, and the flight control surfaces (the wings, tail rudder, elevons, etc) start to influence its motion. The large angle of attack creates a significant aerodynamic drag force, along with a correspondingly large lift force. The drag reduces the high reentry velocity to speeds

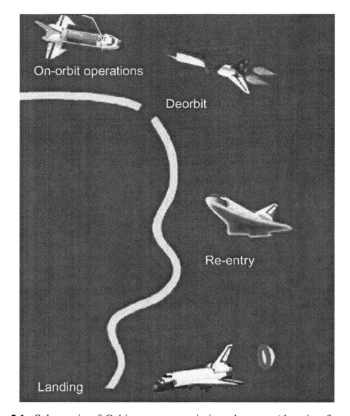

Figure 5.1. Schematic of Orbiter reentry mission elements (drawing from [2]).

low enough to initiate additional aerodynamic maneuvers designed to reduce the accumulative *heat load* during reentry. As the vehicle drops below the speed of sound (Mach 1) at 50,000 feet or 15 km above sea level, the Shuttle commander takes over manual control to execute the final approach and landing maneuvers. Nominal touchdown speeds are about 190 knots or about 100 m/s. Thus, during reentry, the Orbiter's velocity decreases from 7,700 m/s to 100 m/s, and this factor of 77 change in speed translates to a factor of about 6,000 decrease in kinetic energy.

Where did all that kinetic energy go? Basically, the dissipated kinetic energy heated the air around the Orbiter, which in-turn transferred a fraction of it to the Orbiter's thermally protected surfaces via convective, conductive, and radiative heat transfer. During the high-drag, high-lift reentry period the thermal protection system provides the heat transfer protection to enable the vehicle to survive the enormous rates of convective heating created by the high air temperatures that envelop the hypersonic vehicle.

The Orbiter's thermal protection system consists primarily of tiles on the wings and on the top and bottom of the fuselage. Critical to vehicle survivability are the surfaces constructed from the very important "heat-resistant reinforced carbon–carbon (RCC) materials" which cover the tip of its nose and the leading edges of its wings. These are heated to extremely high temperatures midway through the atmospheric reentry period. This high temperature period lasts for more than 10 minutes. In fact, in this chapter we will estimate the peak nose-tip temperatures and show them to be about 1,400°C or 2,500°F.

5.3 USING QUICK-FIRE METHODS TO CRUDELY ESTIMATE PEAK HEATING RATE AND TOTAL HEAT LOADS FROM THE INITIAL ORBITER KINETIC ENERGY

The primary thermodynamic principle involved in reentry is that almost all of the very high kinetic energy of the orbiting spacecraft, present at the start of reentry ($V_E \approx 7.7$ km/s at an altitude of about 120 km), is converted into raising, to very high temperatures, the internal thermal energy of the gas layer passing through the bow shock wave surrounding the vehicle. A small fraction of the heated gas energy is transferred into the wall of the Orbiter through the thermally conducting boundary layer. In basic thermodynamic terms, the total change in kinetic energy is approximately equal to the negative work produced by the non-conservative aerodynamic drag force, D, acting over a long descent path ℓ_D [3].

In Sections 5.3.1 and 5.3.2 we will utilize the Quick-Fire approach presented in Chapter 1 to crudely estimate some of the key reentry parameters, such as the total reentry time, the maximum rate of heat transfer to the Orbiter's nose, and the total amount of energy transferred to the surface over the full trajectory (Q_w, in joules) that heats up the thermal protection system. After we have introduced more detailed BotE models for both the early-entry region (Section 5.5.1) and the lift-assisted glide portion of the reentry trajectory (Section 5.5.2) we repeat these calculations with the aim of substantially improving our estimates. In Section 5.6, we introduce an improved model for estimating the altitude-dependent heat transfer rate, $\dot{q}_w$, along the entry and glide portions of the reentry, and use this to determine the total energy transferred to the vehicle as a function of its lift to drag ratio primarily during the glide portion of the descent.

5.3.1 Quick-Fire problem definition and sketch

We set the following three problems aimed at crudely estimating the magnitude of certain key dimensions and heating measures pertinent to Orbiter reentry:

- Estimate the average flight path angle and the total reentry time for the Orbiter.
- Estimate the maximum heat transfer rate to the Orbiter's nose.
- Estimate the total maximum heat load, the total heat transfer per unit surface area, during reentry.

The sketches in Figures 5.2 and 5.3 will help to guide us in the formulation of the simple mathematical models needed for a "first cut" at these problems.

5.3.2 The Quick-Fire baseline mathematical model, initial results, and interpretation

We first develop the Quick-Fire mathematical model for the descent geometry based on Figure 5.2.

5.3.2.1 *Modeling the descent path ℓ_D, and the average flight path angle, θ_{avg}*

For a vehicle of mass m and velocity V_E, the kinetic energy is

$$KE = \tfrac{1}{2} m V_E^2 \tag{5.1}$$

Most of the kinetic energy of a reentry vehicle is converted into internal (i.e thermal) energy, mainly in the vicinity of the detached shock wave close to the surface. If a space probe enters the atmosphere from an interplanetary mission at an entry velocity greater than 10 km/s, some of the KE will also be converted into significant thermal radiation of the hot gases [5].

The kinetic energy per unit mass (the specific KE) is $\tfrac{1}{2} V_E^2$. The magnitude of the specific KE for $V_E = 7{,}700$ m/s is

$$\tfrac{1}{2} V_E^2 = 3.0 \times 10^7 \text{ joules/kg} = 12{,}900 \text{ BTU/lbm} \tag{5.2}$$

This level of energy is sufficient to dissociate almost all of the molecular nitrogen and oxygen to individual atoms in the air that passes through the shock wave in front of the Orbiter, since it is of the same order of magnitude as the heat required to fully dissociate gaseous nitrogen, 3.4×10^7 J/kg. [6].

Assuming that the total initial vehicle KE (about 3×10^{12} J for an Orbiter mass of 10^5 kg) is reduced to approximately zero by aerodynamic drag (before landing), we can equate all of the KE to the negative work which is produced by the aerodynamic drag force D acting over the distance ℓ_D. This energy/work relationship allows us to determine that distance using the energy balance equation

$$\int_0^{\ell_D} D\, dx = KE_{\text{total}} = \tfrac{1}{2} m V_E^2 \tag{5.3}$$

The gravitational potential energy, $PE = mgh_0$, for a body of mass m at an entry altitude $h_0 = 120$ km, is estimated at about 1×10^{11} J. This is only a small fraction of the initial KE; about 3%. To first order we can neglect PE in the analysis that follows.

If we define an average drag force $\bar{D}$ over the full reentry distance ℓ_D by

$$\bar{D} = \frac{1}{\ell_D} \int_0^{\ell_D} D\, dx,$$

then Equation 5.3 takes the form

$$\bar{D} \cdot \ell_D = \{\bar{q} C_D A\} \cdot \ell_D \approx \{[\tfrac{1}{2}(\bar{\rho} \bar{V}^2)] C_D A\} \cdot \ell_D = \tfrac{1}{2} m V_E^2 \tag{5.4}$$

where $\bar{\rho}$ is in effect some average air density over the full reentry path, $\bar{q}$ is the average dynamic pressure defined above, and $\bar{V}$ is the corresponding average vehicle velocity. C_D is the average effective drag coefficient (order 1) and A is the fixed reference drag area for the Orbiter (≈ 250 m^2).

Solving Equation 5.4 for the drag length ℓ_D

$$\ell_D \approx \left(\frac{m}{\bar{\rho} C_D A}\right) \left[\frac{1}{(\bar{V}/V_E)^2}\right] \tag{5.5}$$

5.3.2.2 Physical parameters and data for the entry-distance problem

We now estimate ℓ_D based on the following crude assumptions and input parameters

- Average altitude $= \tfrac{1}{2}$ entry altitude of 120 km; i.e. $\bar{h} = 60$ km
- Average density (at $\bar{h}$) for an atmospheric scale height $H_{\text{atm}} = 7$ km is given by

$$\bar{\rho} \approx \rho_0 \exp(-h/H_{\text{atm}}) = 1.2 \exp(-60\text{ km}/7\text{ km}) = 2.3 \times 10^{-4} \text{ kg/m}^3$$

- $C_d = 0.8$
- $V_E = 7.70$ km/s
- $\bar{V}/V_E \approx 0.5$; a simple guess of the mean velocity ratio
- $A = 250$ m^2
- $m_{\text{orbiter}} = 1 \times 10^5$ kg

5.3.2.3 Numerical estimate for the entry-distance and average flight-path angle

Substituting the above values into Equation 5.5, we estimate the entry distance

$$\boxed{\ell_D \approx \left(\frac{10^5}{(2.3 \times 10^{-4})(0.8)(250)}\right) \left[\frac{1}{(1/2)^2}\right] = 8.7 \times 10^6 \text{ m} = 8{,}700 \text{ km}} \tag{5.6}$$

5.3.2.4 Interpretation of geometrical results for reentry

If the average flight path angle is very small (as will be shown), then the horizontal distance or range is approximately equal to the length of the hypotenuse ℓ_D in Figure 5.2. From Equation 5.6 we see that our estimate of horizontal range is fairly close to the typical NASA values of between 8,100 to 8,500 km reported for flight range.

The approximate average flight path angle, θ_{avg} for a straight-line descent trajectory from 120 km initial altitude is estimated to be

$$\bar{\theta} \approx \arcsin(120 \text{ km}/8{,}700 \text{ km}) = 0.0138 \text{ radians} \approx 0.8° \qquad (5.7)$$

NASA's records indicate that the initial entry angle for the Orbiter is typically around 1.2°, so our estimate of that angle is reasonable. Also, data show that the flight-path angle θ is maintained, on average, below 1° for the first 1,100 s after entry. This is during the time when the Orbiter's angle of attack is set to 40°. The flight-path angle, however, does increase substantially, to above $\theta = 10°$, about 5 minutes prior to landing.

The estimated total reentry time is approximately the total estimated range divided by the average velocity during reentry

$$t_{max} \approx \frac{\ell_D}{\bar{V}} = \frac{8{,}700 \text{ km}}{3.85 \text{ km/s}} \approx 2{,}300 \text{ s} \qquad (5.8)$$

The total reentry time listed for one of the early Shuttle missions, STS-5, is approximately 1,820 s. So our crude BotE estimate for t_{max} is about 25% greater than the typical Orbiter reentry time.

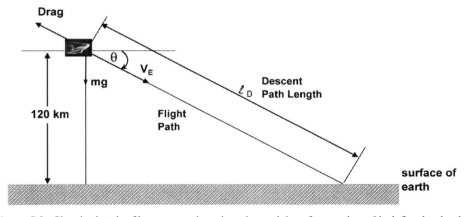

Figure 5.2. Simple sketch of key reentry length scales and drag force, where θ is defined to be the average flight-path angle.

Sec. 5.3]   Using Quick-Fire methods to crudely estimate peak heating rate 227

5.3.2.5 Modeling the maximum heat transfer rate during reentry

We can now develop a crude model for the maximum heat transfer rate, $\dot{q}_w$, (in Watts/m^2) during reentry. Figure 5.3 serves as a reference for the hypersonic heat transfer problem.

The peak heating rate at the blunt Orbiter nose-tip depends upon the specific reentry flight trajectory and the lift and drag forces set up by the Orbiter during its high Mach number reentry. Figure 5.3 can be thought of as a sketch of the high Mach number air flow over a blunted hemisphere fixed in a *wind tunnel*. Let's say that the wind tunnel is a square duct with dimensions $A_{\text{tunnel}} = 1 \times 1$ m. If the air is moving at a speed of V_∞, we can readily calculate the air mass passing through a given cross-section per unit time

$$\dot{m} = \rho_\infty V_\infty A_{\text{tunnel}} \tag{5.9}$$

which is the mass flow rate in kg/s, where ρ_∞ is the density of the gas upstream of the reentry model. The mass flow rate per unit area is then $\rho_\infty V_\infty$ with units of kg/s/m^2.

Every kilogram of air carries with it a certain amount of total energy (made up of both internal thermal energy per unit mass ($\approx C_P T$) and kinetic energy per unit mass ($V^2/2$). At very high Mach numbers, the kinetic energy is the dominant term. Hence the incident energy flux passing through our hypothetical wind tunnel per unit area per unit time ($\dot{Q}_\infty$) is approximately equal to the product of KE/mass and the mass flow rate per unit area

$$\dot{Q}_\infty = (\tfrac{1}{2} V_\infty^2)(\rho_\infty V_\infty) = \tfrac{1}{2} \rho_\infty V_\infty^3 \quad \text{(joule/s/m}^2\text{)} \text{ or (watts/m}^2\text{)} \tag{5.10}$$

The heat flux per unit surface area transferred to the body in our hypothetical wind tunnel, $\dot{q}_w$, is also measured in these same units. This implies that $\dot{q}_w$ may also

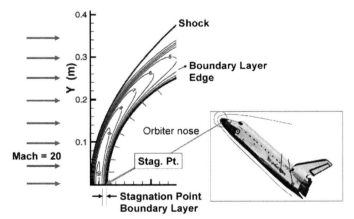

Figure 5.3. Schematic of the Orbiter blunt nose stagnation point shock layer, temperature field, and viscous boundary layer edge for high Mach number reentry conditions. The plotted thermal field shows the line of constant temperature calculated for an equivalent axisymmetric stagnation flow based on the computational fluid dynamics results presented in [4]. Note that the thickness of the thin boundary layer at the stagnation point is crucial in determining the peak nose-cap heat flux.

be proportional to the oncoming incident energy flux $\dot{Q}_\infty$. The proportionality constant (the Stanton number, designated St), is a non-dimensional parameter used to scale heat transfer rates in a wide range of convective heating problems. (It is sometimes also written $\equiv C_h$). The Stanton number for high speed flows is defined as

$$St = \frac{\dot{q}_w}{(\rho_\infty V_\infty)[(C_p T_\infty + \tfrac{1}{2} V_\infty^2) - C_p T_{wall}]} \quad (5.11)$$

As stated, for very high Mach numbers, the kinetic energy term dominates over the free stream and wall temperature (or *enthalpy*) terms. This allows us to approximate St by the expression

$$St \approx \frac{\dot{q}_w}{(\tfrac{1}{2}\rho_\infty V_\infty^3)} = \frac{\dot{q}_w}{\dot{Q}_\infty} \quad \text{for Mach. no.} \gg 1 \quad (5.12)$$

5.3.2.6 Physical parameters and data for estimating peak heat flux

If we are given the magnitude of St, or can employ a model viscous boundary layer equation to calculate St (appropriate to the stagnation point region as in Figure 5.3), then the wall heat transfer rate, $\dot{q}_w$, can readily be calculated. This calculation can be performed if we know both ρ_∞ and V_∞, or if we have a reentry trajectory model that determines V_∞ as a function of ρ_∞.

As a "zeroth-order" cut at the problem, let's grossly assume that 10% of the incident energy flux is heated by the strong shock wave and is transported by conduction through the stagnation point boundary layer into the wall. That is

$$St \approx 0.10 \quad (5.13)$$

At this point we do not have any justification for assuming that St is of order 0.1, but later we will see from analytical models culled from the reentry literature that St is approximately in the range of 0.02 to 0.04 for the peak heating region of the reentry trajectory.

The remainder of the available air energy flux near the vehicle moves along streamlines close to the body of the Orbiter, heating other areas of the craft but not nearly as intensely as at the stagnation point. Almost all of the kinetic energy flowing close to the body is converted into internal thermal energy in the shock layer. In fact, this layer is heated to such a high temperature that the molecules of gas dissociate. Other portions of the heated flow field pass around the body into the turbulent wake and are carried far downstream of the Orbiter.

We start by estimating the magnitude of the incident energy flux $\dot{Q}_w$. Let's crudely adopt the same "average" values of velocity and density that we used to estimate the descent path ℓ_D, and average flight path angle, θ_{avg}, i.e.

$$V_\infty \approx \bar{V} = 0.5 V_E \approx 3{,}850 \text{ m/s}$$

$$\rho_\infty \approx \bar{\rho} \approx 2.3 \times 10^{-4} \text{ kg/m}^3$$

5.3.2.7 Numerical estimate for the peak heat flux

Using the above input data, the corresponding magnitude for $\dot{Q}_\infty$ is

$$\dot{Q}_\infty = \tfrac{1}{2}\rho_\infty V_\infty^3 = (0.5)(2.3 \times 10^{-4})(3{,}850)^3 \approx 6.6 \times 10^6 = 6.6 \text{ MW/m}^2$$

The magnitude of $\dot{Q}_\infty$ is quite sensitive to both velocity (cubic dependence) and to density (exponential in altitude). If our initial assumed velocity was $0.7 V_E$ instead of $0.5 V_E$, then the energy flux would be larger by a factor of 2.75, giving $\dot{Q}_\infty = 18 \text{ MW/m}^2$. If the fractional heat transferred to the wall is assumed to be 10% of $\dot{Q}_\infty$ (an order of magnitude guess as noted above) then the peak wall heat flux is approximately

$$\dot{q}_w \approx (0.1) \cdot (6.6) = 0.66 \text{ MW/m}^2 \tag{5.14}$$

5.3.2.8 Interpretation of the Quick-Fire peak heat flux estimate

For the blunt-nose region of the Orbiter, several different NASA data sources and predictions give values of the maximum heat flux in the range

$$0.31 < \dot{q}_w < 0.52 \text{ MW/m}^2 \tag{5.15}$$

Hence our zeroth-order estimate for heat flux, $\dot{q}_w \approx 0.66 \text{ MW/m}^2$, is close to the upper end of this "experimental ballpark".

In Section 5.6.5 we will review the data represented by Equation 5.15 and cite the appropriate literature sources, where we use a more physically-based boundary layer model to calculate stagnation-point heat transfer rates.

5.3.2.9 Modeling the total maximum heat load during reentry

The estimated total heat load (in J/m^2) transferred through the stagnation boundary layer to the nose region of the Orbiter, Q_w/A_w, is the time integral of the maximum heat flux during the entire flight

$$\text{Total heat load} \equiv Q_w/A_w = \int_0^{t_{max}} \dot{q}_w \, dt \tag{5.16}$$

The quantity Q_w/A_w should be called "total heat load per unit area" but the common name used in the literature is "total heat load" or sometimes just "heat load".

5.3.2.10 Numerical estimate for the maximum heat load

If the effective integration time is based on the duration of the peak heating time period, let's assume it is 50% of the approximate total flight time calculated from Equation 5.8, then $t_{\text{effective}} \approx 1{,}100$ s. An estimate for the total heat load is

$$\text{Total heat load} \equiv Q_w/A_w \approx [(\dot{q}_w)_{\max}] t_{\text{effective}} \tag{5.17}$$

$$Q_w/A_w \approx (0.66 \text{ MW/m}^2)(1{,}150 \text{ s}) = 760 \text{ MJ/m}^2 \tag{5.18}$$

5.3.2.11 Interpretation of the Quick-Fire maximum heat load estimate

The corresponding average heat load for the *entire Orbiter body* is considerably smaller (about a factor of six lower than that given by Equation 5.18) at approximately $100 \, \text{MJ/m}^2$. This reduction in the average heat load is due, in large part, to the fact that the heating on the top side of the Orbiter is quite low; nearly two orders of magnitude smaller than on the underside or windward side of the Orbiter.

For the blunt-nose region of the Orbiter, several different NASA sources and predictions (see Section 5.6.5) give values of total heat load in the range

$$260 < Q_w/A_w < 320 \, \text{MJ/m}^2 \tag{5.19}$$

Our crude, zeroth-order prediction (Equation 5.18) is about a factor of 2.4 greater than the upper end of this data-driven total heat-load range.

The total heat load expressed as a fraction of the Orbiter's initial KE
Another way of writing Equation 5.17 is

$$Q_w/A_w \approx [(\dot{q}_w)_{\max}] t_{\text{effective}} = (\tfrac{1}{2}\rho_\infty V_\infty^3) \cdot (St) \cdot \left(\frac{\ell_D}{V_\infty}\right) \tag{5.20}$$

where

$$\ell_D = \frac{\tfrac{1}{2}mV_E^2}{\{[\tfrac{1}{2}(\rho_\infty V_\infty^2)] C_D A\}} \tag{5.21}$$

Substituting Equation 5.21 into Equation 5.20 and canceling the common terms yields

$$Q_w/A_w = \frac{St}{(C_D A)} (\tfrac{1}{2}mV_E^2) \tag{5.22}$$

Normalizing Q_w by the initial vehicle kinetic energy, gives us

$$Q_w/(\tfrac{1}{2}mV_E^2) = \left[\frac{St \cdot A_w}{C_D A_{\text{drag}}}\right] \tag{5.23}$$

This simple result was originally derived in the 1950s by Allen and Eggers [7]. In their specific equation, they also substituted the approximate low-speed boundary layer estimate for Stanton number as equal to one-half the boundary layer skin friction coefficient, C_f; that is, $St \approx C_f/2$.

Equation 5.23 essentially states that the ratio of the total body heat, Q_w, to the initial kinetic energy is of the order of the Stanton number multiplied by the ratio of the wetted area to the drag area. A designer seeking to reduce the heat load should try to reduce the wetted area relative to the frontal or drag area of the vehicle.

The initial kinetic energy for the Orbiter on entering the atmosphere is about 3×10^{12} J. For the peak heating region we "guessed" a Stanton number of order 0.1. On average, it is about one order of magnitude *smaller* for the entire high temperature *lower side* of the Orbiter. So the effective Stanton number for the whole body is perhaps about 0.01. The wetted area of all the heated surfaces is of order $1{,}000 \, \text{m}^2$; the drag area, A_{drag}, is about $200 \, \text{m}^2$; and C_D is order 1. Hence, the ratio of total heat for the entire Orbiter body to the initial kinetic energy of the Orbiter (based on Equation 5.23) is of order

$$Q_w/(\tfrac{1}{2}mV_E^2) \approx 0.05 = 5\% \tag{5.24}$$

The remaining 95% of the initial kinetic energy goes into heating the air that surrounds the vehicle during the approximately 2,000 s of reentry. Much of this heated air eventually is entrained into the turbulent wake of the vehicle.

In absolute terms, the total heat to the Orbiter body, Q_w, is approximately

$$Q_w = 0.05(3 \times 10^{12}) = 1.5 \times 10^{11} \text{ J} = 1.5 \times 10^5 \text{ MJ} \tag{5.25}$$

That is quite a bit of heat going into the Orbiter, and is the primary reason for the extensive engineering effort by engineers to develop the ceramic tiles for its thermal protection system.

We note that a good fraction of this total heat must be absorbed primarily by the Orbiter's TPS tiles. The quasi steady-state temperature increase of the TPS material during reentry is approximately $\Delta T_{\text{tile}} = Q_{\text{per tile}}/C_{p_{\text{tile}}} m_{\text{tile}}$. It follows that to maintain N tiles below the maximum allowable fracture temperature, say some $\Delta T_{\max}$ for a given material, the total heat load during reentry, Q_w, must be kept below a given threshold (see [7]) set approximately by the expression

$$Q_w < (N_{\text{tiles}} \cdot m_{\text{per tile}}) C_{p_t} \Delta T_{\max} \tag{5.26}$$

5.3.2.12 From Quick-Fire estimates to more accurate results

In the more detailed analysis that follows, we upgrade these Quick-Fire models with more accurate models for the reentry flight dynamics and for the stagnation-point convective heat transfer rates. We employ these more physically-based BotE models in order to calculate better estimates for Orbiter altitude and velocity as a function of time, the peak heat flux to its nose, and the total heat load to its thermal protection system in the vicinity of the stagnation point.

We start by examining a well-known fluid-physics based model for the stagnation point heat transfer process.

5.4 A LOOK AT HEAT FLUX PREDICTION LEVELS BASED ON AN ANALYTICAL MODEL FOR BLUNT-BODY HEATING

In the following discussion we utilize a specific fluid-mechanics-based equation for the rate of heat transfer $\dot{q}_w$ to the wall of the Orbiter's spherical nose within a stagnation point flow field. We follow the 1958 approach and derivation of Lees [6] and Allen and Eggers [7]. From this model heat transfer equation we calculate the stagnation-flow Stanton number for typical reentry conditions as defined in Equation 5.12 for high Mach number flows. This expression allows us to quantify the Stanton number to within about 20%, rather than (as we did above) simply apply an order-of-magnitude guess.

For a number of models in the literature, $\dot{q}_w$ is approximately proportional to the product of the cube of the free stream velocity V_∞ and the square root of the free stream density ρ_∞. In some more-refined models, the velocity exponent is determined to be a little larger than 3.0.

The well-known studies by Lees [6], Allen and Eggers [7] and their colleagues in the 1950s also showed that for high speed stagnation flows, as with their low speed counterparts, $\dot{q}_w$ is inversely proportional to the square root of the nose radius r_{nose}. Lees' interpretation of this dependence on blunt-nose radius is that "at hypersonic speeds all reentry bodies must be blunt-nosed to some extent in order to reduce reentry heat transfer rates to manageable proportions". In fact, his students demonstrated this overall conclusion by placing a relatively sharp conical nose-cone made from dense paraffin wax in the Caltech Mach 6 wind tunnel. When the hypersonic flow was established, the stagnation-point heating melted the wax there, transforming the cone into a more stable, well-rounded, blunt-nosed body [personal communication from Dr. James Wu].

The general dependences of peak heat transfer rates on speed, altitude, and nose radius are summarized by the following approximate expression

$$\dot{q}_w = \frac{A}{\sqrt{r_{nose}}} (\rho_\infty)^{1/2} (V_\infty)^3 \tag{5.27}$$

where A is a constant determined from the stagnation flow field geometry and the thermal and viscous properties of air.

The corresponding expression for high speed Stanton number is

$$St_{high\ speed} = \frac{\dot{q}_w}{(\frac{1}{2}\rho_\infty V_\infty^3)} = \frac{2A}{\sqrt{r_{nose} \cdot \rho_\infty}} \tag{5.28}$$

One can calculate an approximate estimate for A based on Allen and Eggers' and Lees' extrapolations to hypersonic speeds and high temperatures of the low-speed formula for St obtained from the well-known laminar stagnation-flow thermal-boundary-layer solution of Sibulkin for flow around a low-speed *blunt body* of revolution [8]

$$St_{low\text{-}speed} \equiv \frac{\dot{q}_w}{[\rho_\infty V_\infty C_p (T_\infty - T_{wall})]} = \frac{0.763}{\sqrt{Re_{r_{nose}}}} (Pr^{-0.6}) \sqrt{\frac{r_{nose} \cdot \bar{K}}{V_\infty}} \tag{5.29}$$

where $Re_{r_{nose}} \equiv \frac{\rho_\infty V_\infty r_{nose}}{\mu_\infty}$ and Pr is the Prandtl number for air. In this low-speed case Pr is a constant fluid parameter of order 1.0. $Re_{r_{nose}}$ is defined as the Reynolds number based upon the *radius* of the equivalent stagnation-point sphere, r_{nose}, and also on the upstream velocity V_∞. For low speeds, the density ρ_∞, and the viscosity μ_∞ are assumed constants of the incompressible fluid.

The important flow parameter $\bar{K}$, defined equal to the local inviscid flow velocity gradient at the stagnation point, is

$$\bar{K} = \left(\frac{dV_e}{dx}\right)$$

As indicated in the sketch below, by definition, the x direction is perpendicular to the principal axis of the oncoming flow at the stagnation point.

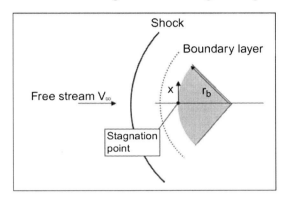

The subscript e in the expression for $\bar{K}$ designates flow variables at the outer edge of the stagnation point boundary layer.

In Equation 5.29, the *grouped* parameter, $\left(\frac{r_{\text{nose}} \cdot \bar{K}}{V_\infty}\right)$, is a *non-dimensional* measure of the stagnation point velocity gradient. For low-speed flow over a spherical body, basic potential flow theory determines that this non-dimensional velocity gradient parameter is $\left(\frac{r_{\text{nose}} \cdot \bar{K}}{V_\infty}\right) = 1.5$.

For a fixed difference between the free stream temperature and the wall temperature of the sphere, $(T_\infty - T_w)$, Equation 5.29 indicates that the heat flux $\dot{q}_w$ is proportional to the square root of the free stream velocity, $\dot{q}_w \sim V_\infty^{1/2}$.

Also $\dot{q}_w$ is inversely proportional to the square root of the nose radius, r_{nose}, i.e. $\dot{q}_w \sim \sqrt{V_\infty/r_{\text{nose}}}$ based on the definition of the nose Reynolds number.

In the high Mach number counterpart to this low-speed stagnation-flow problem, the velocities and densities that determine the wall heat flux occur within the high-temperature *subsonic* flow region immediately *downstream of the bow shock wave*, referred to as region 2, just outside the stagnation point.

Let's define the high-speed Stanton number in terms of the inviscid "post-shock" region 2 variables ρ_2, V_2, μ_2, and T_2 (or T_{recovery}), i.e.

$$St_{\text{high-Mach no.}} \equiv \frac{\dot{q}_w}{[\rho_2 V_2 C_{p_2}(T_{\text{recovery}} - T_{\text{wall}})]} \approx \frac{0.763}{\sqrt{Re_{r_{\text{nose}}}}}(Pr_2^{-0.6})\sqrt{\frac{r_{\text{nose}} \cdot \bar{K}_{\text{high}}}{V_2}} \quad (5.30)$$

where for the post-shock region 2, $Re_{r_{\text{nose}}} \equiv \frac{\rho_2 V_2 r_{\text{nose}}}{\mu_2}$.

The velocity gradient, $\bar{K}_{high}$ will be shown to be directly related to the inviscid hypersonic pressure gradient, dP_e/dx, evaluated at the stagnation point, i.e. the rate of change of the boundary layer *edge* pressure with distance x along the blunt nose.

As implied by Figure 5.3, the recovery or stagnation temperature behind the shock wave is approximately equal to the very high temperature of the shocked gas upon being brought to rest at the stagnation point, namely $T_{recovery}$. Based on the conservation of energy relationship across the shock, the recovery *enthalpy* of an ideal gas, the ($Cp_2 T_{recovery}$) term, is approximately equal to the initial kinetic energy per unit mass. We also assume that the wall temperature, T_w, is very low in comparison to the recovery temperature. Under these assumptions

$$C_p(T_{recovery} - T_w) \approx \frac{V_\infty^2}{2}$$

Solving Equation 5.30 for $\dot{q}_w$, we find that the heat flux is then given by

$$\dot{q}_w = 0.763(Pr_2^{-0.6})\sqrt{\rho_2 \mu_2}\sqrt{\bar{K}_{high}}\left(\frac{V_\infty^2}{2}\right)$$

Regrouping the terms

$$\dot{q}_w = \frac{0.763(Pr_2^{-0.6})}{2\sqrt{r_{nose}}}\sqrt{\frac{r_{nose} \cdot \bar{K}_{high}}{V_\infty}}\sqrt{\frac{\rho_2}{\rho_\infty}}\sqrt{\mu_2}[\sqrt{\rho_\infty}V_\infty^{5/2}] \qquad (5.31)$$

In Equation 5.31, the density ratio ρ_2/ρ_∞ across the shock wave for an ideal gas is given by $\rho_2/\rho_\infty = (\gamma_2 + 1)/(\gamma_2 - 1)$, where γ_2 is the ratio of specific heats for the post-shock environment for which the gas is fully dissociated. For that very high-temperature region, Lees [6] indicates that $\gamma_2 \approx 1.2$ (or perhaps 1.1) based on dissociation calculations from several sources.

We note that the viscosity μ of a gas, like air, increases significantly due to temperature. Viscosity μ_2 is approximately proportional to the square root of the local recovery temperature, $\mu \sim T_{recovery}^{1/2}$, hence $\mu_2 \sim V_\infty$. From Equation 5.31, this high temperature viscosity relationship introduces an extra velocity dependent term to the equation for the high speed heat transfer rate that is proportional to $V_\infty^{1/2}$. The product of all terms in Equation 5.31 that contribute to the heat flux, $\dot{q}_w$, are then seen to be proportional to $\sqrt{\rho_\infty} \cdot V_\infty^3$.

The only remaining constant is the high-speed non-dimensional velocity gradient parameter, $\left(\dfrac{r_{nose} \cdot \bar{K}_{high}}{V_\infty}\right)$, and this can be estimated for hypersonic flows, assuming that surface pressures on the sphere are given by the hypersonic "Newtonian" flow model [6]

$$\left(\frac{r_{nose} \cdot \bar{K}_{high}}{V_\infty}\right) = \sqrt{\frac{2}{\rho_2/\rho_\infty}} = \sqrt{\frac{2(\gamma_2 - 1)}{(\gamma_2 + 1)}}$$

Following Lees we approximate the effects of gas dissociation in the shock layer by setting $\gamma_2 \approx 1.2$ in the above expressions in order to scale the constants in Equation 5.31.

Sec. 5.4] A look at heat flux prediction levels 235

With these assumptions for the gas properties behind the shock, we obtain the following equation for the high speed temperature heat flux, previously written down in Equation 5.27

$$\dot{q}_w = \frac{A}{\sqrt{r_{\text{nose}}}} (\rho_\infty)^{1/2} (V_\infty)^3 \tag{5.32}$$

Based on Lees' *hypersonic* extrapolation of the low-speed stagnation point model, we estimate that the magnitude of the constant A is

$$A_{\text{estimate}} = 1.36 \times 10^{-4}$$

in units of $\sqrt{\dfrac{\text{kg}}{\text{m}}}$.

Employing a similar set of underlying boundary layer assumptions, Sutton and Graves [9] calculated the constant A for the equivalent stagnation-point heat transfer problem, but did so by calculating *all* of the gas properties based on the assumption that the air downstream of the shock wave is *fully dissociated*. They also included in their calculations the Newtonian model for the stagnation point velocity gradient under hypersonic flow conditions. These model changes have the basic form of Equation 5.32, but the constant A calculated for air is somewhat larger than our preliminary estimate of 1.36×10^{-4} due to Sutton and Graves' detailed numerical evaluation and the utilization of the thermodynamic properties of high temperature *dissociated* air. They also took into account, in their calculations, the *diffusion* of the various dissociated air species to the cooled wall of the spherical nose. We will not develop a model for the diffusion processes within the stagnation point boundary layer. From their *equilibrium* dissociated air and diffusion calculations, the recommended constant of Sutton and Graves for reentry into the earth's atmosphere, as translated from the initial form of their model, is

$$A_{\text{Sutton-Graves}} \approx 1.75 \times 10^{-4} \frac{\sqrt{\text{kg}}}{\text{m}} \tag{5.33}$$

Our model estimate for this constant, $A \approx 1.36 \times 10^{-4}$, is reasonably close to the Sutton–Graves constant; we under-predict its magnitude by only about 20%.

Using the Sutton–Graves constant, the corresponding expression for the hypersonic Stanton number (accounting for dissociation) is

$$St_{\text{Sutton-Graves}} = \frac{3.5 \times 10^{-4}}{\sqrt{r_{\text{nose}} \cdot \rho_\infty}} \tag{5.34}$$

Equation 5.34 can be rewritten in the familiar form which is dependent on the free-stream (or infinity) Reynolds number scaled by the nose radius. In order to obtain this form from Equation 5.34 we again assume that the gas viscosity in the shock layer is proportional to the *square root* of the recovery temperature. The equivalent expression for St is

$$St_{\text{high-speed}} \approx 1.57 \sqrt{\frac{M_\infty}{Re_{\infty(r_{\text{nose}})}}} \qquad (5.35)$$

where $Re_{\infty(r_{\text{nose}})} \equiv \dfrac{\rho_\infty V_\infty r_{\text{nose}}}{\mu_\infty}$.

Note that the free stream Mach number M_∞ is now a scaling parameter in Equation 5.35 due to the dependence of viscosity on temperature. By definition the Mach number is $M_\infty = V_\infty/(\text{speed of sound}) = V_\infty/\sqrt{\gamma R T_\infty}$, where γ is the ratio of specific heats for air, and R is the gas constant in the perfect gas equation of state, $p = \rho R T$.

Either Equation 5.34 or Equation 5.35 is an acceptable expression for estimating peak blunt-nose heat flux. The choice of the form of the equation depends on the input data provided. With our simplified models for trajectory density and velocity (derived in the following section) Equation 5.34 is the first choice. Often, however, NASA provides information for both Reynolds numbers and Mach numbers as a result of full-scale flight tests or wind-tunnel model tests, in which case we would use Equation 5.35.

5.4.1 Numerical estimates of Stanton number using the Sutton–Graves constant

In our first estimate of the magnitude of the incident energy flux, $\dot{Q}_\infty$, we crudely used the following "average" values of reentry velocity and density

$$V_\infty \approx \bar{V} = 0.5 V_E \approx 3{,}850 \text{ m/s}$$

$$\rho_\infty \approx 2.3 \times 10^{-4} \text{ kg/m}^3 \quad (\text{altitude} \approx 60 \text{ km})$$

and showed that the corresponding magnitude for $\dot{Q}_\infty$ is

$$\dot{Q}_\infty = 6.6 \text{ MW/m}^2$$

We arbitrarily assumed that the Stanton number was 0.10 or 10%. So now let's estimate St from Equation 5.34 with an assumed nose radius of 1 m, a typical value for the full-scale Orbiter

$$St_{\text{Sutton-Graves}} = \frac{3.5 \times 10^{-4}}{\sqrt{r_{\text{nose}} \cdot \rho_\infty}} = \frac{3.5 \times 10^{-4}}{\sqrt{(1.0 \text{ m}) \cdot (2.3 \times 10^{-4})}} = 0.023 \quad \text{or } 2.3\% \qquad (5.36)$$

This is a factor of 4 smaller than our original 10% guess for Stanton number. Our prediction for the corresponding heat flux at this altitude and flight speed is $\dot{q}_w = St \cdot \dot{Q}_\infty = (0.023) \cdot (6.6) = 0.15 \text{ MW/m}^2$, which is about 50% below the NASA data range for maximum blunt-nose heat flux

$$0.31 \text{ MW/m}^2 < \dot{q}_{w_{\text{NASA}}} < 0.52 \text{ MW/m}^2$$

The reason for the discrepancy between our BotE estimate for $\dot{q}_w$ and NASA data is our initial crude guess of the Orbiter's velocity at the specified altitude of 60 km. We had guessed the velocity to be about 50% of the entry velocity of 7.7 km/s. A more realistic value for a gliding reentry (based on the STS-5 trajectory data set discussed below) would be

$$V_{\infty_{STS-5}} \approx 0.68 V_E \approx 5{,}200 \text{ m/s} \quad \text{at altitude} = 60 \text{ km} \tag{5.37}$$

Using this velocity, along with $\rho_\infty(60 \text{ km}) \approx 2.3 \times 10^{-4} \text{ kg/m}^3$, as input we obtain the following estimate for the incident energy flux $\dot{Q}_\infty$

$$\dot{Q}_{\infty_{STS-5}} = \tfrac{1}{2}\rho_\infty V_\infty^3 = (0.5)(2.3 \times 10^{-4})(5{,}200)^3 \approx 16.2 \times 10^6 = 16.2 \text{ MW/m}^2 \tag{5.38}$$

Our revised prediction for heat flux at the vehicle speed of 5.2 km/s at $h = 60$ km now rises to

$$\dot{q}_w = St \cdot \dot{Q}_\infty = 0.023(16.2 \text{ MW/m}^2) = 0.37 \text{ MW/m}^2 \tag{5.39}$$

This heat flux is 2.3 times larger than our previous estimate, which was based on a lower guess for the Orbiter's velocity at 60 km altitude. The revised heat flux given by Equation 5.39, based on a Stanton number of 0.023, now falls within the NASA data range for peak nose heat flux of $0.31 < \dot{q}_w < 0.52 \text{ MW/m}^2$. In fact, it is typical of many of the maximum heat flux data points listed reported by NASA for the early Shuttle missions.

Assuming that this peak heating rate is maintained for about 1,150 s, as per our initial crude estimate for heat load, we now find that the total heat load per unit area for the stagnation region (based on the higher velocity assumption of 5,200 m/s) is approximately

$$Q_w/A_w \approx (0.37 \text{ MW/m}^2)(1{,}150 \text{ s}) = 425 \text{ MJ/m}^2$$

This new heat load estimate falls about 33% beyond the upper end of the heat load range given by NASA's initial design study and for STS-1 to STS-5 [10]

$$260 < Q_w/A_w < 320 \text{ MJ/m}^2$$

We will estimate the maximum heat flux and maximum heat load again in Section 5.6, after developing a more detailed flight trajectory model for reentry in Section 5.5.

5.5 SIMPLE FLIGHT TRAJECTORY MODEL

The equations of motion for a body of mass m moving with velocity $v(t)$ subject to the forces of gravity (mg), aerodynamic drag (D), and aerodynamic lift (L) are written below. The equations are derived for a coordinate system *along* and *transverse* to the trajectory (Figure 5.4). Note that we model reentry motions relative to a non-rotating earth.

238 Estimating the Orbiter reentry trajectory and the associated peak heating rates [Ch. 5

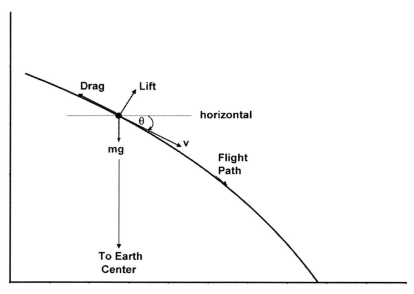

Figure 5.4. Force balance on a reentry vehicle with mass m moving on a curved trajectory. At time t, the velocity vector $\vec{v}$ tangent to the trajectory is inclined at angle θ (the flight-path angle) relative to the local horizontal. The gravity vector always points to the center of the earth. The direction of the drag force is taken to be opposite that of the velocity vector $\vec{v}$.

The equation of motion relating the forces along the curved flight path to the acceleration dv/dt, is

$$m\left(\frac{dv}{dt}\right) = -D + mg \sin\theta \qquad (5.40)$$

where θ is the flight-path angle measured relative to the local horizon, as defined in Figure 5.4.

The second equation of motion, written for the direction perpendicular to the curved flight path, is

$$-mv\left(\frac{d\theta}{dt}\right) = L - m\left(g - \frac{v^2}{r}\right)\cos\theta \qquad (5.41)$$

which relates the rate of change of the flight-path angle, $\frac{d\theta}{dt}$, to the aerodynamic lift force L, the gravitational constant g, and the opposing effective centrifugal force mv^2/r, where r is the distance to the center of the earth, h is the vehicle's altitude, and $r = R_{\text{earth}} + h$.

It is evident that the rate of change of $h(t)$, the time varying altitude of the reentry body, is equal to the vertical component of velocity along the flight path

$$\frac{dh}{dt} = -v \sin\theta \qquad (5.42)$$

Sec. 5.5] Simple flight trajectory model 239

We will approximate Equations 5.40, 5.41, and 5.42, and obtain simple solutions in two separate flight domains.

The first trajectory model and solution is for the *initial descent period*, roughly the first 275–300 seconds after entering the atmosphere at $h = 120$ km. For this domain we show that the altitude $h(t)$ decreased approximately linearly with time. We will also estimate the small change in Orbiter velocity during this early period due to aerodynamic drag. In this early reentry region we initially assume that the aerodynamic lift is negligible, and the flight path angle θ is more or less constant and equal to the shallow entry-angle θ_E, which is approximately 1° [4, p. 284]. We then perturb this zeroth-order solution to determine the approximate change in flight path angle with time due to lift.

In the second model, appropriate to the next 900 seconds, we assume that the vehicle moves along a steady or equilibrium *glide* trajectory, where the lift force per unit vehicle mass is exactly balanced by the gravitational and centrifugal force components normal to the flight path. We will show that the equilibrium flight path angle θ is determined by the lift coefficient, C_L, the mass m, and the local air density ρ.

The simple BotE models outlined above directly address the *Orbiter* reentry trajectory problem. The approach, where an engineer estimates the relative contributions of the dominant forces in different flight regimes and then simplifies and solves the equations of motion, can be, and has been, applied to a number of *non-Orbiter* entry problems; see Appendix (Section 5.7). We emphasize that, even though we have not developed the specific models and worked out the solutions to these other entry problems, the basic BotE approach developed here can often provide a valuable first-cut engineering estimate for a range of flight dynamic and aerothermodynamic variables.

5.5.1 A simple BotE model for the initial entry period; the entry solution

Let's first consider the Orbiter entering the atmosphere at an altitude $h = 120$ km as a result of a deorbit burn from a circular orbit. The initial flight-path angle $\theta_E \approx 1.20°$, is typical of published entry angles [11]. The initial entry velocity is estimated to be $v_E \approx 99\%$ of the circular velocity for a typical Shuttle mission ($v_{circular} \approx 7.78$ km/s for a 200 km high circular orbit). Thus we set $v_E = 7.70$ km/s.

5.5.1.1 Altitude model

We now calculate a simple solution for Orbiter altitude h as a function of time using Equation 5.42. We assume, for this zeroth-order solution, that the Orbiter velocity and entry angle remain approximately constant for about the first five minutes of the reentry period, i.e.

$$v = \text{constant} = v_E = 7.70 \text{ km/s}$$

$$\theta \approx \theta_E = 1.20° = 0.0209 \text{ radians}$$

Towards the end of the initial 275 to 300 s period we observe (from Orbiter altitude data) that the flight-path angle rapidly decreases from 1.2° to about 0.1°

as the effect of aerodynamic lift on the trajectory becomes significant. In this lift-significant domain ($t > 275$ s), known as the "steady equilibrium glide" region, we will use a second solution that takes into account both lift and drag (see Section 5.52).

From Equation 5.42, we approximate the right-hand side of the altitude-rate equation as follows

$$\frac{dh}{dt} = -v \sin \theta \approx -v_E \theta_E \quad \text{assuming } \theta \text{ (in radians) is small} \ll 1 \qquad (5.43)$$

Integrating Equation 5.43 over time from the initial height h_0 at $t = 0$ (marking the initial entry time) yields the simple linear expression

$$h(t) = h_0 - v_E \theta_E t \qquad (5.44)$$

Substituting our estimated values for the entry constants

$$h(t) = 120 \text{ km} - (0.161 \text{ km/s})t \qquad (5.45)$$

From Equation 5.45 we see that at $t = 100$ s, the altitude of the Orbiter drops to about 104 km, and that by 300 s it has decreased to 71.7 km.

5.5.1.2 Velocity change model

Let's now estimate the change in the Orbiter's velocity with time, assuming that the drag force (proportional to $\frac{1}{2}\rho v^2$, where $\rho = $ air density) is dominant over the small component of gravity in the flight direction, $mg \sin \theta$.

Although this assumption is not actually true at the very highest entry altitude of about 120 km where the air density is quite low (approximately 10^{-7} kg/m^3), the drag force becomes dominant over the flight-direction-component of gravity soon afterwards, as the vehicle descends below 100 km. Basically, drag increasingly decelerates the vehicle as a result of the exponential increase in air density with decreasing altitude.

The approximate small flight-path angle form of the equation of motion along the flight path, from Equation 5.40, is

$$\frac{dv}{dt} = -\frac{D}{m} \qquad (5.46)$$

Drag increases as the square of the relative velocity (between the vehicle and the surrounding air) based on the well-known equation

$$D = (\tfrac{1}{2}\rho v^2) C_d A \qquad (5.47)$$

We assume that wind and thermally-driven air motions (i.e. convection) are negligible at altitudes above 20 to 30 km, so that the relative velocity used in the above expression is equal to the vehicle velocity, $v(t)$.

In Equation 5.47, C_d is defined to be the drag coefficient (an aerodynamic constant of order 1.0 for blunt bodies or for the Orbiter at large angles of attack),

Sec. 5.5]   **Simple flight trajectory model**   241

and A is the reference drag area for the Orbiter, which NASA lists for both drag and lift as equal to the area of the Orbiter's delta-shaped wing $\approx 250\,\text{m}^2$.

To solve for vehicle velocity as a function of time, we integrate Equation 5.46 using an empirical expression for the atmosphere relating density to altitude.

A typical empirical model for atmospheric density is the *exponential model* where

$$\rho = \rho_0 \exp(-h/H_{\text{atm}}) \tag{5.48}$$

and ρ_0 is approximately equal to the atmospheric density at sea level ($\approx 1.225\,\text{kg/m}^3$) and H_{atm} is an empirically derived exponential *scale height* for the earth's atmosphere ($\approx 7.16\,\text{km}$, or $23{,}500\,\text{ft}$). These constants are based on semi-log curve fits to atmospheric density data by Dean Chapman and his colleagues at the NACA, dating back to 1958 [12].

Substituting the linear time-varying altitude expression, Equation 5.44, into Equation 5.48 we obtain the following equation for density as a function of time

$$\rho(t) = \rho_0 \exp[-(h_0 - v_E \theta_E t)/H_{\text{atm}}] \tag{5.49}$$

Entering this into the equation of motion along the flight path (Equation 5.46) gives

$$-\frac{dv}{dt} = \frac{D}{m} = \frac{(\tfrac{1}{2}\rho v^2) C_d A}{m} = \left(\frac{\rho_0 C_d A}{2m}\right) v^2 \exp[-(h_0 - v_E \theta_E t)/H_{\text{atm}}] \tag{5.50}$$

Let's assume that v varies slowly with flight time in the early entry period and decreases only a small amount from its initial magnitude. We also acknowledge that density increases exponentially with time. With these assumptions we can integrate Equation 5.50 to find the incremental change in velocity, $\Delta v(t) = v_E - v(t)$ in the following approximate form

$$\frac{\Delta v(t)}{\left(\frac{\rho_0 C_d A}{2m}\right) v_E^2} \approx \exp\left(-\frac{h_0}{H_{\text{atm}}}\right) \int_0^t (\exp(t/\tau))\,dt = \tau \cdot \exp\left(-\frac{h_0}{H_{\text{atm}}}\right) [\exp(t/\tau) - 1] \tag{5.51}$$

with the time constant for velocity change, $\tau \equiv \dfrac{H_{\text{atm}}}{v_E \cdot \theta_E} \approx 44\,\text{sec}$.

Recognizing that $\exp\left(-\dfrac{h_0}{H_{\text{atm}}}\right)[\exp(t/\tau)] = \rho(t)$ we can also write Equation 5.51 in the following form

$$\frac{\Delta v(t)}{v_E} \approx \left(\frac{\rho_0 C_d A}{2m}\right)\left(\frac{H_{\text{atm}}}{\theta_E}\right)\left(\frac{\rho(t)}{\rho_0} - \frac{\rho(0)}{\rho_0}\right) \tag{5.52}$$

To estimate the magnitude of the relative velocity change at $t = 300\,\text{s}$ after entry, we substitute the following constants characterizing the Orbiter and the atmosphere into Equations 5.49 and 5.52

$$\rho_0 = 1.225 \text{ kg/m}^3$$
$$H_{atm} = 7.16 \text{ km}$$
$$C_d = 0.8$$
$$\theta_E = 0.021 \text{ radians } (1.2°)$$
$$A = 250 \text{ m}^2$$
$$v_E = 7.70 \text{ km/s}$$
$$m_{orbiter} = 1 \times 10^5 \text{ kg}.$$

The density ratio at $t = 300$ s, calculated from Equation 5.49 at an altitude of $h \approx 71.6$ km is

$$\left(\frac{\rho(t=300)}{\rho_0}\right) = 4.54 \times 10^{-5}$$

From Equation 5.52 we find that the fractional change in velocity over 300 s is quite small

$$\frac{[v_E - v(t=300)]}{v_E} = 0.019, \quad \text{a 1.9\% reduction in velocity.} \tag{5.53}$$

Thus we estimate that the entry velocity of 7.70 km/s decreases a modest 0.15 km/s to 7.55 km/s in the first five minutes of entry as the Orbiter descends from 120 km to about 72 km, i.e.

$$v(t=300) \approx 7.55 \text{ km/s} \tag{5.54}$$

This small decrease in velocity during the initial reentry period justifies our use of the small perturbation approximation in the derivation of Equation 5.22.

The ratio of the "initial glide" velocity to the velocity of a hypothetical body orbiting the earth at zero altitude, $\frac{v_{ig}}{v_0}$, is a key parameter that sets the initial velocity ratio for the glide trajectory solution for $250\text{ s} < t < 300\text{ s}$; i.e. approximately when the Orbiter can be modeled as a gliding reentry body. $\frac{v_{ig}}{v_0}$ is defined and estimated as follows

$$\frac{v_{ig}}{v_0} \equiv \frac{v_{initial\text{-}glide}}{v_0} \approx \frac{v_{initial\text{-}entry}(300\text{ s})}{\sqrt{g_0 r_{earth}}} = \frac{7{,}550 \text{ km/s}}{\sqrt{(9.81 \text{ m/s}^2)(6{,}378 \times 10^3 \text{ m})}} = 0.954 \approx 0.95 \tag{5.55}$$

As the Orbiter descends to lower altitudes where the air density and dynamic pressure $q \equiv \frac{1}{2}\rho v^2$ rise rapidly, the drag force increases significantly. This higher level of drag greatly decelerates the vehicle. At altitudes below about 80 km, the effects of lift also become important and need to be incorporated into our glide dynamics model for this region. Section 5.52 gives our model for this "equilibrium glide" period.

5.5.1.3 Flight-path angle change due to lift

As will be shown by the "equilibrium glide" model of Section 5.2, at an altitude of about 78 km the glide path angle, θ_{glide}, is quite small, estimated at about 1/8th of the entry angle, $\theta_E = 1.2°$. The reduction of θ over a short period of time in the initial entry region is due to the increasing positive lift force, which is in turn proportional to the increasing dynamic pressure q encountered during this initial entry period. We develop here a simple estimate for the change in θ with time from the equations of motion.

Equation 5.41 relates the rate of change of the flight-path angle, $\frac{d\theta}{dt}$, to the aerodynamic lift force L, the gravitational constant g, and the opposing effective centrifugal force mv^2/r. In the early entrance region the trajectory is like that of a body in a circular orbit, since there is a close balance between the gravitational and effective centrifugal forces. Following the arguments of Lees [13], a good first approximation to the vertical equation of motion is to assume that the *difference* between the gravitational and effective centrifugal forces is quite small compared with the lift force. Under this approximation, and assuming that θ is a very small angle, Equation 5.41 reduces to the simple form

$$mv\left(\frac{d\theta}{dt}\right) = -L \tag{5.56}$$

which clearly shows that as lift L increases in magnitude, the flight-path angle θ decreases.

We previously showed that the approximate form for the equation of motion along the flight path (Equation 5.46) is

$$m\frac{dv}{dt} = -D \tag{5.57}$$

Dividing Equation 5.56 by Equation 5.57, yields the differential expression

$$v\frac{d\theta}{dv} = \frac{L}{D} = \frac{C_L}{C_D} \tag{5.58}$$

or

$$d\theta = (C_L/C_D)\left(\frac{dv}{v}\right) \tag{5.59}$$

Therefore the incremental change in flight-path angle is equal to the relative change in flight velocity multiplied by an aerodynamic lift to drag constant of order 1.

Equation 5.52 is a simple expression for the relative change in velocity during the initial entry period

$$\frac{\Delta v(t)}{v_E} = \frac{v_E - v(t)}{v_E} \approx \left(\frac{\rho_0 C_d A}{2m}\right)\left(\frac{H_{\text{atm}}}{\theta_E}\right)\left(\frac{\rho(t)}{\rho_0} - \frac{\rho(0)}{\rho_0}\right)$$

where
$$\rho(t) = \rho_0 \exp[-(h_0 - v_E \theta_E t)/H_{atm}]$$

From Equation 5.59, the corresponding expression for $\Delta\theta(t)$ is

$$\Delta\theta(t) = (\theta_E - \theta(t)) \approx (C_L/C_D)\left(\frac{\rho_0 C_d A}{2m}\right)\left(\frac{H_{atm}}{\theta_E}\right)\left(\frac{\rho(t)}{\rho_0} - \frac{\rho(0)}{\rho_0}\right) \quad (5.60)$$

Solving for $\theta(t)$

$$\theta(t) \approx \theta_E - (C_L/C_D)\left(\frac{\rho_0 C_d A}{2m}\right)\left(\frac{H_{atm}}{\theta_E}\right)\left(\frac{\rho(t)}{\rho_0} - \frac{\rho(0)}{\rho_0}\right) \quad (5.61)$$

Equation 5.61 indicates that the flight-path angle drops with time from an initial value of $\theta_E = 1.2°$ towards 0° as the density increases. For large times, according to Equation 5.61, $\theta(t)$ *surprisingly* becomes negative. If this were true it would mean that the Orbiter would stop descending and begin to rise. However, this rise in the Orbiter's trajectory doesn't actually occur, because when θ is still slightly positive, the original assumption of an identical balance between gravity and centrifugal forces underpinning Equation 5.31, breaks down and distorts any subsequent dynamics predictions.

It is at approximately that altitude where θ is still slightly > 0 that a balance is ultimately achieved between three principal forces: the centrifugal force, gravity, and the lift force. With this balance, the lifting reentry vehicle enters the so-called "equilibrium glide" region of the flight. At $t \approx 275$ s we would then transition from our initial entry model, valid for times $0 < t < 300$ s, to the "equilibrium glide" model. The initial conditions for the glide model are based on the magnitudes of the altitude and velocity derived from our simple *initial entry solution* taken when θ drops approximately to zero.

Figure 5.5 shows the flight-path angle trends with time during the initial entry period based on Equation 5.61. In the next section we will compare these results to flight-path angles inferred from STS-5 altitude data, $h(t)$.

5.5.1.4 Comparison of initial entry model with STS-5 data

Plots of our model estimates for altitude, velocity, and flight path angle (as a function of time for the initial entry period) are shown respectively in Figures 5.6, 5.7, and 5.8 for $C_L/C_D = 1.0$, where comparisons are made with the data from STS-5, which was the first "operational" mission of the Space Transportation System [14].

In Figure 5.6 two model calculations for vehicle altitude are shown. The blue curve shows a linearly descending altitude with time. This zeroth-order solution (Equation 5.44) assumes a constant flight-path angle with θ set equal to $\theta_E = 1.2°$. The red curve depicts the first-order solution for the perturbed flight-path angle (Equation 5.61). The magnitude of the altitude rate prediction, dh/dt, diminishes significantly as the initial entry period nears completion at 275 to 300 s. This is as expected since $-dh/dt \sim \theta$, and Figure 5.5 shows that θ drops rapidly towards a

Sec. 5.5]	**Simple flight trajectory model**	245

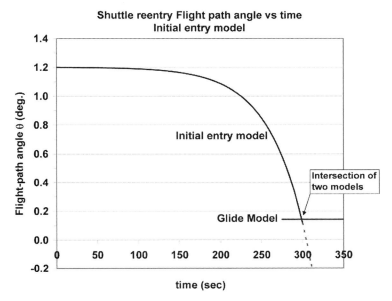

Figure 5.5. Flight path angle $\theta(t)$ during the early entry period (Equation 5.61) for $C_L/C_D = 1.0$ and $C_D = 0.8$. For reference, the estimated flight path angle for the equilibrium glide model is shown as a short horizontal line with glide angle $\theta_{\text{Glide}} = 0.14°$. The intersection point for the two models is indicated. This is the point in time where the flight-path angles from the two solutions become equal.

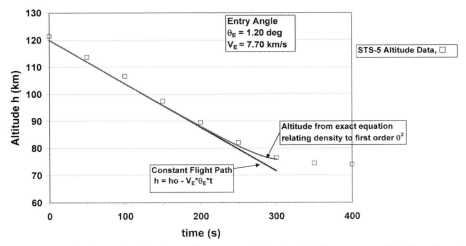

Figure 5.6. Altitude vs time for the early entry period using BotE reentry models based on: (a) the zeroth-order constant flight-path model (Equation 5.44), and (b) the first-order model for flight-path angle (Equation 5.61) with altitude from Lees' "exact" equation [13] relating density to the first-order model for $\theta(t)^2$. Comparisons are made with the Orbiter altitude data from the Shuttle's STS-5 mission [14].

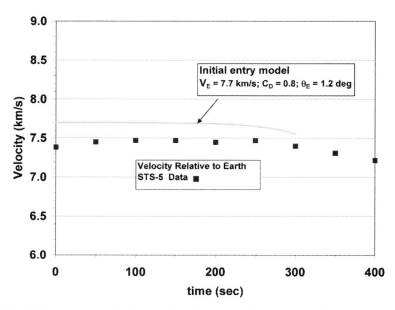

Figure 5.7. Orbiter reentry velocity vs time for the early entry period, $0 < t < 300$ s. Initial entry "constant flight-path-angle drag model" compared to STS-5 data. Velocity data measured relative to rotating earth. Simple constant-flight-path angle model velocities, $v(t)$, are defined for a non-rotating two-dimensional inertial coordinate system with its origin at the center of the earth.

value significantly below the entrance angle of 1.2° prior to the end of the initial entry time domain.

The predicted altitude as a function of time in Figure 5.6 can be estimated with either of two different procedures.

One way numerically integrates the product of modeled flight-path angle $\theta(t)$ from Equation 5.61 and entry velocity v_E, using dh/dt from Equation 5.42

$$h = h_0 - \int_0^t v(t)\theta(t)\,dt \approx \left[h_0 - v_E \int_0^t \theta(t)\,dt\right] \qquad (5.62)$$

and the other way uses Equation 5.63, the following closed-form non-linear solution of Lees [13] for altitude h (or indeed air density) as a function of a given flight-path angle $\theta(t)$

$$\frac{h(t)}{H_{\text{atm}}} = -\ln\left\{\exp(-h_0/H_{\text{atm}}) + \left(\frac{m}{\rho_0 C_L A H_{\text{atm}}}\right)(\theta_E^2 - \theta^2(t))\right\} \qquad (5.63)$$

This non-linear solution for the time-varying altitude can readily be derived by combining the small-angle version of Equation 5.42 for dh/dt with Equation 5.56 for $d\theta/dt$. We leave its derivation as an exercise for the reader.

Figure 5.6 displays the calculated values of $h(t)$ evaluated from Equation 5.63 using as input the values for $\theta(t)$ from our first-order model for time-varying flight-path angle (Equation 5.61). The real reentry data included for comparison were obtained by digitizing plots of measured altitude and velocity as a function of time for the first complete mission, STS-5. Those individual graphs were presented, in a single multi-variable composite graph in the 1986 technical paper by Ko and colleagues [14].

The *linear* initial entry model for altitude shown in Figure 5.6 was calculated from Equation 5.44, which is based on the constant flight-path angle assumption. This model altitude is in good agreement with the STS-5 data for times $t < 200$ s. But for $200\,\text{s} < t < 300\,\text{s}$, the first-order model based on the integrated form of Equation 5.61 is a better predictor of the change in slope. This is the critical region where lift becomes large enough to transition the trajectory of the Orbiter onto an equilibrium glide path. Agreement between the first-order model and the STS-5 data for altitude is quite good in this time interval, both with respect to the shape and magnitude of the time-varying trajectory curve.

We also test the model predictions against STS-5 for each of the two other principal variables, velocity and path angle, in the initial entry region: Figure 5.7 presents model and data for velocity, $v(t)$, and Figure 5.8 compares model, Equation 5.61, and data for the initial entry flight-path angle, $\theta(t)$.

Note from Figure 5.7 that the model-generated velocity estimates are nearly constant at 7.70 km/s for the first 200 s. Thereafter, for $200 < t < 300$, velocity v gradually drops as the drag becomes more significant at altitudes below 90 km, about 300,000 ft. The STS-5 data are also nearly constant in the first 200 s and begin to diminish at about the 240 s mark. However, the STS-5 velocity data are about 0.25 km/s below the predictions. This difference is primarily because the NASA velocity data, which was obtained from ground-based radars, measures vehicle velocities *relative to a rotating earth*, whereas our model was based on a *non-rotating inertial* coordinate system. The estimated velocity correction needed to reconcile the difference in coordinate systems is, of course, the magnitude of the component of rotational velocity of the earth *in the direction along the flight path*. If, hypothetically, the flight path followed a great circle in a plane coincident with the earth's *equator*, the velocity correction to NASA's relative velocity data would be +0.46 km/s. For an unperturbed orbit inclined 28° to the equator, with the Orbiter flying west to east (as in the case of a minimum-energy launch from the Kennedy Space Center), the correction would be about $0.46\,\text{km/s} \times \cos(28°) = 0.41\,\text{km/s}$.

For the first engineering test flight of the Shuttle, STS-1, data records indicate that at reentry the Orbiter was traveling northeast on a heading about 40° (counterclockwise) from the equator, for which the correction is approximately $0.46 \times \cos 40° = 0.35\,\text{km/s}$. Since we estimate the difference between model velocity predictions and data to be about 0.25 km/s, in translating to inertial velocities our estimate would over-shift the NASA data by about 0.1 km/s during the early phase of reentry. A possible explanation for this 0.1 km/s bias is the acceleration of gravity, g of our simple model, because we assumed it to be constant at its ground level value

g_0 of 9.81 m/s² throughout our altitude-varying reentry calculations, whereas in fact "g" is *not* constant with altitude; it increases slightly as the Orbiter descends from about 200 km to about 100 km. Effective "g" increases about 3% over this range of altitude, about enough to account for a 0.1 km/s increase in predicted Orbiter velocity as it descends. If such a correction was made, it would raise the model predicted inertial velocities by that amount and therefore cancel, approximately, the difference between model and data when viewed in an inertial system.

This estimate of a 0.1 km/s correction due to the height or radial dependence of "g" can be inferred from the *exact* expression for the velocity of a body in a circular orbit

$$v_{\text{circular}} = \sqrt{\frac{\mu_e}{r}} = \sqrt{\frac{g_0 r_e^2}{r_e + h}} \qquad (5.64)$$

where g_0 is the acceleration of gravity at the earth's surface. The exact expression for gravitational acceleration is not included in our approximate closed-form reentry dynamics model.

From Equation 5.64, the ratio of circular orbital velocities at altitudes differing by 100 km is

$$\frac{v_{100\,\text{km}}}{v_{200\,\text{km}}} = \sqrt{\frac{6{,}378 + 200}{6{,}378 + 100}} = 1.016$$

i.e. the fractional velocity difference is $(v_{100} - v_{200})/v_{200} \approx 0.016 = 1.6\%$.

If $v_{200} = 7.8$ km/s, then the difference in velocity of two orbiting bodies at altitudes of 200 and 100 km would be approximately $\Delta v \approx 0.016(7.8) = 0.12$ km/s.

This "g" correction to the near-circular path of the Orbiter as it descends, in combination with the correction for the earth's rotational component in the flight direction, may thus crudely explain most, but perhaps not all, of the differences between our modeled flight-path velocities in inertial space and NASA's earth-relative velocities.

We now move on to compare our model against real data for an even more sensitive reentry parameter, the flight-path angle θ as a function of time in the initial period.

Figure 5.8 shows remarkable agreement between the predicted flight-path angle $\theta(t)$ based on our first-order model (Equation 5.61) with $C_L/C_D = 1.0$ and the measured flight-path angle estimated from the STS-5 altitude vs time data, $h(t)$. We note that the initial flight-path angle is nearly constant with time in the first 200 s but diminishes rapidly in the interval 200 s $< t <$ 300 s as the effects of lift become pronounced. It is seen that the modeled flight-path angle for the *glide period*, sets an effective *lower limit* of about 0.14° after $t = 300$ s for this initial reentry period. The STS-5 data lie close to this model "floor" with flight-path angle values in the range of 0.16° to 0.19° for the start of the glide regime.

Now we will model the *equilibrium* glide angle and the corresponding Orbiter velocity, density, and altitude as a function of time.

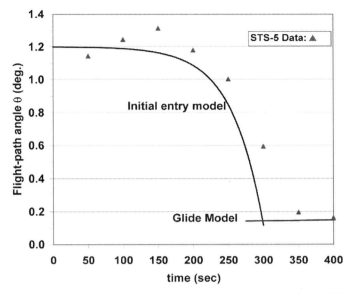

Figure 5.8. Shuttle reentry Flight path angle vs time. Initial entry model and glide model flight-path angles compared to STS-5 flight path angle data estimated by numerical differentiation of the measured data for $h(t)$ from [14]. The numerical value of the derivative $\Delta h/\Delta t$, divided by the measured local velocity v, is used to estimate the experimental path angle θ shown for STS-5 (the green triangles).

5.5.2 The equilibrium glide model

A second major period of reentry begins when the Orbiter, at a pre-set angle of attack of 40°, descends to altitudes where the air density becomes large enough for aerodynamic lift to play a major part in the balance of forces perpendicular to the flight path.

A *subsonic* aircraft can be maintained at a fixed altitude in a steady level flight equilibrium condition provided that the lift force on the aircraft exactly balances the force of gravity (i.e. the weight of the aircraft)

$$\text{Lift}_{\text{aircraft}} = mg \tag{5.65}$$

If gliding *without* thrust, an aircraft will descend at an almost constant speed v and a constant angle θ_{glide}. For low speeds, this *glide angle* can be calculated by ignoring centrifugal effects. With $dv/dt = 0$ and $d\theta/dt = 0$ in Equations 5.40 and 5.41, one can easily show that for small angles the low-speed aircraft glide angle θ_{glide} is given by the well-known expression

$$\theta_{\text{glide-aircraft}} = \text{drag/lift} = \frac{1}{C_L/C_D} \tag{5.66}$$

where C_L/C_D is the ratio of the lift to drag coefficients.

For a spacecraft in a steady-state circular orbit there is no lift force to balance gravity, but there is the so-called centrifugal force. Actually, by Newton's laws of motion for a curved trajectory, the force of gravity is equal to the object mass m multiplied by the angular acceleration v^2/r, where r is the radius of the earth plus the orbital altitude h_0. We can write this fictitious "force" balance as

$$\left.\begin{array}{c}(mv^2/r)_{\text{orbit}} = mg \\ \text{where } g = g_0 r_E^2/r^2 \text{ and } r = r_E + h_0\end{array}\right\} \quad (5.67)$$

At high altitudes on reentry, the Orbiter is traveling at a very high speed with Mach number in the range of 10 to 20. In this regime the vehicle's lift to drag ratio ≈ 1.0, as calculated for a flat-plate-like wing at a large angle of attack and hypersonic speeds. It is these conditions that enable the Orbiter to fly long distances at an approximately constant, very small glide angle, $\theta_{\text{glide-reentry}}$. For this near-orbital part of the flight, the centrifugal force cannot be ignored in Equation 5.41. For an "equilibrium glide", $d\theta/dt = 0$ and gravity, centrifugal force, and lift must be in balance. We then have from Equation 5.41 that

$$L - m\left(g - \frac{v^2}{r}\right)\cos\theta = 0 \quad (5.68)$$

5.5.2.1 Glide velocity as a function of density

For very small flight-path angles, $\cos\theta \approx 1$. Also, the radius r to the Orbiter is approximately equal to the radius to the surface of the earth, $r_e = 6{,}378$ km for altitudes $h \ll r_e$. With these approximations we can solve Equation 5.68 for v^2 using the standard expression for lift, $L = 1/2(\rho v^2)C_L A$

$$\frac{v^2}{gr_e} \approx \frac{1}{\left[1 + \left(\dfrac{C_L A r_E}{2m}\right)\rho\right]} \quad (5.69)$$

Taking the square root, and defining the velocity of a hypothetical body orbiting the earth at zero altitude by the expression $v_0 \equiv \sqrt{g_0 r_e}$, the equation for the Orbiter velocity in the glide regime becomes

$$v \approx \frac{v_0}{\left[1 + \left(\dfrac{C_L}{C_D}\right)\left(\dfrac{C_D A r_E}{2mg}\right)\rho g\right]^{1/2}} = \frac{v_0}{\left[1 + \left(\dfrac{C_L}{C_D}\right)\left(\dfrac{\rho g r_E}{2 \cdot BN}\right)\right]^{1/2}} \quad (5.70)$$

Note that we have separated out, in the denominator of Equation 5.70, a term inversely proportional to the ballistic number, which is defined as follows,

$$BN \equiv \frac{mg}{C_D A} = \frac{\text{weight}}{\text{(effective drag area)}} \quad (5.71)$$

The ballistic number has units of Force/Area, newtons per square meter, or pascals (Pa). BN often appears in aerodynamic calculations where both drag and gravity forces are prominent. (Indeed, we encountered it in Chapter 4 when discussing the foam impact problem.)

From Equation 5.70 we observe that during the "equilibrium glide" the velocity decreases as the atmospheric density increases. Using Orbiter values for BN and a lift to drag ratio $C_L/C_D \approx 1$, it can be shown that the second term in the denominator of Equation 5.70 that contains the lift coefficient becomes of order unity for altitudes below about 65 km. At this altitude, the glide velocity drops to about $\sqrt{1/2}$ of the orbital velocity or approximately 5.4 km/s. So gravity and lift play nearly equal roles in determining the magnitude of the Orbiter's velocity in the mid-stages of flight when the altitude is approximately 1/2 the entry altitude. In contrast, during the initial entry period ($t < 300$ s) we found that there is very little change in flight velocity at these higher altitudes, since air density and the effects of drag are quite low.

In the analysis that follows, all of the derived equations for the engineering quantities of interest (altitude, glide angle, reentry time, distance traveled) can readily be expressed as functions of the velocity ratio, v/v_0. This ratio can be thought of as *the independent variable* for our problem.

We can also write density ρ as a function of the velocity ratio v/v_0. Solving either Equation 5.69 or Equation 5.70 for density, ρ, gives us the following useful expression

$$\rho = \frac{2 \cdot BN}{(C_L/C_D)v_0^2}[(v/v_0)^{-2} - 1] \qquad (5.72)$$

Note that as v/v_0 drops from a value slightly less than unity to a value close to zero (near touchdown), air density increases greatly, which is as expected. We can use Equation 5.72 to determine reentry altitude as a function of v/v_0, i.e. $h(v)$, in conjunction with the empirical air density model (Equation 5.48)

$$h(v) = -(H_{atm})\ln(\rho/\rho_0) = -(H_{atm})\ln\left(\frac{2 \cdot BN}{(C_L/C_D)(\rho_0 v_0^2)}[(v/v_0)^{-2} - 1]\right) \qquad (5.73)$$

where H_{atm} is the empirical exponential scale height for the atmosphere, with a value ≈ 7.16 km.

5.5.2.2 Glide flight-path angle as a function of velocity

We now find a rather simple algebraic expression for the flight-path angle θ as a function of velocity v by relating the rate of change of altitude with time (valid for small θ)

$$dh/dt \approx -v \cdot \theta \qquad (5.74)$$

to the rate of change of velocity with time along the flight path (Equation 5.46). We can relate dh/dt to dv/dt can by differentiating Equation 5.73. However, in order to simplify the algebra we use the following chain-rule property for derivatives

$$\frac{dv}{dt} = \left(\frac{dv}{d\rho}\right)\left(\frac{d\rho}{dh}\right)\left(\frac{dh}{dt}\right) = \left(\frac{dv}{d\rho}\right)\left(\frac{d\rho}{dh}\right)(-v \cdot \theta) \qquad (5.75)$$

Differentiating v in Equation 5.69 with respect to ρ

$$\left(\frac{dv}{d\rho}\right) = -\frac{\left(\frac{C_L A r_e}{2m}\right) \cdot v^3}{2 g r_e} \tag{5.76}$$

From the empirical exponential formula (Equation 5.48)

$$\left(\frac{d\rho}{dh}\right) = -\frac{\rho}{H_{\text{atm}}} \tag{5.77}$$

Substituting Equations 5.76 and 5.77 into Equation 5.75 yields

$$\frac{dv}{dt} = -\left[\frac{\left(\frac{C_L A r_e}{2m}\right) \cdot v^3}{2 g r_e}\right] \left[\frac{\rho}{H_{\text{atm}}}\right] (v \cdot \theta) \tag{5.78}$$

From the approximate equation of motion along the flight path (Equation 5.46), it is apparent that for very small glide angles, the rate of change of velocity (i.e. the deceleration of the vehicle) is set by the drag force

$$\left(\frac{dv}{dt}\right) \approx -\frac{D}{m} = -\left(\frac{C_D A}{2m}\right)(\rho v^2) = -\frac{g}{2 \cdot BN}(\rho v^2) \tag{5.79}$$

Setting the expressions for dv/dt given by Equations 5.78 and 5.79 equal to one another, we arrive at the following equation relating the equilibrium glide angle θ to the velocity v

$$\left(\frac{C_D A}{2m}\right)(\rho v^2) = \left[\frac{\left(\frac{C_L A r_e}{2m}\right) \cdot v^3}{2 g r_e}\right] \left[\frac{\rho}{H_{\text{atm}}}\right] (v \cdot \theta) \tag{5.80}$$

Solving Equation 5.80 for $\theta = \theta_{\text{glide}}$, we obtain the following very simple formula for the glide angle

$$\theta_{\text{glide}} = \frac{2}{\left[\left(\frac{C_L}{C_D}\right)\left(\frac{r_e}{H_{\text{atm}}}\right)(v/v_0)^2\right]} \tag{5.81}$$

From Equation 5.81, the magnitude of the glide angle at the earliest "glide" time (where the above solution becomes applicable with $v/v_0 \approx 0.95$ per Equation 5.55) is estimated to be

$$\theta_{\text{glide}/275\,\text{s}} = \frac{2}{\left[(1.0)\left(\frac{6{,}378}{7.16}\right)(0.95)^2\right]} = 2.5 \times 10^{-3} = 0.14°$$

Observe that this very small glide angle is a small fraction (about 1/8th) of the initial entry angle $\approx 1.2°$.

Sec. 5.5]   **Simple flight trajectory model**   253

As the Orbiter drops deeper into the atmosphere, drag slows it as air density increases. Equation 5.81 shows that the glide angle significantly increases from 0.14° as v is reduced. At some point the original assumption of a small flight-path angle is violated and the approximate glide solution is no longer valid. But as we will see when we compare our model predictions to STS-5 data, the "equilibrium glide" region of the trajectory is reasonably well modeled.

Given a model solution for density as a function of velocity, we can estimate the maximum rate of heat transfer to the rounded nose of the Orbiter and to the leading edges of the wings. We will employ the basic hypersonic boundary layer model for the rate of heat transfer to the surface of a blunt body in a hypersonic flow field (Equation 5.32). To determine the variation in Orbiter heat flux with time, an equation needs to be constructed that links time to velocity during the glide period. In order to obtain an expression for time t as a function of vehicle velocity for the glide regime we will integrate the reciprocal of Equation 5.79, as shown in the following subsection.

5.5.2.3 Flight time as a function of velocity

Up to this point we have not related velocity, density, or flight path angle directly to flight time, t. This connection can readily be made by integrating Equation 5.79 to solve for t as a function of v.

From Equation 5.79 the reciprocal of the acceleration equation for dv/dt is

$$\frac{dt}{dv} = -\frac{2 \cdot BN}{g}\left(\frac{1}{\rho v^2}\right) \tag{5.82}$$

Integrating this expression for time t

$$t - t_i = -\left(\frac{2 \cdot BN}{g}\right) \int_{v_i}^{v} \frac{dv}{\rho v^2} \tag{5.83}$$

We set the initial glide velocity to be $v = v_i$ at the initial glide time $t = t_i$. Using Equation 5.72 we determined that the term ρv^2 is a function of v only

$$\rho v^2 = \frac{2 \cdot BN}{(C_L/C_D)}\left[1 - \frac{v^2}{gr_e}\right] \tag{5.84}$$

Substituting Equation 5.84 into Equation 5.83

$$t - t_i = -\left(\frac{v_0}{g}\right)(C_L/C_D) \int_{v_i/v_0}^{v/v_0} \frac{d(v/v_0)}{[1 - (v/v_0)^2]} \tag{5.85}$$

where the constant $v_0 \equiv \sqrt{gr_e} = 7.91$ km/s.

The basic form of the integral in Equation 5.85 is known to be

$$\int \frac{dx}{(1 - x^2)} = \tfrac{1}{2} \ln\left|\frac{1 + x}{1 - x}\right|$$

so the final expression for $t(v)$ is

$$t - t_i = \left(\frac{v_0}{2g}\right)(C_L/C_D) \ln\left|\frac{(1 - v/v_0)(1 + v_i/v_0)}{(1 + v/v_0)(1 - v_i/v_0)}\right| \quad (5.86)$$

As the logarithmic solution is only valid for velocity ratios $v_i/v_0 < 1$, the initial glide velocity must be less than the effective orbital velocity for a hypothetical body at the surface of the earth, v_0.

5.5.2.4 Estimating maximum flight time

Let's set $v = 0$ in Equation 5.86 in order to estimate the time t_{max} that it takes for a gliding vehicle to come to rest at zero velocity

$$t_{max} - t_i = \left(\frac{v_0}{2g}\right)(C_L/C_D) \ln\left|\frac{(1 + v_i/v_0)}{(1 - v_i/v_0)}\right| \quad (5.87)$$

Note that t_{max} is sensitive to the ratio of initial glide velocity to orbital velocity, v_i/v_0, since this ratio is only slightly less than unity, and the quantity $(1 - v_i/v_0)$ appears directly in the denominator of the logarithmic function.

We now quantitatively estimate t_{max} using the following input values:

- v_0 = orbital velocity for a body at the earth's surface = $\sqrt{g_0 r_e}$ = 7.91 km/s = 7,910 m/s
- v_i = initial vehicle velocity at the start of the equilibrium glide trajectory = the vehicle velocity at the end of the entry period (Equation 5.54) evaluated at approximately 300 seconds, i.e. $v_{end\ of\ entry} \approx v(300\ s) \approx 7.55$ km/s
- t_0 = initial time after the Orbiter begins its entry into the atmosphere. We will show graphically that the steep entry solution for altitude as a function of time, $h_{entry}(t)$ intersects the more gradually sloped "glide" solution at approximately $t \approx 275$ s
- $C_L/C_D \approx 1.0$

C_L/C_D is the lift to drag ratio for the Orbiter at a fixed angle of attack. Under hypersonic conditions with a high angle of attack of 40°, both aerodynamic theory and experimental data indicate that the lift to drag ratio is of order 1.0; some sources, and a simple Newtonian model for a flat-plate, valid in the case of high Mach numbers, give a ratio in the range ≈ 1.1 to 1.2.

From Equation 5.87

$$t_{max} = t_i + \left(\frac{v_0}{2g}\right)(C_L/C_D) \ln\left|\frac{(1 + v_i/v_0)}{(1 - v_i/v_0)}\right| \approx 275\ s + \left(\frac{7910}{2 \cdot 9.81}\right)(1.0) \ln\left|\frac{(1 + 0.95)}{(1 - 0.95)}\right|$$

$$\boxed{t_{max} \approx 275\ s + 1{,}477\ s \approx 1{,}750\ s} \quad (5.88)$$

NASA's measurement of the touchdown time for STS-5 is 1,821 s [14]. Remarkably, and perhaps fortuitously, our model estimate for t_{max} is only 4% below the NASA value.

Of course, there are a number of other factors that one should consider when trying to estimate maximum flight time, including the fact that the angle of attack is held at 40° for only about 60% of the flight period after the time of atmospheric entry. The Orbiter's angle of attack drops to about 5° just prior to landing; during the final approach the subsonic lift to drag ratio increases to about 4.5. In addition, the Orbiter undertakes five separate *roll maneuvers* during reentry. These maneuvers, called *S turns*, are designed to further slow the vehicle prior to landing. And then there is a low altitude circle maneuver used to line up with the runway, followed by the final flare maneuver as the Orbiter lands subsonically.

5.5.2.5 Estimating maximum horizontal distance traveled after "entry interface"

We can also readily estimate the maximum horizontal distance, s, traveled by the Orbiter during reentry. The origin for horizontal distance traveled along the surface of the earth is set $= 0$ at the "entry interface" arbitrarily defined as being 400,000 feet, some 120 km.

Assuming that the vehicle motion lies in a plane, then the rate of change of horizontal distance with time is given by

$$\frac{ds}{dt} = v \cos \theta \tag{5.89}$$

If we further assume that the flight path angle is quite small, $\cos \theta \approx 1$, it is then possible to integrate Equation 5.89, using the small angle form of Equation 5.89

$$s = \int v \, dt = \int \frac{v \, dv}{(dv/dt)} = \int \frac{v \, dv}{-\frac{g}{2 \cdot BN}(\rho v^2)} \tag{5.90}$$

Since $\rho v^2 = \frac{2 \cdot BN}{(C_L/C_D)} \left[1 - \frac{v^2}{gr_e} \right]$ from Equation 5.84, Equation 5.90 can readily be integrated in closed-form to yield the following expression for distance, s, as a function of velocity v during reentry

$$s = \frac{r_e}{2}(C_L/C_D) \ln \left| \frac{(1 - (v/v_0)^2)}{(1 - (v_i/v_0)^2)} \right| \tag{5.91}$$

Note that the distance, s, scales with the radius of the earth multiplied by the lift to drag ratio, C_L/C_D, and that s (and its maximum value s_{max}) is an even more sensitive function of the ratio of initial glide velocity to orbital velocity, v_i/v_0, than is the estimated flight time t_{max} (Equation 5.87). This is because the denominator of the logarithm is $1 - (v_i/v_0)^2$ and the initial velocity ratio, v_i/v_0, is ≈ 0.95, which is quite close to 1.0. If the initial velocity ratio is set identically equal to 1.0 in this simple approximation (Equation 5.91) then s clearly becomes unbounded.

We now quantitatively estimate the distance traveled during the equilibrium glide portion of the trajectory $s_{max\text{-glide}}$, by setting the final velocity equal to zero,

256 Estimating the Orbiter reentry trajectory and the associated peak heating rates [Ch. 5

i.e. $v = v_{\text{final}} = 0$ in Equation 5.91 and using the following input values:

v_0 = reference orbital velocity ≈ 7.91 km/s $= 7,910$ m/s

v_i = initial vehicle velocity for the equilibrium glide trajectory ≈ 7.52 km/s.

The lift to drag ratio, $C_L/C_D \approx 1.0$ because this is the nominal value for a hypersonic vehicle at an angle of attack of $40°$.
The maximum glide distance is then given by

$$s_{\text{max-glide}} = \frac{r_e}{2}(C_L/C_D) \ln \left| \frac{1}{(1-(v_i/v_0)^2)} \right| \quad (5.92)$$

Quantitatively

$$s_{\text{max-glide}} = \frac{6{,}378 \text{ km}}{2}(1.0) \ln \left| \frac{1}{(1-(0.95)^2)} \right| = 7{,}424 \text{ km} \quad (5.93)$$

In the initial "steep entry" region that precedes the equilibrium glide portion of the flight, the Orbiter travels an additional horizontal distance Δs_{entry}. The added distance flown during this portion of the flight (flight-path angle $\approx 1.2°$) can readily be estimated. The distance Δs_{entry} is proportional to the estimated time interval $\Delta t_{\text{entry}} \approx 275$ s according to the simple model

$$\Delta s_{\text{entry}} \approx (v_{\text{initial}})\Delta t_{\text{entry}} = 7.70 \text{ km/s}(275 \text{ s}) = 2{,}118 \text{ km} \quad (5.94)$$

To increase the fidelity of our calculation, we should actually use the initial "relative velocity" in this calculation, rather than the inertial velocity.
Adding the two portions of the flight together (Equations 5.93 and 5.94) gives the estimated maximum total distance traveled during reentry, $\Delta s_{\text{max-total}}$

$$\Delta s_{\text{max-total}} = \Delta s_{\text{entry}} + s_{\text{max-glide}} = 7{,}424 + 2{,}118 \approx 9{,}540 \text{ km} \quad (5.95)$$

NASA's measure of the total distance traveled for STS-5 $\approx 8{,}500$ km [4, p. 284]. Our estimate (Equation 5.95) is some 12% greater. As an additional data reference, STS-1 traveled $\approx 8{,}100$ km during reentry [11]. We overestimate this by 18%. However, as noted above, an Orbiter performs banking, or S-roll, maneuvers to reduce speed along the entry trajectory, with the result that the total flight distance is reduced relative to that of a hypothetical vehicle landing without these maneuvers.

5.5.2.6 Comparisons of the glide model calculations with STS-5 data

Plots of our glide model estimates for altitude, velocity, and flight-path angle as a function of time for the initial entry period ($0 < t < 275$ s) and for the main portion of the glide period ($275 \text{ s} < t < 1{,}000 \text{ s}$) are shown in Figures 5.9, 5.10, and 5.11 respectively. The angle of attack is maintained at $40°$ for $t < 1{,}000$ s. This is the equilibrium glide region of nearly constant lift to drag ratio, where $C_L/C_D \approx 1.0$. It is here that our simplified glide solution provides a good approximation. With a reasonable flight-trajectory prediction, we will subsequently show that the peak heat

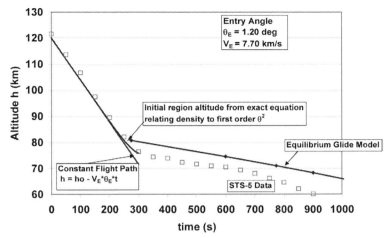

Figure 5.9. Altitude vs time. Initial entry period through the "equilibrium glide" period from BotE reentry models (Equation 5.44 initial entry region, and Equations 5.73 and 5.86 for the glide region). Comparisons are made with data from STS-5 [14].

transfer rates to the Orbiter nose-tip and the leading edge of the wing can be fairly well estimated. For $t > 1,000$ s, the angle of attack is reduced rapidly, dropping to about $10°$ at $t \approx 1,500$ s. This change in angle of attack cannot be easily modeled assuming a simple constant glide condition.

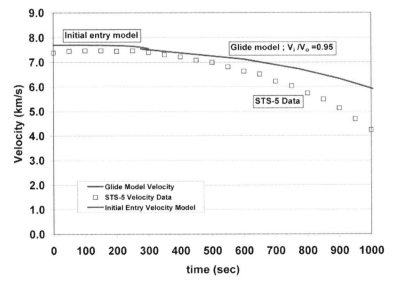

Figure 5.10. Velocity vs time. Initial entry period through the "equilibrium glide" period from BotE reentry models. Comparison of initial entry model (Equation 5.22) and glide model velocity calculations (Equation 5.86) with STS-5 data.

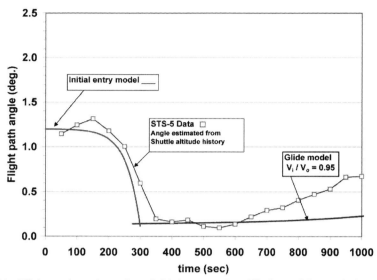

Figure 5.11. Flight-path angle vs time. Initial entry + "equilibrium glide" period. Comparison of entry model (Equation 5.61) and glide model (Equation 5.81) with STS-5 data.

In this subsection, we examine the overall model trends with time for the three flight variables h, v, and θ in the equilibrium glide regime and compare these calculations with the Orbiter data from STS-5 [14].

Finally, we examine the entire flight dynamics range, covering the period from $t = 0$ at the entry interface through to touchdown at about 1,820 s. Our basic glide solution is extended to cover the time domain from $t = 275$ s to 1,820 s. Figure 5.12 compares our complete reentry results with the STS-5 data for altitude and velocity.

In the period from $t = 275$ s to $t = 1,000$ s, Figures 5.9, 5.10, and 5.11 show good agreement between model and data for h, v, and θ for $t < 600$ s. After that, the STS-5 data display a faster reduction in altitude with time, dh/dt, than does the glide model prediction (Figure 5.9) and the flight-path angle shows a more rapid growth with time than does our model calculation. In addition, the real data show a higher deceleration rate, dv/dt, for time $t > 600$ s than is predicted by the glide model (Figure 5.10). This greater rate of change of velocity with time leads to an increasing divergence between the model and data velocities. At $t \approx 650$ s, model velocities over-predict the true velocities by only about 8%. However, by $t = 1,000$ s the over-prediction increases to about 40% ($\Delta v \approx 1.8$ km/s), which is quite significant. Reduced velocities lead to greater rates of change of flight-path angle, $d\theta/dt$ (Figure 5.11).

As can be observed from the equation of motion for forces and accelerations normal to the flight path (Equation 5.41)

$$v\left(\frac{d\theta}{dt}\right) = -\frac{L}{m} + \left(g - \frac{v^2}{r}\right)\cos\theta$$

Sec. 5.5]   Simple flight trajectory model   259

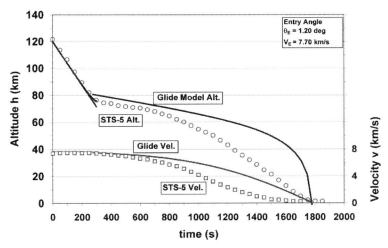

Figure 5.12. Altitude and velocity vs time, from entry interface to touchdown at $t = 1,820$ s. Initial entry period through the equilibrium glide period from BotE reentry models. Comparisons are made with STS-5 data (circles and squares).

lower v magnitudes not only reduce the impact of lift L (which is proportional to v^2) but also, perhaps more importantly, reduce the centrifugal force, v^2/r which, due to the minus sign, acts to reduce $d\theta/dt$. The term which acts to dramatically increase $d\theta/dt$ is, of course, the force of gravity g. We decided to model g as a constant, even though it actually increases approximately 1% in magnitude for every 100,000 ft reduction in altitude. With lower than expected velocities the Orbiter descends more rapidly and the air density rapidly increases, which causes the drag to increase, and that in turn slows the vehicle more than expected.

In Figure 5.12 we compare our solution to the data for the total flight. Clearly the glide model gives *poor* results in the last 1,000 s of flight, but surprisingly the predicted altitude h drops to zero at a time close to the correct touchdown time around 1,800 s.

Next we introduce a new aerodynamic flight dynamics parameter, the Orbiter bank or roll angle. After we had completed the glide calculations above, it came to our attention that NASA employs a particular flight-control maneuver in which the vehicle banks or rolls as an active means to increase its deceleration during the glide period. As a result of merely lowering the vertical lift to drag ratio in a revised set of model calculations to capture the effect of banking, our predicted equilibrium glide altitudes and velocities (as a function of time) move much closer to the NASA data.

5.5.2.7 *The effect of bank maneuvers and S-turns on the glide model calculations*

An important overall Orbiter system constraint that helps set the trajectory of the Orbiter is the total amount of heat (in joules) transferred to the vehicle from the hot reentry gases. We follow the basic discussion of this issue by Hirschel and Weilend in [15], who discuss how this constraint can be satisfied simply by banking the Orbiter. In principle, this can be applied to any other winged reentry vehicle during descent.

In Section 5.3.2.11 we noted that the total heat, Q_w, caused by convective heat transfer must be "soaked-up" by the thermal protection system, which is mostly composed of tiles. Since the temperature increase of the TPS material during reentry is given approximately by $\Delta T_{tile} = Q_{per\ tile}/C_{p_{tile}} m_{tile}$, it follows that to keep the tile temperature below its maximum allowable fracture temperature, the total heat load during reentry, Q_w, must be kept below a given threshold, and this becomes a constraint on the Orbiter's altitude and speed trajectory relationship.

The total heat load per unit area can be approximated by the product of the maximum heat flux in the trajectory and the effective time interval $\Delta t_{effective}$, which is about 50% of the total flight time

$$Q_w/A_w \approx [\dot{q}_w]_{max} \cdot \Delta t_{effective}$$

where $(\dot{q}_w)_{max}$ is the maximum heat flux per unit area (W/m^2) due to aerodynamic heating (Equation 5.32). For the glide flight regime, both terms are functions of the lift to drag ratio, C_L/C_D. From Equation 5.32, heat flux is proportional to $\sqrt{\rho v^2} v^2$ and $\rho v^2 \sim 1/(C_L/C_D)$, hence $(\dot{q}_w)_{max} \sim (1/\sqrt{C_L/C_D})$. Equation 5.87 tells us that the time interval is proportional to (C_L/C_D). The total heat load to the vehicle is therefore proportional to

$$Q_w \sim (1/\sqrt{C_L/C_D}) \cdot (C_L/C_D) \sim \sqrt{C_L/C_D} \qquad (5.96)$$

Thus the heat load can be reduced below some given upper-bound if the lift to drag ratio (C_L/C_D) is reduced. It appears that reduction of C_L/C_D (based on the system heating limits) became one of several constraint objectives in NASA's effort to optimize the Orbiter reentry trajectory. In fact, NASA introduced five or more separate bank maneuvers during the reentry period. They are of varying duration, and are active for about 85% of the reentry time period, as shown in Figure 5.13 from [16].

The change of lift to drag ratio with bank angle

When the Orbiter is banked either to the right or to the left, the component of the lift in the direction perpendicular to the flight path is given approximately by

$$\text{Lift}(\phi) \approx (\text{Lift}_{\phi=0}) \cos \phi \qquad (5.97)$$

where ϕ is the bank angle.

Since the angle of attack is not changed during the bank maneuver, the drag remains approximately constant, i.e. $D \approx$ constant. Therefore we model the lift to drag ratio dependence on bank angle as

$$(C_L/C_D) \approx \lfloor (C_L/C_D)_{\phi=0} \rfloor \cdot \cos \phi \approx (1.0) \cdot \cos \phi \qquad (5.98)$$

for a fixed 40° angle of attack.

Sec. 5.5] Simple flight trajectory model 261

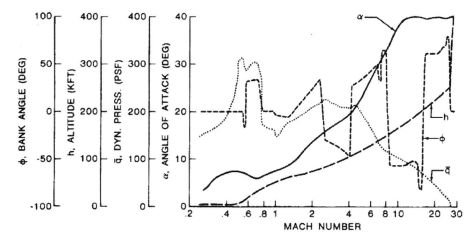

Figure 5.13. Typical Orbiter entry trajectory. Note the curve of bank angle (ϕ) in degrees as a function of flight Mach number [16].

There are four major maneuvers for Mach numbers $M > 5$, as shown in Figure 5.13. The first, at $M = 25$, shows a roll of $+70°$ in bank angle, subsequently easing off gradually to $+60°$. The second maneuver, at $M = 18$, rolls the vehicle back to a $-80°$ bank angle, then lowers the bank almost immediately to $-50°$. Maneuvers 3 and 4 roll to $+40°$ and $-40°$ bank angles respectively at lower Mach numbers.

We now attempt to re-evaluate our trajectory calculations using only a single roll angle magnitude in our simple glide model. We select

$$\phi \approx \pm 45° \quad (5.99)$$

as our "typical" bank angle. This angle is applicable only during the high Mach number portion of the flight, before there is any significant reduction in angle of attack α. That large angle of attack reduction starts at about $M = 10$ or at about 1,100 s after the "start" of reentry.

The revised lift to drag ratio for our equilibrium glide model calculation is estimated to be

$$(C_L/C_D) \approx [(C_L/C_D)_{\phi=0}] \cdot \cos 45° \approx 0.707 \quad (5.100)$$

This lift to drag ratio magnitude gives an approximate "best fit" between the model and data for $h(t)$ and $v(t)$. The revised plots at $(C_L/C_D) \approx 0.707$ for a $45°$ bank angle are compared in Figure 5.14 to both the $(C_L/C_D) \approx 1.0$ for a $0°$ bank angle calculations and the STS-5 data.

The revised altitude and velocity curves, based on a modeled constant bank angle of $45°$ show a marked improvement, relative to the STS-5 data, for flight times in the main part of the glide region; $275\,\text{s} < t < 1,100\,\text{s}$. As the glide path angle θ begins to exceed $5°$ after $t = 1,100\,\text{s}$, our model predictions for altitude diminish rapidly with time.

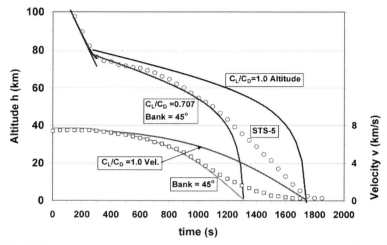

Figure 5.14. Altitude and velocity vs time, from entry interface to zero altitude and velocity. Glide model calculations at (a) $C_L/C_D = 1.0$ with zero bank angle, and (b) $C_L/C_D = 0.707$ with bank angle $\phi = 45$ degrees, are compared with data from the STS-5 mission.

5.5.2.8 Extending the glide model calculations a little further

At about 1,100 s, the flight-control system reduces the Orbiter angle of attack α from 40° to about 10° over a 5 minute period (Figure 5.13). At smaller angles of attack α, the lift to drag ratio increases quite significantly. Without banking, C_L/C_D nearly doubles from 1.0 at $\alpha = 40°$ to about 2.0 at $\alpha = 10°$ (with $C_D \approx 0.3$). With the continuation of S-turn banking in the 5 minute period after 1,100 s, we estimate that the maximum level of C_L/C_D will be about $C_L/C_D \approx 1.5$.

With these parameters, we look at the benefit of appending a small additional segment to our trajectory model that starts at 1,100 s and ends at about 1,500 s. In this range we set $C_L/C_D = 1.5$ with $C_D = 0.3$ for the lower angle of attack.

Figure 5.15 shows that the altitude and velocity predictions of the appended glide segment initiated at $t = 1,100$ s, agree reasonably well with the STS-5 data to about 1,400 s. Clearly the shift to a higher lift to drag ratio, with $C_L/C_D = 1.5$, has *prevented* our "baseline" glide model calculation, with $C_L/C_D = 0.707$, from dropping too rapidly in altitude. After 1,400 s, our appended solution falters badly. Hence our simple analysis appears to have reached its useful limit.

Some further flight control changes are initiated at this point which allow the real Orbiter to fly a circuitous path to a safe landing. Actually, when the vehicle's speed drops below the speed of sound, at an altitude of 15 km, the commander takes over manual control of the approach and landing process from the flight-computer system. He then maneuvers the Orbiter, using its various flight control surfaces, and heads to the runway in order to land. These terminal maneuvers occur about 30 minutes after the time of entry interface.

In summary, we have successfully "flown" our glide model (with a constant C_L/C_D lowered with S-turn bank maneuvering) down to altitudes of about 40 km

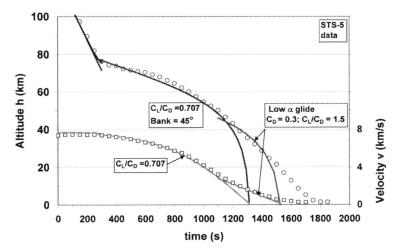

Figure 5.15. Altitude and velocity as a function of time. STS-5 data are represented by circles and squares. A small additional glide segment is added to the baseline glide model to account for the rapid drop in angle of attack that starts at 1,100 s. For this region, we set $C_L/C_D = 1.5$ and $C_D = 0.3$. Initial velocity = 2.77 km/s to match the high angle of attack model at this time.

where the Orbiter's speed drops to about 2 km/s (Mach ≈ 6). With an added high lift to drag model segment, we can push our modeling a little further, down to an altitude of 30 km. However, at some time the simplicity of BotE modeling is forced to yield to a more complex form of analysis.

Given the confidence in our model gained through comparisons with Orbiter reentry data, from the entry interface down to $h = 40$ km and $v = 2$ km/s, we will now revisit an earlier challenge. We seek to predict the maximum rates of heat transfer, $\dot{q}_{w_{max}}$, to the surface of the nose (and the leading edge of the wings). In the next section we use Equation 5.32 to show that we can readily use our bank-angle-modified glide-trajectory solution in order to estimate the time dependent history of the heat flux, $\dot{q}_w(t)$. We basically take the glide trajectory and heating model developed by Eggers and Allen in the 1950s [17], and employ it to predict the performance of the present day real-world case of a Shuttle reentry.

Finally we calculate the time dependent heat load, $Q_w(t)$, and determine by integration its total value from the time of entry interface ($t = 0$) through to when the Orbiter ends its initial high angle of attack glide phase, at a flight time of about $t = 1,100$ s.

5.6 CALCULATING HEAT TRANSFER RATES IN THE PEAK HEATING REGION

The first two questions we address here are important in setting a bound on the heat transfer rate:

- At what velocity, density, altitude, and time is the heat transfer rate, $\dot{q}_w(t)$, a maximum during reentry?
- What is the maximum magnitude of the heat transfer rate $\dot{q}_w$ [W/m^2]?

The answers to these questions are readily obtained using the equilibrium glide model for the trajectory in conjunction with the Sutton–Graves heat transfer equation. For our glide model, the magnitude of the lift to drag ratio C_L/C_D is the lift to drag ratio at a 40° angle of attack, modified by the "effective bank angle" of the Orbiter during S-turn maneuvers.

The Sutton–Graves equation for heat flux (Equation 5.32) is

$$\dot{q}_w = \frac{A}{\sqrt{r_{\text{nose}}}} (\rho_\infty)^{1/2} (V_\infty)^3 \qquad (5.32)$$

with the constant $A = 1.75 \times 10^{-4}$ measured in units of $\sqrt{\frac{\text{kg}}{\text{m}}}$. We see that $\dot{q}_w$ is proportional to the square root of density multiplied by the cube of velocity. Let's define a function $Z(\rho, v)$ that is proportional to the square of $\dot{q}_w$, i.e.

$$Z(\rho, v) \equiv (\rho v^2) v^4 \qquad (5.101)$$

From the equilibrium glide model we utilize the equation for ρv^2 as a function of v only (Equation 5.84)

$$\rho v^2 = \frac{2 \cdot BN}{(C_L/C_D)} \left[1 - \frac{v^2}{gr_e} \right] \quad \text{where} \quad BN = \frac{mg}{C_D A} \qquad (5.102)$$

If we define the ratio $\dfrac{v^2}{gr_e} = \left(\dfrac{v}{v_0}\right)^2 \equiv \bar{v}^2$, then Equation 5.102 becomes

$$\rho v^2 = \frac{2 \cdot BN}{(C_L/C_D)} [1 - \bar{v}^2] \qquad (5.103)$$

In this form, the squared heat flux function Z is then given by

$$Z(\bar{v}) \equiv C(1 - \bar{v}^2)\bar{v}^4 \qquad (5.104)$$

where C is a constant.

Differentiating $Z(\bar{v})$ with respect to $\bar{v}$ and setting the derivative to zero gives us the value of $\bar{v}$ that maximizes the heat flux

$$4\bar{v}^3 - 6\bar{v}^5 = 0 \qquad (5.105)$$

The Eggers–Allen glide solution of Equation 5.105, from their 1955 paper on the performance of long-range hypervelocity vehicles [17], is

$$\bar{v}_{\text{max-}\dot{q}} = \sqrt{2/3} = 0.816 \qquad (5.106)$$

As $v_0 = \sqrt{gr_e} = 7.92$ km/s, the velocity at the maximum heat transfer rate is

$$v_{\text{max-}\dot{q}} = 6.46 \text{ km/s} \qquad (5.107)$$

From Equation 5.103, we solve for the density at this particular velocity during the equilibrium glide trajectory

$$\rho_{\text{max-}\dot{q}} = \frac{2m}{(C_L A r_E)}\left[\frac{1-\bar{v}^2}{\bar{v}^2}\right] = \frac{2m}{(C_L A r_E)}\left[\frac{1}{2}\right] = \frac{m}{(C_L A r_E)} = \frac{mg}{(C_D A v_0^2)} \cdot \frac{1}{(C_L/C_D)}$$

$$\rho_{\text{max-}\dot{q}} = \frac{mg}{(C_D A v_0^2)} \cdot \frac{1}{(C_L/C_D)} \qquad (5.108)$$

With $(C_L/C_D) = 0.707$ for $45°$ bank maneuvers and $C_D = 0.8$, we determine the magnitude of $\rho_{\text{max-}\dot{q}}$

$$\rho_{\text{max-}\dot{q}} = \frac{10^5 (9.81)}{(0.8 \cdot 250 \cdot (7{,}920)^2)} \cdot \frac{1}{0.707} = 1.11 \times 10^{-4} \text{ kg/m}^3 \qquad (5.109)$$

We now calculate the altitude corresponding to this density

$$h = -H_{\text{atm}} \ln\left(\frac{\rho}{\rho_0}\right) \quad \text{where } \rho_0 = 1.225 \text{ kg/m}^3$$

$$h_{\text{max-}\dot{q}} = -(7.16 \text{ km}) \ln(0.906 \times 10^{-6}) = 66.7 \text{ km} \qquad (5.110)$$

From our plot of altitude vs time (Figure 5.14) we see that the time t when the Orbiter descends to 66.7 km, which we have determined is the time of max $\dot{q}$, is approximately

$$t_{\text{max-}\dot{q}} = 665 \text{ s} \qquad (5.111)$$

The maximum value of the Orbiter's nose heat flux, $\dot{q}_w$, can be estimated from Equation 5.32 by using the calculated density (Equation 5.109) and velocity, (Equation 5.107) at this time. It is important to recall that $\dot{q}_w$ is inversely proportional to the radius of the RCC nose cap.

5.6.1 Selecting the nose radius

The nose radius of an idealized hemisphere, r_{nose}, is a very important geometrical parameter that helps to determine the magnitude of the heat transfer rate. This radius determines the inviscid and viscous flow fields in the vicinity of the stagnation point. Equation 5.32 clearly shows that the magnitude of $\dot{q}_w$ is *inversely* proportional to $\sqrt{r_{\text{nose}}}$. When comparing two different calculations, ostensibly of the same geometry but differing by a factor of two in assumed radius, the heat flux equation clearly shows that the calculation with the larger radius yields a value for $\dot{q}_w$ approximately 30% below that of the calculation with the smaller radius. This is the primary reason why both the Orbiter's nose *and* the swept-wing leading edges each are designed *not* to be sharp, but rather to have a substantial radius of curvature.

Table 5.1. Equivalent nose radius for Orbiter heat transfer calculations based on equivalent axisymmetric body [19, 20].

Angle of attack, α, for Orbiter	Equivalent nose radius, r_{nose}, for Orbiter
22°	0.49 m
40°	1.29 m
42°	1.37 m

We raise the issue here because when one reviews a number of papers in the reentry literature, they all have the inverse square root dependence of nose radius in their heat flux models, but they often differ substantially in the assumed radius for the Orbiter nose. For example, the 1977 paper by Goodrich et al. [18], lists a value of 28.33 inches ($r_{\text{nose}} = 0.72$ m) for the nose radius based on the Rockwell's Orbiter 140-B drawings. However their calculations show an extended subsonic region near the nose whose distance from the leading edge is more than the nose radius of 0.72 m. This led the authors to state that "the flow in the stagnation-point region is actually influenced by a *much larger effective body* with a larger effective radius of curvature than the actual geometry". They also conclude that the effective radius is influenced by both gas chemistry and the Orbiter angle of attack. Zoby [19], and Moss and Bird [20], calculate the geometry for an *equivalent axisymmetric* body at zero angle of attack that models the centerline flowfield over the Orbiter at a given angle of attack. The result of their efforts allowed them to determine the *equivalent* nose radius as a function of angle of attack. Some of the *equivalent* nose radius-of-curvature values are listed here in Table 5.1.

5.6.2 Comparing the model maximum rate of heat transfer, $\dot{q}_{w_{\text{max}}}$, with data

For our subsequent calculations at $\alpha = 40°$, we select $r_{\text{nose}} = 1.29$ m based on Table 5.1.

Using Equations 5.107 and 5.109 the maximum rate of heat transfer in the glide trajectory is

$$\dot{q}_{w_{\text{max}}} = \frac{1.75 \times 10^{-4}}{\sqrt{r_{\text{nose}}}} (\rho_{\infty_{\text{max}}})^{1/2} (V_{\infty_{\text{max}}})^3 = \frac{1.75 \times 10^{-4}}{\sqrt{1.29}} (1.11 \times 10^{-4})^{1/2} (6{,}460)^3$$

$$\boxed{\dot{q}_{w_{\text{max}}} = 0.437 \frac{\text{MW}}{\text{m}^2} = 437 \frac{\text{KW}}{\text{m}^2} \quad (h = 66.7 \text{ km}; v = 6.46 \text{ km/s})} \quad (5.112)$$

Williams and Curry [1] generated an estimate of the peak heat flux based on STS-5 measurements of nose surface-temperatures as a function of time, and their estimate for peak heat flux was approximately 40 BTU/ft^2-s, which in MKS units is

$$\boxed{\dot{q}_{w_{\text{max}}} \approx 0.46 \frac{\text{MW}}{\text{m}^2} \quad t \approx 600 \text{ s}, h \approx 70 \text{ km}, \text{STS-5 nose data [1]}} \quad (5.113)$$

Zoby [19] compared several detailed Orbiter heating studies with his approximate method for the windward symmetry plane of the Shuttle, and his prediction of maximum heat flux was in accord with his extrapolation of *wind-tunnel data* to full-scale flight conditions, 46 BTU/ft²-s, or

$$\dot{q}_{w_{max}} \approx 0.52 \frac{MW}{m^2} \quad \text{equivalent full-scale nose data;} \; h \approx 68.9 \, \text{km [19]} \quad (5.114)$$

At the leading edge of the Orbiter wing, nearly identical experimental heat flux levels were reported on by Cunningham [21] using radiometer data from within the body of the Orbiter wing. The maximum heat flux was determined to be approximately 45 BTU/ft²-s, or

$$\dot{q}_{w_{max}} \approx 0.51 \frac{MW}{m^2} \quad \text{STS-1,2 wing leading edge data;} \; t \approx 600 \, \text{s [21]} \quad (5.115)$$

We conclude that our BotE estimate (Equation 5.112) is within about 15% of both the reported STS data and the above mélange of aerothermodynamic calculations.

5.6.3 Model estimate for nose radiation equilibrium temperature, T_{max}

As the Orbiter begins to heat up on entering the atmosphere, high rates of heat transfer to the nose region cause the surface (or wall) temperature to increase rapidly with time. The rate of change of temperature with time, dT/dt, is inversely proportional to the product of the mass and specific heat of the wall material. At sufficiently high wall temperatures, the wall starts to significantly radiate heat to space, particularly if the nose surface has a high emissivity close to 1. The 0.25 to 0.5 inch thick reinforced carbon-carbon (RCC) material that thermally protects the nose and the leading edges of the Orbiter's wings has an emissivity $\varepsilon \approx 0.9$. RCC is the thermal protection material that is used on the Orbiter when surface temperatures are expected to exceed 2,300°F or 1,260°C. If the reusable tiles were to be utilized for the nose, they would fail at such temperatures. When the rate of convective heating to the wall, $\dot{q}_w$, exactly balances the rate of radiative cooling from its surface ($\dot{q}_{radiation}$), the two processes are in aerothermodynamic equilibrium and $dT/dt = 0$, i.e.

$$\dot{q}_{radiation} = \dot{q}_w \quad \text{(for aerothermodynamic equilibrium)} \quad (5.116)$$

If we assume that the nose of the Orbiter is in this state of equilibrium when wall temperatures are above 1,000°F (540°C or 813°K) we can directly calculate the corresponding equilibrium wall temperature, T_w (in degrees Kelvin), using the well-known Stefan–Boltzmann radiation law

$$\dot{q}_{radiation} = \varepsilon \sigma T_w^4 \quad (5.117)$$

where ε is the emissivity of the RCC material ≈ 0.9 and σ is the Stefan–Boltzmann constant $= 5.67 \times 10^{-8} \dfrac{\text{W}}{\text{m}^2(^\circ\text{K}^4)}$. Substituting Equation 5.117 into Equation 5.116, and setting $\dot{q}_w =$ the heat flux to the nose, yields an expression for the equilibrium wall temperature in the nose region

$$T_w = \left(\frac{\dot{q}_w}{\varepsilon\sigma}\right)^{1/4} \quad \text{equilibrium radiation temperature} \quad (5.118)$$

At the maximum heat flux conditions (Equation 5.112), we obtain the following estimate of the wall temperature for the RCC nose

$$T_{w_{\max}} = \left(\frac{0.44 \times 10^6}{(0.9)(5.67 \times 10^{-8})}\right)^{1/4} = 1{,}714^\circ\text{K} = 1{,}441^\circ\text{C} = 2{,}626^\circ\text{F} \quad (5.119)$$

Williams and Curry [1] give a maximum stagnation point "inner moldline" temperature $T_{\text{nose-IM}}$ for the STS-5 data of $2{,}605^\circ\text{F}$ or $1{,}429^\circ\text{C}$. They used optical radiometer measurements, along with supporting thermal models, to estimate the surface temperature as a function of time during reentry. Temperatures deduced from radiometers installed within the wing leading-edge showed comparable maximum temperatures during reentry.

Our simple aerothermodynamic equilibrium relation (Equation 5.118) provides a good first-order estimate of the peak nose and leading edge temperatures within about a 100°C uncertainty range.

In the following section, we not only estimate heat fluxes at their maximum values, but calculate them as a function of time based on our equilibrium glide model. With $\dot{q}_w(t)$ as input, the corresponding temperature time history can be calculated directly from Equation 5.118.

5.6.4 Model calculations of $\dot{q}_w$ as a function of time

Figure 5.16 is a plot of the wall heat flux at the stagnation point of the Orbiter's nose calculated from the Sutton–Graves model for the equilibrium glide region. Here, we let the air density and vehicle velocity evolve with time as previously presented in Figure 5.14 for the equilibrium glide model.

We also separately estimate the stagnation point heat flux in the *initial entry* region using the initial entry time dependent altitudes and velocities shown earlier in Figures 5.6 and 5.7.

These two heat transfer solutions are combined in Figure 5.17, which shows that the predicted heat flux ramps up rapidly in the early entry region that starts at an altitude of 120 km. At 275 s the early entry model for $\dot{q}_w$ reaches about 70% of the peak heat flux estimated for the equilibrium glide phase. The entry solution and the glide solution for $\dot{q}_w$ are about equal in magnitude at this time; i.e. at about the time where we transition between the two models. However, the slopes of the two heat

Sec. 5.6]   Calculating heat transfer rates in the peak heating region   269

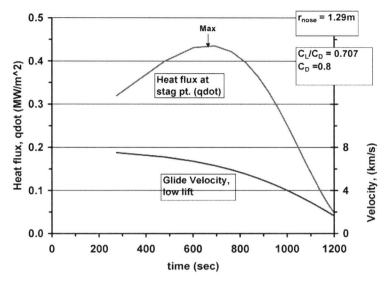

Figure 5.16. Heat flux $\dot{q}_w$ calculated from the Sutton–Graves model with nose radius = 1.29 m. Atmospheric density and Orbiter velocity from the baseline glide model. Equilibrium glide model valid in region $275 < t < 1{,}100$ s. $C_L/C_D = 0.707$, $C_D = 0.8$. Initial velocity ratio, $v_i/v_0 = 0.95$.

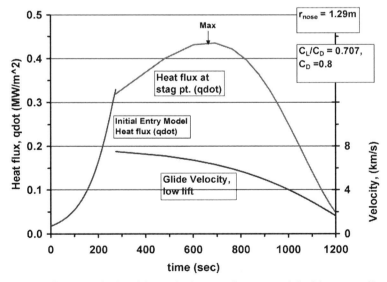

Figure 5.17. Heat flux $\dot{q}_w$ calculated from the Sutton–Graves model with nose radius = 1.29 m. Atmospheric density and Orbiter velocity from the initial entry model $0 < t < 275$ s appended to the baseline glide model. The initial entry lift to drag ratio = 1.0 and the velocity is 7.70 km/s at $t = 0$. The equilibrium glide model, with lift to drag ratio based on banking is used in the region $275 < t < 1{,}100$ s.

flux solutions are *discontinuous* at the match point. This slope discontinuity is to be expected, given the fact that the two flight dynamics models are based on different equations of motion that were separately chosen for the simplicity of their mathematical and physical representations.

In Figure 5.18, we compare our model results to the time-varying stagnation-point heat flux data derived from the optical nose-cap radiometer measurements of wall temperature, collected during Shuttle mission STS-5 [1]. These synthetic $\dot{q}_w$ flight data were calculated from our equilibrium radiation model, Equation 5.117, using the reported "inner moldline" flight temperatures (as a function of time) presented in Figure 5 of [1], as noted in Section 5.6.3. STS-5 radiometer temperatures were measured *inside* the Orbiter nose cap (the thickness of the reinforced carbon-carbon wall is quite thin, approximately $\frac{1}{4}$ inch). For simplicity we assume in our estimation of the STS-5 heat flux history that the gas-side exterior wall temperatures at the stagnation point are exactly equal to the measured inner-moldline temperatures on the opposite side of the wall.

Note in Figure 5.18 that our model's temporal shape and predicted heat transfer magnitudes are in reasonably good agreement with the flight data, both maximums are of the order of $0.4\,\mathrm{MW/m^2}$. However, the glide portion of our model does show a more sharply peaked $\dot{q}_w$ profile near $t = 700\,\mathrm{s}$, in comparison to the correspondingly flatter STS-5 plot at this time. We also observe that the STS-5 heat flux history *temporally lags* our model results by about 100 seconds. We cannot fully explain

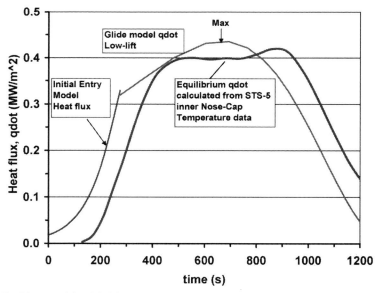

Figure 5.18. The combined initial entry model $0 < t < 275\,\mathrm{s}$ and the baseline glide model, $275 < t < 1{,}100\,\mathrm{s}$ are compared to the stagnation point heat flux derived from the optical nose-cap radiometer measurements of wall temperature, collected during STS-5 [1].

this difference. However, it appears that part of the answer is related to the thermal inertia of the supporting structure, i.e. we expect that temperatures *inside* the nose cap will not instantaneously respond to rapid changes in temperature (or heat flux) on the gas side of the nose-cap due to the mass of the nose-cap walls and the supporting interior structural elements. Overall, it appears from an examination of Figure 5.18 that the heating rates calculated from our simple glide and entry models are comparable to the rates measured by NASA for the STS-5 trajectory. (Lower peak temperatures and heat fluxes were reported on earlier flights, STS-1 to STS-4, than for STS-5. These earlier temperature data were, however, believed to have been biased low as a result of the heavy black deposits subsequently found on the radiometer lens inside the nose cap [1]).

In Section 5.3.2.9 we pointed out that the total heat load during reentry, Q_w (per unit area) is given by the integral of the heat flux over time. In the next section, we integrate our model results for the time-dependent heat fluxes (Figure 5.18) over time to calculate the estimated total heat load.

5.6.5 Model calculations for total heat load at the stagnation point

The total heat load, or heat per unit area convected to the walls at the stagnation point, is defined as

$$\text{Total heat load} \equiv Q_w / A_w = \int_0^{t_{max}} \dot{q}_w \, dt \quad (5.120)$$

The stagnation heat flux can be calculated from the Sutton–Graves equation (Equation 5.32) as

$$\dot{q}_w = \frac{\text{Const}}{\sqrt{r_{nose}}} (\rho)^{1/2} (v)^3 \quad (5.121)$$

where the "Constant" $= A = 1.75 \times 10^{-4} \frac{\sqrt{kg}}{m}$. For convenience we have dropped the subscript ∞. Substituting Equation 5.121 into Equation 5.120 and then changing the variable of integration to velocity v, yields

$$Q_w / A_w = \frac{\text{Const}}{\sqrt{r_{nose}}} \int_0^{t_{max}} (\rho)^{1/2} v^3 \, dt = \frac{\text{Const}}{\sqrt{r_{nose}}} \int_{v_i}^{0} \frac{(\rho)^{1/2} v^3}{(dv/dt)} dv \quad (5.122)$$

We now evaluate this integral using the expressions for density, velocity, and acceleration based on the equilibrium glide model. In Equation 5.82, we showed for the glide model that the reciprocal of the acceleration is

$$\frac{dt}{dv} = -\frac{2 \cdot BN}{g} \left(\frac{1}{\rho v^2} \right) \quad (5.123)$$

Substituting Equation 5.123 into Equation 5.122 gives

$$Q_w / A_w = \frac{\text{Const}}{\sqrt{r_{nose}}} \left(-\frac{2 \cdot BN}{g} \right) \int_{v_i}^{0} \frac{v \, dv}{(\rho)^{1/2}} \quad (5.124)$$

Equation 5.54 gives the following expression for ρv^2 based on the glide model

$$\rho v^2 = \frac{2 \cdot BN}{(C_L/C_D)} \left[1 - \frac{v^2}{gr_e}\right] \quad (5.125)$$

Solving for $1/\sqrt{\rho}$

$$\frac{1}{\sqrt{\rho}} = \frac{(C_L/C_D)^{1/2}}{(2 \cdot BN)^{1/2}} \frac{\bar{v}\sqrt{gr_e}}{\sqrt{1-\bar{v}^2}} = \left(\frac{C_L A r_e}{2m}\right)^{1/2} \left(\frac{\bar{v}}{\sqrt{1-\bar{v}^2}}\right) \quad (5.126)$$

where the non-dimensional velocity ratio is defined as $\bar{v} \equiv v/v_0 = v/\sqrt{gr_e}$.

Substituting Equation 5.126 into the integral in Equation 5.124 and normalizing the velocity in the limits of integration gives

$$Q_w/A_w = \frac{\text{Const}}{\sqrt{r_{\text{nose}}}} \left(\frac{2 \cdot BN}{g}\right) \left(\frac{C_L A r_e}{2m}\right)^{1/2} (v_0^2) \int_0^{\bar{v}_i} \left(\frac{\bar{v}^2 \, d\bar{v}}{\sqrt{1-\bar{v}^2}}\right) \quad (5.127)$$

Simplifying the expression using $BN \equiv mg/C_D A$

$$Q_w/A_w = (\text{Const} \cdot v_0^2) \sqrt{\frac{r_e}{r_{\text{nose}}}} \left(\frac{2m}{C_D A}\right)^{1/2} \left(\frac{C_L}{C_D}\right)^{1/2} \cdot I \quad (5.128)$$

The integral I can be written down in closed form as

$$I = \int_0^{\bar{v}_i} \left(\frac{\bar{v}^2 \, d\bar{v}}{\sqrt{1-\bar{v}^2}}\right) = \tfrac{1}{2}\left[\sin^{-1} \bar{v}_i - \bar{v}_i \sqrt{1-\bar{v}_i^2}\right] \quad (5.129)$$

For our glide calculations we nominally set $\bar{v}_i = 0.95$. The magnitude of the integral is then

$$I = 0.4783 \quad (5.130)$$

From Equation 5.128, we see that the heat load is proportional to $\sqrt{C_L/C_D}$ so that lower lift to drag ratios reduce the total heat load as previously discussed. This is the reason why bank maneuvers were introduced during the equilibrium glide phase. With m, C_D, A, r_{nose}, and $C_L/C_D = 0.707$ as defined, we calculate the total heat load from Equation 5.128 to be

$$Q_w/A_w = (1.75 \times 10^{-4} \cdot (7{,}920)^2) \sqrt{\frac{6{,}378 \times 10^3}{1.29}} \left(\frac{2 \cdot 10^5}{0.8(250)}\right)^{1/2} (0.707)^{1/2} \cdot (0.4783)$$

$$Q_w/A_w = 310 \text{ MJ/m}^2 \quad \text{glide model with bank correction} \quad (5.131)$$

The added heat load from the initial entry solution, the area under the initial entry portion of the heat flux curve in Figure 5.17 is approximately

$$Q_w/A_w \approx 27 \text{ MJ/m}^2 \quad \text{initial entry model} \quad (5.132)$$

The combined total stagnation point heat load for both the entry region and the glide region is estimated to be

$$(Q_w/A_w)_{\text{combined}} = 27 + 310 = 337 \text{ MJ/m}^2 \quad \text{"entry model + glide model"} \quad (5.133)$$

NASA values for heat load when scaled from a reference 1 foot nose radius [22] to a 1.29 m nose radius (Figure 5.19) lie in the range

$$260 < Q_w/A_w < 320 \text{ MJ/m}^2 \quad \text{NASA data} \quad (5.134)$$

We conclude that our glide-modeled heat-load estimate, $Q_w/A_w = 337 \text{ MJ/m}^2$, slightly exceeds the upper end of the NASA estimated stagnation-point total heat load range [23].

Our model peak heat flux estimate, 437 KW/m^2 is plotted as a coordinate pair with the total heat load, 337 MJ/m^2 on the x axis in Figure 5.19 (the diamond data symbol). Also included are a number of other experimental data and numerical model estimates for heat load and peak heat flux from various sources.

In summary, our BotE model calculations are able to reasonably predict both the qualitative and quantitative trends in the reentry trajectory, heat transfer rate, and total heat load. These simplified calculations, based on models developed more than half a century ago, can effectively set the stage for the much more detailed

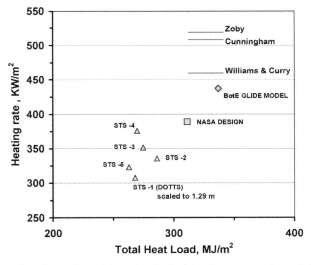

Figure 5.19. Heat flux $\dot{q}_w$ and total heat load, Q_w/A_w, from a number of different data and numerically-modeled references compared to BotE glide model results. Only peak heating rates are listed in the studies of Zoby [19], Williams and Curry [1], and Cunningham [21]. Representative heat flux and heat load values from STS-1 to STS-5 are given by Dotts [23]. The reference heating rate was initially based on a theoretical 1 ft radius reference sphere. The results are rescaled here to a 1.29 m radius spherical Orbiter nose.

calculations and measurements that are often required by engineering teams as they work to more accurately calculate the thermal performance of present and future spacecraft during reentry.

5.7 APPENDIX: BotE MODELING OF *NON-ORBITER* ENTRY PROBLEMS

The simple BotE models, outlined in Section 5.5, were tailored to characterize the major forces that specifically determine the *Orbiter* reentry trajectory. To develop such models, it is necessary for an engineer to comprehend the particular design limits set for the Space Shuttle program. For example, in order to reduce the overall deceleration forces on the vehicle and also help constrain the peak heat transfer rates to the vehicle, the Orbiter s initial entry angle was chosen by NASA to be quite small, of order 1°. In addition, rotating the vehicle to a high angle of attack just prior to entering the atmosphere produces the strong lift and drag forces that also make it possible to control and limit both the maximum rate of descent and the convective heat fluxes to the Orbiter's thermal protection system. Taking account of these flight trajectory and aerodynamic force constraints, an engineer should be able to make certain estimates as to the magnitudes of the accelerations and dominant forces that appear in the equations of motion, Equations 5.40 to 5.42. On the basis of such estimates one can then develop a simplified set of dynamics equations that can be used, in many instances, to obtain several well-known mathematical solutions (e.g. the equilibrium *glide* trajectory solution). These simple analytical results make it possible to quickly calculate Orbiter flight path angles, velocities, and altitudes as a function of time in various flight regimes.

This approach, where an engineer estimates the dominant forces and accelerations of the vehicle and then simplifies the equations of motion, can be, and has been, utilized in a number of *non-Orbiter* entry problems (e.g. the reentry of the manned Mercury and Apollo capsules into the earth's atmosphere and the entry of an unmanned vehicle into a Martian atmosphere where a variety of aerodynamic and propulsion devices are used to safely land the spacecraft on the surface of the planet). This is not to say that detailed numerical solutions of the equations of motion are not possible or are not needed; quite the contrary. When fully mature these numerical calculations often provide important benchmarks and system trends for the engineer over a wide range of conditions. However, as illustrated throughout this book, simple approximate mathematical solutions and quick quantitative estimates of the key variables can often lead to a broad understanding of the interactions in a complex system and can also quickly establish the dependence of the engineering quantities of interest on the principal system variables and initial conditions. Such BotE solutions and estimates often lead to the development of basic rules of thumb for the system and subsequently translate into improved design tradeoff decisions.

Some of the early entry/reentry engineers who first simplified the full non-linear equations of motion and then developed useful approximate solutions to a variety of atmospheric entry problems were Julian Allen, Alfred Eggers [7], and Dean

Chapman [12]. The following are examples of the classical problems they addressed in their seminal papers: (a) *ballistic* entry at moderate to large entry angles (It is shown for these conditions that the gravitational and centrifugal forces can be neglected compared to the drag force and that the flight-path angle remains effectively unchanged during most of the entry trajectory), (b) *skip* entry at relatively small entry angles (If we assume that the *difference* between the vertical gravitational force and the centrifugal force is small compared with the lift force, then one can readily calculate the changes in flight-path angle with time and determine the conditions under which a spacecraft will *skip* one or more times out of, and then back into the atmosphere) and (c) *glide* entry, a generalization of the Orbiter glide problem (If we assume that the flight path angle is constant for a vehicle with large aerodynamic lift and drag and travels on a small glide-path angle within the atmosphere, then from Equation 5.41 a balance is achieved between the lift, gravity and centrifugal force terms. This balance leads directly to a simple closed-form solution to the glide problem that provides a simple means for calculating altitude as a function of velocity and time.

In addition to these well-known problems and solutions, a whole host of new problems associated with previous or future planetary entry missions challenge present-day flight engineers and academics. Many of these problems are being, or have been, tackled using multidimensional numerical flight dynamics methods and software. However, we assert that BotE techniques should be employed, where possible, to calculate first-order estimates of mission performance for those complex systems that are designed to take advantage of various "aero-assist" devices to slow a vehicle when entering a planet's atmosphere.

5.8 REFERENCES

[1] Williams, S.D. and D. M. Curry, "Assessing the Orbiter Thermal Environment Using Flight Data," *J. Spacecraft*, Vol. 21, No. 6, November, 1984.
[2] Wikipedia "Space Shuttle" (STS mission profile graphic). This work is in the public domain in the United States because it is a work of the United States Federal Government under the terms of Title 17, Chapter 1, Section 105 of the US Code. http://en.wikipedia.org/wiki/Space_Shuttle
[3] Fortescue, P. and J. Stark (Eds.), *Spacecraft Systems Engineering*, Wiley,1992.
[4] Sketch from the "Columbia sacrifice" website: http://www.columbiassacrifice.com/$C_hypersonic.htm
[5] Griffin, M. D. and J. R. French, *Space Vehicle Design*, 2nd edition, AIAA education series, 2004
[6] Lees, Lester, "Laminar heat transfer over blunt-nosed bodies at hypersonic flight speeds, pp. 259–269, *Jet Propulsion Journal*, April, 1956.
[7] Allen, H.J. and A.J. Eggers, "A study of the motion and aerodynamic heating of ballistic missiles entering the earth's atmosphere at high supersonic speeds," NACA TR 1381, 1958.
[8] Sibulkin, M.J., "Heat transfer near the forward stagnation point of a body of revolution," *Journal of the Aeronautical Sciences*, vol. 19, no. 8, August 1952.

[9] Sutton, K. and Graves, R. A., "A general stagnation-point convective-heating equation for arbitrary gas mixtures," NASA TR R-376, 1971.
[10] Reid, R. "Orbiter Entry Aerothermodynamics," Space Shuttle Technical Conference, Houston, NASA CP-2342, part 2, page 1051, June, 1983.
[11] Johnson Space Flight Center/DM5 Flight Design and Dynamics Division, "Entry Aerodynamics" tables. Descent data post flight summaries, 1998.
[12] Chapman, D.R., "An Approximate Analytical Method for Studying Entry into Planetary Atmospheres," NACA Technical Note 4276, May 1958.
[13] Lees, Lester, F. Hartwig, C.B. Cohen, "Use of Aerodynamic Lift During Entry into the Earth"s Atmosphere," *ARS Journal*, September 1959.
[14] Ko, W., R. Quinn, L. Gong, "Finite-Element Reentry Heat-Transfer Analysis of Space Shuttle Orbiter", NASA Technical Paper 2657, December 1986.
[15] Hirschel, E. H. and C. Weilend, *Selected Aerothermodynamic Design Problems of Flight Vehicles*, jointly published by AIAA and Springer, 2009.
[16] Young, J. and J. Underwood et al., "The aerodynamic challenges of the design and development of the Space Shuttle", Space Shuttle Technical Conference, Part 1, NASA CP 2342.
[17] Eggers, A.J., H.J. Allen, and S.E. Neice, "A comparative analysis of the performance of long-range hypervelocity vehicles," NACA RM A54L10, 1955.
[18] Goodrich, W.D. and C. Lit, C. Houston, P. Chiu, L. Olmedo, "Numerical Computations of Orbiter Flowfields and Laminar Heating Rates", *J. Spacecraft*, Vol. 14, No. 5, pp. 257–263, May 1977.
[19] Zoby, E.V., "Approximate Heating Analysis for the Windward Symmetry Plane of Shuttle-like Bodies at Large Angle of Attack," published in *Thermophysics of Atmospheric Entry*, edited by T.E. Horton, Vol. 82, *Progress in Astronautics and Aeronautics*, 1981.
[20] Moss, J.N. and Bird, G.A., "Direct simulation of translational flow for hypersonic reentry conditions," AIAA paper 84-0223, 1984.
[21] Cunningham, J. and J.W. Haney, Jr., "Space Shuttle wing leading edge heating environment prediction derived from development flight data," published in "Shuttle Performance: Lessons Learned," NASA CP-2283, pp. 1083-1110, 1983.
[22] Edwards, C. and S. Cole, "Predictions of Entry Heating for lower Surface of Shuttle Orbiter," NASA TM 84624, July 1983.
[23] Dotts, R.L., J. Smith, and D. Tillian, "Space Shuttle Orbiter Reusable Surface Insulation Flight Test Results," in Shuttle Performance Lessons Learned NASA CP 2283, part 2, NASA Langley Conference, March 1983.

6

Estimating the dimensions and performance of the Hubble Space Telescope

6.1 THE HUBBLE SPACE TELESCOPE

The Hubble Space Telescope (HST) has revolutionized modern astronomy since it was first launched into orbit by the Space Shuttle Discovery on April 24, 1990. Not since the invention of the telescope in the 17th century, have the mysteries of the universe been uncovered at such a rapid pace. The most detailed looks at the farthest galaxies in the universe have been obtained using images taken by the HST. It has made some extraordinary discoveries, the most notable being the confirmation of dark matter, observations supporting the accelerating universe theory, and studies of newly discovered planets outside our solar system [1].

6.1.1 HST system requirements

The Hubble was designed to meet three major system requirements:

(1) High angular resolution—the ability to image fine detail
The HST was initially designed in the 1980s to be the first large space telescope with a resolution capability an order of magnitude better than what was then possible with ground-based telescopes in the visible and near ultraviolet. The resolution was designed to be diffraction-limited at visible wavelengths, with angular resolutions on the order of less than 0.1 arcseconds [2].

In Section 6.2, we will use BotE methods to estimate the diffraction-limited angular resolution for the HST and calculate the equivalent system focal length, f_{eq}, needed to achieve that resolution. We will use the magnitude of f_{eq} to determine the resulting secondary mirror magnification, m, for a Cassegrain telescope design.

I. E. Alber, *Aerospace Engineering on the Back of an Envelope.*
© Springer-Verlag Berlin Heidelberg 2012.

(2) High sensitivity—the ability to detect very faint objects

This goal for the HST was to be able to observe objects at least 50 times fainter than was possible from the ground. Much of this sensitivity was achieved due to the simple fact that the HST was imaging stars from space. The HST doesn't suffer from the losses in signal strength due to the earth's atmosphere, particularly in those electromagnetic bands where certain spectral lines from distant stars are strongly absorbed (primarily in the infrared and ultraviolet).

(3) Ultraviolet performance—the ability to produce ultraviolet images and spectra at high spatial resolution

The HST is designed to see celestial objects in the ultraviolet down to wavelengths of 120 nanometers and has the ability to measure the positions of these objects to within 0.003 arcseconds [2].

6.1.2 HST engineering systems

The Hubble Space Telescope is about the size and weight of a city bus that must fit within the Space Shuttle's cargo bay [3]. It is moderately sized compared to other much larger ground-based telescopes. Figure 6.1 is a NASA contractor

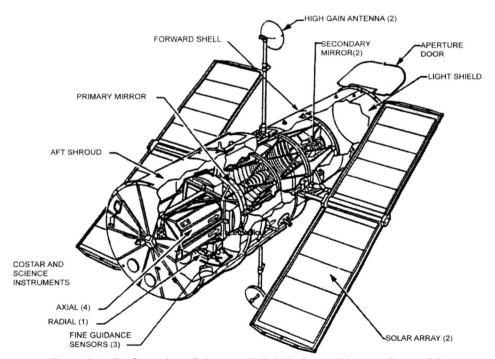

Figure 6.1. Configuration of the overall Hubble Space Telescope System [4].

Sec. 6.1]   The Hubble Space Telescope   279

drawing of its major components. Note the callouts for the primary and secondary mirrors. The light shield tube and its viewing aperture door are placed in front of the optics section that houses the 2.4 meter diameter primary mirror and a much smaller secondary mirror.

The HST was designed to be launched into a low earth orbit (600 km orbital altitude) and subsequently serviced by astronauts using NASA's Space Shuttle. Design tradeoffs were made to keep the HST within the constraints imposed by the length and width of the Shuttle cargo bay. In addition, the mass of the entire HST satellite was required to be considerably less than the maximum possible cargo mass that the Shuttle can deliver to a given orbit. Fundamental to the strategy for maintaining the full scientific potential of the HST, it was designed from the start to accommodate orbital maintenance and refurbishment by the Shuttle's astronauts. During a maintenance mission, a robotic arm aboard the Orbiter vehicle allows the HST to be captured from orbit and then berthed in the cargo bay for stability and access to its electronic and optical systems. This set of operations makes it possible to install various spectral sensors and new cameras. The Optical Telescope Assembly for the HST did not achieve its intended design performance until the first Shuttle servicing in December 1993 (STS-61) when vital "corrective optics" were installed by the crew of the Shuttle Endeavour.

6.1.3 Requirements for fitting the HST into the Orbiter

Once the decision was made to design, develop, build, and maintain the HST in conjunction with the Shuttle, a number of specific design requirements followed. Here we mention one requirement pertaining to the maximum size of the HST.

The HST could not exceed the Orbiter's cargo bay dimensions of 18.3 m in length and 4.6 m in width. Taking into account the other satellite deployment equipment carried in the cargo bay, the length of the HST payload was limited to about 16 m or about 52 feet. In addition, the mass of the HST had to be less than the maximum cargo-carrying capability for the Shuttle when placing a satellite in an orbit at 600 km altitude, which meant that the HST mass could be no greater than about 15,000 kg.

Some important system design questions that the early HST designers must have addressed are:

(a) *What are the basic mirror dimensions and optical telescope positions necessary for the HST to reach its design objectives?*
(b) *What HST design tradeoffs must be made to keep the Hubble payload within the total cargo bay length and cargo mass constraints imposed by the Shuttle design?*

In addition there is the very important HST system requirement for optical resolution given in Section 6.1.1, namely that the telescope's optical resolution be diffraction-limited at visible wavelengths, with angular resolutions on the order of less than 0.1 arcseconds.

280 Estimating the dimensions and performance of the Hubble Space Telescope [Ch. 6]

The corresponding design question is:

(c) *What is the equivalent optical system focal length needed for the HST to satisfy the diffraction-limited resolution requirement?*

We generate a sketch of the basic Cassegrain HST optical telescope design and then first answer question (c) by estimating the overall equivalent system focal length, f_{eq}, necessary for the HST to generate high quality diffraction-limited astronomical images. With f_{eq} and the secondary magnification, m_2, both defined, we then turn our attention to answering the telescope size and mass questions: (a) and (b).

6.2 THE HST OPTICAL TELESCOPE DESIGN

The Optical Telescope Assembly operates on much the same principles as the reflecting telescopes invented in the 17th century by Newton, Cassegrain, and Gregory. Reflecting telescopes gather starlight using mirrors that concentrate it onto a detector which may be either a human eye or, in modern telescopes, an array of CCD pixels. The optical design of the HST is a Ritchey-Chrétien form of a Cassegrain telescope with a large concave 2.4 meter diameter primary mirror and a much smaller convex secondary mirror.

A schematic of a classical Cassegrain telescope is shown in Figure 6.2. This is a cleaner version of the sketch presented in Chapter 1, where we took a first-cut look at the telescope sizing problem using the Quick-Fire BotE approach. In this figure the incoming light rays from a distant light source are nearly parallel because celestial objects are very far away.

We assume that the diameters of all of the mirrors are substantially larger than the wavelength of the incident light; in fact, this ratio is of the order of 10^7. This

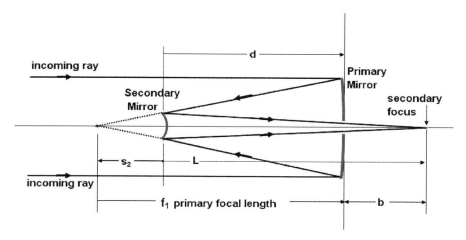

Figure 6.2. Two-mirror Cassegrain optics for the HST with principal dimensions and mirrors identified. Note the reflection of the incoming rays from the primary and secondary mirrors.

allows us to use the *geometric optics* approximation where light paths, or light rays, are taken to be straight lines perpendicular to the wave fronts of the light [5, Halliday and Resnick]. A "wave front" is the locus (or surface) of points that have the same phase. This is true for any wave field. For example, when we look at the water waves created when a stone is dropped into a still pond, the expanding two-dimensional wave fronts (on the surface of the water) are concentric circles with their centers at the splash point. The corresponding "rays", normal to these circles, are straight radial lines which emanate from that splash point.

When a light ray strikes a mirror, its path changes immediately according to the specular law of reflection where *the angle of reflection equals the angle of incidence*. As a historical note, the law of reflection was known to Euclid more than two thousand years ago.

The incoming, approximately parallel light rays (emanating from an object at infinity) reflect off the *concave* primary mirror in Figure 6.2 and are directed (as a consequence of the primary mirror's curvature and the law of reflection) towards the on-axis primary mirror focal point at a distance f_1 from the primary mirror. But just before they would reach the primary focal point these rays change direction as they reflect from the *convex* secondary mirror. They then pass through a hole in the primary mirror and converge at the secondary focus forming an image on the telescope's focal plane. As Figure 6.2 shows, the secondary focus lies at an axial distance, b, behind the primary mirror.

An important focal length scale that facilitates the imaging of fine astronomical details is the equivalent system focal length, f_{eq}. We define and calculate it for the HST in the following section.

6.2.1 The equivalent system focal length

Bely [6] states: It can be shown that, to first order, any system composed of several mirrors is equivalent to a single mirror the focal length of which is called the equivalent, or "effective" focal length, f_{eq}. The non-dimensional focal ratio, or f-ratio of the system, F_{eq}, is the magnitude of f_{eq} normalized by D, the diameter of the primary mirror. Thus F_{eq} is defined as

$$F_{eq} = \frac{|f_{eq}|}{D} \qquad (6.1)$$

To calculate the effective focal length, f_{eq}, we first trace those rays that travel from the secondary mirror to the focal plane backwards to the left. This trace of rays (shown as dotted lines in Figure 6.3) intersects the corresponding incident rays at height y_1, which is the vertical distance from the optical axis to the outer radius of the primary mirror. Schroeder [7] says "the distance between the intersection plane (where the backward projected rays intersect the incoming rays) and the focal point, measured along the axis is the equivalent or effective focal length, f_{eq}".

From the geometry of Figure 6.3, we can determine f_{eq} using the proportionality property of similar triangles constructed from the forward and projected rays in that

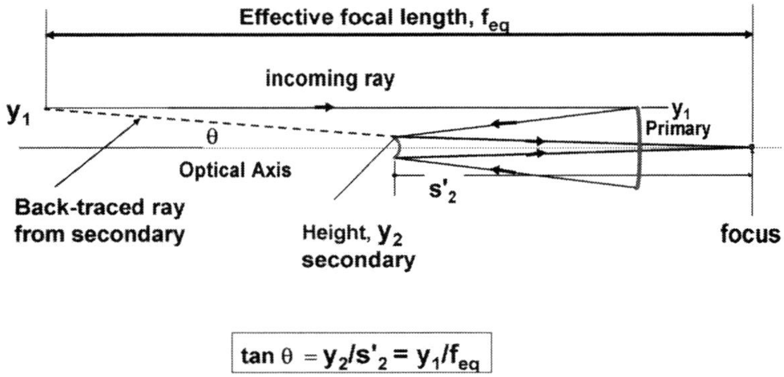

Figure 6.3. Effective or equivalent focal length, f_{eq}, defined graphically. Note that the ray that travels from the upper edge of the secondary mirror to the focal point on the focal plane is back-traced (dashed line) until it intersects a line parallel to the optical axis at the height y_1 of the primary mirror outer edge. y_2 is the half height of the secondary mirror.

figure. We equate the ratios of the altitude to base lengths of two similar triangles to obtain

$$\frac{y_1}{f_{eq}} = \frac{y_2}{s'_2} \qquad (6.2)$$

In Equation 6.2, y_1 is the half-height of the primary mirror, y_2 is the half-height of the secondary mirror, and s'_2 is the image distance from the secondary mirror to the secondary focal point (recall from Chapter 1 that s'_2 is sometimes referred to as the constructional length L for a Cassegrain telescope, Figure 6.2).

Using the equation for the lateral magnification of the secondary mirror [5], written as either m_2 or simply m, we express m as the ratio of the image distance to the negative of the object distance

$$m \equiv m_2 = -\frac{\text{image distance}}{\text{object distance}} = \frac{s'_2}{(-s_2)} \qquad (6.3)$$

In Equation 6.3 the secondary object distance, $-s_2$, is the distance (behind the secondary mirror) to the *virtual* image provided by the primary mirror (Figure 6.2).

Solving for the equivalent focal length f_{eq} from Equation 6.2 gives

$$f_{eq} = s'_2 \left(\frac{y_1}{y_2}\right) = \left(\frac{s'_2}{-s_2}\right)\left(\frac{-s_2}{k}\right) = m\left(\frac{-s_2}{k}\right) \qquad (6.4)$$

The parameter k is defined as the ratio of the secondary to primary mirror heights, $k \equiv y_2/y_1$.

We can also use the geometric properties of similar triangles in order to write an alternate expression for the ratio $\left(\frac{-s_2}{k}\right)$ based on the following ray drawing, Figure 6.4.

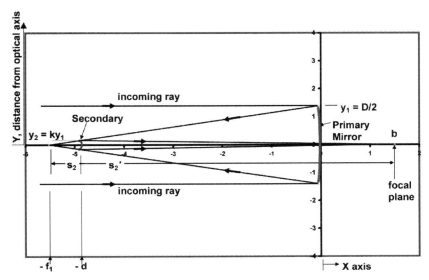

Figure 6.4. Ray geometry for two-mirror Cassegrain telescope calculated with the paraxial approximation. All relative scales are based on our calculated HST telescope dimensions.

Note the large right triangle in Figure 6.4 with base f_1 whose hypotenuse starts at the upper edge of the primary mirror ($x = 0, y = y_1$) and ends on the optical axis at the primary focal point ($x = -f_1, y = 0$). Also observe the small right triangle of height y_2 ($= ky_1$) and base of length $-s_2$ behind the secondary mirror that is geometrically similar to the larger triangle. Since the height to base ratios for each of the similar triangles are equal

$$\frac{y_1}{f_1} = \frac{y_2}{(-s_2)}$$

then

$$\frac{(-s_2)}{k} = f_1 \qquad (6.5)$$

Substituting Equation 6.5 into Equation 6.4 yields the following expression for the equivalent focal length given m and f_1

$$f_{eq} = mf_1 \qquad (6.6)$$

If we are given both f_{eq} and f_1 we can calculate the secondary magnification, m, from the ratio of these two focal lengths, i.e.

$$m = f_{eq}/f_1 \qquad (6.7)$$

Dividing each focal length by the primary diameter D, to form the equivalent and primary focal ratios, the resulting expression for m is

$$m = F_{eq}/F_1 \qquad (6.8)$$

In the next section we estimate the focal ratio F_{eq} for an HST that satisfies the important diffraction-limited resolution requirement. If the primary focal ratio is specified (as below) as half the radius of curvature of the primary mirror divided by D, then we can directly calculate the secondary magnification, m, from Equation 6.8. As we will see, "m" appears in all the expressions for the Cassegrain telescope dimensions.

6.2.2 How do designers determine the required system focal ratio, F_{eq}?

One of the HST's three major system requirements (Section 6.1.1) was to achieve optical angular resolution at the diffraction limit, defined below, for visible wavelengths.

6.2.2.1 Defining angular resolution

The resolution of any telescope is defined as the smallest angular separation of two point sources of light (e.g. two stars) that will still allow them to be resolved as individual point sources, despite their having overlapping diffraction patterns. The exact value of the angular separation at which a pair of adjacent diffraction patterns overlap to such an extent that they become indistinguishable depends upon the particular angular criterion that is chosen [8].

One criterion, defined by Rayleigh, states that the smallest angular separation, $\Delta\theta$, that two stars can have and remain resolvable as individual stars is

$$\Delta\theta = 1.22 \frac{\lambda}{D} \quad \text{(Rayleigh criterion)} \tag{6.9}$$

where λ is the wavelength of the light being considered and D is the diameter of the telescope's circular aperture.

The Rayleigh limit in Equation 6.9 is measured by the angular distance from the center of a single Airy diffraction disc to its first dark diffraction ring. In Fraunhofer diffraction, light collected by the circular primary mirror, of diameter D, interferes with itself creating a ring-shaped diffraction, or Airy pattern, that blurs the image.

Figure 6.5 shows an ideal Fraunhofer diffraction pattern from a point source in the "computer-drawn" circular light-intensity image.

When two adjacent stars of equal brightness are separated by this Rayleigh angular spacing, the diffraction intensity pattern from the first star adds to the offset intensity pattern of the second star. The result is a double-hump in which the peak of each hump marks the point location of each star, with a minimum, or dip, in the combined intensity pattern located midway between the two humps. For the Rayleigh criterion spacing, the intensity "dip" is 27% below the peak intensity of either star [10].

Another criterion for resolving two point sources is the "Sparrow criterion". In this case, there is a smaller dip in intensity of less than 2% for the combined two star diffraction pattern when the angular separation $\Delta\theta$ of the peaks is set to

$$\Delta\theta = 1.0 \frac{\lambda}{D} \quad \text{(Sparrow criterion)} \tag{6.10}$$

as shown in Figure 6.6.

Sec. 6.2] The HST Optical Telescope design 285

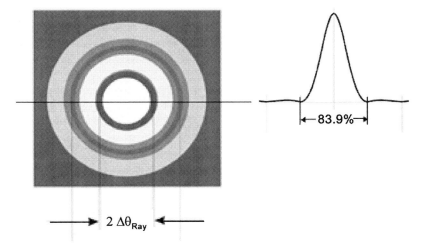

Figure 6.5. Airy diffraction disk pattern. The angle on the sky $\Delta\theta_{Ray} = 1.22\lambda/D$ is the angular distance from the center of the disk to the first dark ring for the Rayleigh angular criterion. The corresponding angular diameter $= 2\Delta\theta_{Ray}$ is shown. A circle of this diameter encloses 83.9% of the total light intensity energy in the disk as shown in the centerline intensity plot displayed to the right. This plot is extracted from [9].

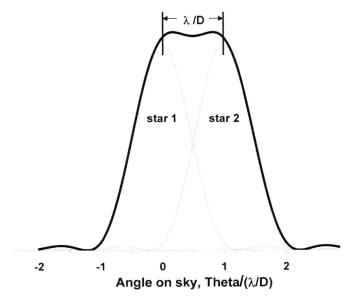

Figure 6.6. Combined Airy diffraction spot intensity pattern for two nearby stars separated by the angular distance $\Delta\theta = 1.0\lambda/D$ for the Sparrow resolution criterion. Note the small 2% dip in the combined pattern midway between each of the two hypothetical diffraction patterns.

We will use this Sparrow criterion in our subsequent calculations. In practice, either the Sparrow or Rayleigh criterion can be used. It is the scale of the angular spacing required to achieve diffraction-limited resolution ($\Delta\theta \sim \lambda/D$), rather than the particular constant (1.0 or 1.22 or ...) that is important in forming an estimate of the effective HST system focal length.

To carry our analysis forward, we need to relate the separation angle on the sky (say from the Sparrow criterion) to a linear distance on the focal plane. This is derived in the following section. We will see that the constant of proportionality is simply $1/f_{eq}$.

6.2.2.2 Relating the position on the focal plane to a measured sky angle

Consider that the light from a given star (call it star #1) generates an Airy disk pattern precisely at the center of the HST's focal plane. We might state that the star, as viewed on the sky, has zero angular displacement from the telescope's optical axis, i.e. $\theta = 0$. Now let's say that the telescope concurrently images star #2, which is equally bright and offset in angle from star #1 by an amount equal to the Sparrow diffraction-limited resolution, $\Delta\theta = \lambda/D$. What, then, is the distance of the peak intensity of star #2 (in millimeters) from the center of the focal plane? We use simple ray geometry to obtain a geometrical equation that relates the linear displacement distance Δ of a given spot or image on the focal plane to the angle θ separating two points in the sky.

Figure 6.7 depicts a very distant object (an upright arrow) that is reflected by a concave mirror and imaged at its primary focal point. A ray from the arrow's tail (think of it as star #1 in our previous discussion of resolution, since it lies on the optical axis) is reflected at the vertex of the mirror. The red and green lines represent two distinct rays that emanate from the head of the object (consider the head of the arrow to be star #2, located at a given angular displacement from star #1). Let's assume the object is located at an *axial* distance z which is extremely far from the center of curvature of the mirror. In that case the rays from the tail region are brought to a focus at the mirror's focal distance f on the optical axis. The

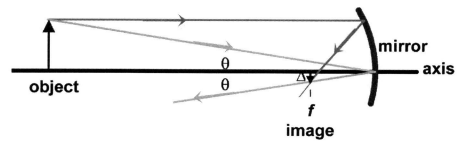

Figure 6.7. Reflection of a distant object (in the sky) by a convex mirror with primary mirror focal length f. The head of the image is displaced vertically a distance Δ below the principal optical axis at an axial distance $z = f$ from the vertex of the mirror at $z = 0$. Note that the image is inverted and diminished in size relative to the object (the magnification $\ll 1$).

"green" ray from the arrow's head strikes the vertex of the mirror at an *angle of incidence* θ relative to the optical axis. This ray reflects off the mirror with an *angle of reflection* exactly equal to θ. It is brought to a focus at an axial distance $z = f$ from the vertex, but at a distance Δ *below* the optical axis. This image position is also where, by geometrical construction, the angled green ray intersects the "red" ray from star #2, initially parallel to the axis. The red ray is specularly reflected downward from the mirror and intersects the reflected green ray at the focal point as shown in Figure 6.7.

From our sketch we see that the displacement Δ of the head from the axis is simply equal to the original incidence angle θ multiplied by the concave mirror's focal length, f, i.e.

$$\Delta = \theta \cdot f \tag{6.11}$$

For a more complex mirror system like a Cassegrain dual mirror system, we can evaluate its final optical angle using the calculated equivalent focal length, f_{eq}, as defined in Section 6.2.1.

The equation that relates the linear displacement distance Δ_{image} of a given spot or image on the focal plane of a *complex mirror system* to the angle θ (in radians) separating two points *in the sky* is thus

$$\theta \cdot f_{eq} = \Delta_{image} \tag{6.12}$$

We now use this equation to determine the magnitude of the equivalent focal length, f_{eq}, that would be required to produce a focal plane image for which θ is the angular separation defined for diffraction-limited resolution ($\theta = \lambda/D$) and for which the final diffraction spot image from the two stars is sampled by one or more focal plane CCD pixels of a given width, Δ_{pixel}.

Thus we assume that the resolution that can be achieved is directly related to the size of one or more CCD solid-state detectors located on the focal plane. For the HST's advanced camera for surveys (ACS), the pixel, or detector size Δ_{pixel}, is approximately 15 microns or 15×10^{-6} m [11].

An estimate for f_{eq} can be made if we match the diffraction-limited spot size to the prescribed silicon pixel size Δ_{pixel}, multiplied by a sampling scale factor, n, of order one, i.e.

$$\Delta_{diffraction\ spot} = n\Delta_{pixel} \tag{6.13}$$

In other words, n is the number of pixels that we would like to fall within the Airy diffraction disk to properly *sample* its signal.

Substituting Equation 6.13 into Equation 6.12 and using the Sparrow resolution criterion for diffraction-limited imaging, yields the following equation for f_{eq}

$$\left(\frac{\lambda}{D}\right)f_{eq} = n\Delta_{pixel} \tag{6.14}$$

In this equation, λ is the wavelength of the incident optical light radiation and D is the aperture diameter (which we take to be the diameter of the primary mirror).

For visible-light observations, astronomers typically use $\lambda = 0.633$ microns in their diffraction calculations.

For *critical* sampling, the Nyquist sampling theorem requires, $n = 2$, or two samples per diffraction spot to prevent aliasing noise in the image.* However, Bely [6, p.138] states that at a cost of some small increase in focal plane low spatial frequency noise, due to aliasing, selecting $n = 1$ (rather than 2) will increase the HST's optical sensitivity and also reduce the required total exposure time.

For the HST, the diffraction-limited angular resolution θ_{limit}, expressed in *arcseconds* instead of radians, is calculated here from the Sparrow equation assuming the primary mirror aperture diameter $D = 2.4$ m and the visible wavelength $\lambda = 0.633 \times 10^{-6}$ m,

$$\theta_{limit\text{-}HST} = (2.0626 \times 10^5 \text{ arcsec per radian})(\lambda/D) = 0.054 \text{ arcsec} \qquad (6.15)$$

We calculate the ratio f_{eq}/D (defined as the equivalent system focal ratio F_{eq}) using Equation 6.14. Assuming $n = 1$ (under-sampling by half) and setting the HST detector width to 15 microns, we are able to estimate the system focal ratio, F_{eq}, required for diffraction-limited HST imaging as

$$F_{eq} = \frac{\Delta_{pixel}}{\lambda} = \frac{15 \times 10^{-6} \text{ m}}{0.633 \times 10^{-6} \text{ m}} = 23.7 \approx 24 \qquad (6.16)$$

It turns out that $F_{eq} = 24$ is precisely the optical system focal ratio selected by the telescope designers to satisfy the HST's diffraction-limited *resolution* requirement [12, Soulihac and Billeray].

The corresponding equivalent focal length (expressed in meters) for the HST is quite large

$$f_{eq} = D \cdot F_{eq} = (2.4 \text{ m}) \cdot (24) = 57.6 \text{ m} \qquad (6.17)$$

6.2.3 Telescope plate scale

By substituting $f_{eq} = 57.6$ m into Equation 6.12, we can calculate a characteristic telescope scale factor, P, that measures the ratio of the angle θ in the sky (in arcseconds, denoted by the straight-double-quote symbol) to the given focal plane offset distance Δ_{image} (usually expressed in millimeters). Astronomers call P the plate scale.

The magnitude of the HST plate scale is calculated using Equation 6.12, the expression which relates an angle on the sky θ to its focal plane displacement distance Δ, i.e.

$$P = \frac{\theta}{\Delta} = \frac{(2.0626 \times 10^5 [\text{arcsec}])}{f_{eq}[\text{mm}]} = \frac{2.0626 \times 10^5}{57.6 \times 10^3} = 3.58''/\text{mm} \qquad (6.18)$$

---

* Aliasing is an effect caused by sampling an image with pixels that are too widely spaced (or undersampled). Consider imaging a flag with closely spaced stripes. Undersampling makes the flag appear as if the stripes were more widely spaced. Thus high texture and low texture images become indistinguishable or "aliases" of one another.

Using this value for P, we can simply determine for the HST the angular displacement θ on the sky for any given offset Δ on the focal plane. For example, when $\Delta =$ the width of one focal plane pixel in [mm] $= 15 \times 10^{-3}$ mm, then

$$\theta_{\text{one pixel}} = P \cdot \Delta = (3.58''/\text{mm}) \cdot (15 \times 10^{-3}\,\text{mm}) = 0.054\,\text{arcsec or } 0.054'' \quad (6.19)$$

This, as expected, is the diffraction-limited angular size we previously determined based on $n = 1$ and the size of a single pixel.

If the total displacement distance Δ is much larger than the size of a pixel (for example when Δ equals the distance to the outer radius of the circularly shaped HST focal plane, about 0.23 m from the optical axis as given in [11]), then the angle on the sky is

$$\theta_{\text{focal plane radial angle}} = P \cdot \Delta = (3.58''/\text{mm}) \cdot (230\,\text{mm}) = 823'' \text{ or } 823\,\text{arcsec}$$

Note that the total HST angular field of view is equal to twice the radial angle calculated above or $1{,}646'' = 0.46°$. This total field of view is about equal to the *angular diameter* of either the moon or the sun as seen from the earth. If you hold your thumb at arm's length, then it has an angular width of 2° as observed by your eye. The moon would be about 1/4 of that angular width. This is an example of an astronomical "rule of thumb".

It would require a camera with a CCD array the size of the entire HST focal plane to fully image the moon (the number of pixels across the diameter of the focal plane $= 1{,}646$ arcsec divided by 0.054 arcsec per pixel $= 30{,}500$ pixels in one direction). The total number of pixels required to fill the circular area with diameter 0.46 m on the focal plane $= 730$ megapixels, or about two orders of magnitude greater than the number of pixels in a typical commercial digital camera. However, the individual cameras attached to the optical instruments of the HST don't have a field of view as large as the full focal plane. For example, the HST wide field camera (WFC) has an angular area of the order of 200×200 square arcsec. Our plate scaling indicates that this focal plane would need to be sampled by a total of about 16 megapixels.

6.2.4 Selection of HST's primary mirror focal ratio, $F_1 = |f_1|/D$

Many astronomers state that high primary "speeds" (corresponding to lower primary focal ratios, F_1) are essential in telescopes used for imaging faint nebular objects. Equation 6.8 shows that as F_1 is reduced, the secondary magnification increases, because $m \sim 1/F_1$. For a given camera or mirror focal length f_1, the primary mirror diameter D must be increased to produce a smaller F_1. A larger diameter increases the number of astronomical photons per second that reach the focal plane. And as the photon flux increases, the required exposure time becomes smaller. This is what is meant by a higher "speed". Thus the smaller you can make F_1 the "faster" your optical system will become.

Bely [6] states, "Since the cost of a telescope observatory is strongly affected by the length of a telescope tube, the tendency is to have the primary mirror as fast (i.e. as low a focal ratio, F_1) as is technically possible."

Note that a low value of f_1 means that the mirror radius of curvature R_1 must be small since, as we will show, f_1 is related to R_1 by the relationship $f_1 = -R_1/2$. A low value of F_1 means more lens curvature, but this is more difficult to provide in a large glass mirror structure.

Bely also comments that "Fast optics used to be difficult to fabricate, but, over the last 30 years, enormous progress has been made in construction methods, and the f-ratio of the primary mirror has been coming down dramatically."

We note that primary focal ratios less than 2.0 have readily been achieved in the construction of large telescopes in recent decades. The primary mirror f-ratio selected in the late 1980s for the HST was $F_1 = 2.3$. For the primary diameter of 2.4 m, we can readily calculate the primary focal length, f_1, and the primary mirror concave radius of curvature on the optical axis, R_1:

$$f_1 = F_1 D = (2.3)(2.4 \text{ m}) = 5.52 \text{ m} \tag{6.20}$$

$$R_1 = -2f_1 = -11.04 \text{ m} \tag{6.21}$$

The meaning of the negative value for R_1 is given in Section 6.2.7.2, where we present the sign conventions for optical distances and focal lengths used in our optical analysis.

According to Bely, for a given F_{eq} the mass and size of the secondary mirror will be smaller as F_1 is reduced and a faster primary leads to a shorter telescope *constructional length L* with a corresponding gain in stiffness [6]. We found in Chapter 1 that the constructional optics length, L, is approximately equal to f_1 plus the back distance b (which is 1.5 m for the instrument section of the HST). Thus, using Equation 6.20 we estimate that a crude approximate estimate for the constructional length

$$L \approx f_1 + b \approx 5.52 \text{ m} + 1.5 \text{ m} = 7.03 \text{ m}$$

This estimate is valid for Cassegrain telescopes with HST focal lengths and high levels of secondary magnification, m.

6.2.5 Calculating the magnification m and exact constructional length L

In Section 6.2.2.2, our calculations for the equivalent system focal ratio for the HST gave us the estimate $F_{eq} \approx 24$. Since $F_1 = 2.3$ for the HST, we can directly use Equation 6.8 to calculate the secondary magnification, m

$$m = F_{eq}/F_1 = 24/2.3 = 10.435 \tag{6.22}$$

Equation 1.14 provides the exact expression for the constructional length

$$L = \frac{m}{(m+1)}(f_1 + b) \tag{6.23}$$

In Chapter 1 we did not know a specific value for m, nor how to calculate it, we just assumed that it was large and therefore set the ratio $m/(m+1)$ to 1. Here we

now can calculate the ratio exactly using $m = 10.435$
$$m/(m+1) = 0.913$$

Substituting this constant (about 9% lower than the value of unity previously assumed) into Equation 6.19 gives us a more accurate constructional length

$$L_{\text{exact}} = \frac{m}{(m+1)}(f_1 + b) = (0.913)(5.52 + 1.5) = 6.406 \text{ m} \qquad (6.24)$$

Our original Quick-Fire BotE estimate for the optical constructional length of 7.02 m calculated in Chapter 1 is therefore about 0.6 m greater than this more exact calculation. It is evident that our original estimate was pretty close to the exact result.

While L is only weakly dependent on m, there are two HST dimensions that are more strongly dependent on the magnification, m, namely the diameter of the secondary mirror, D_2, and the radius of curvature of the secondary mirror, R_2. We now determine the basic equations and calculate the magnitudes of these two secondary mirror parameters.

6.2.6 Estimating the secondary mirror diameter

From the geometry of Figure 6.3 and from the similar triangle equality given in Equation 6.2 we previously derived an explicit expression for the ratio y_2/y_1. By definition, the secondary image distance s'_2 equals the constructional length

$$s'_2 \equiv L$$

and because Equation 6.7 gives

$$f_{\text{eq}} = mf_1$$

we can express the ratio of secondary to primary heights as

$$\frac{y_2}{y_1} = \frac{s'_2}{f_{\text{eq}}} = \frac{L}{mf_1} = \frac{L}{mF_1 D} \qquad (6.25)$$

We see that the height of the secondary mirror to the height of the primary mirror y_2/y_1 is proportional to the construction length L and *inversely proportional to the magnification m*.

Substituting the given primary focal ratio, $F_1 = 2.3$, the calculated value for L from Equation 6.24, and the secondary mirror magnification m from Equation 6.22 into Equation 6.25 yields

$$\frac{y_2}{y_1} = \frac{L}{mF_1 D} = \frac{6.406 \text{ m}}{(10.435)(2.3)(2.4 \text{ m})} = 0.1112 \qquad (6.26)$$

Note from Equation 6.26 that the diameter of the secondary mirror, which is twice the half height, is only 11% of the primary mirror diameter. In meters, the estimated secondary mirror diameter is

$$D_2 = 2y_2 = 2\left(\frac{y_2}{y_1}\right)\left(\frac{D_1}{2}\right) = (0.1112)(2.4 \text{ m}) = 0.267 \text{ m} \qquad (6.27)$$

6.2.7 Estimating the radius of curvature of the HST secondary mirror

We calculate the on-axis radius of curvature of the secondary mirror, R_2, using what is referred to as the *law of reflection for spherical mirrors*, or more simply the *spherical mirror equation*. This law, applicable in the small incidence-angle approximation, relates the mirror's image distance, s', to the mirror's object distance, s, and to its radius of curvature, R [5, Halliday and Resnick, Chapter 39].

An object can either be located in front of a given mirror or virtually behind it. In the case of a Cassegrain, like the HST, the image formed by the primary mirror becomes a virtual object for the secondary, and is located behind the secondary at the primary mirror focal point, $-f_1$ (see Figure 6.2 and Figure 6.4). As we have already calculated the distance to the secondary mirror's image point from the secondary mirror (i.e. the constructional length L) we can solve the *spherical mirror equation* for the radius of curvature, R_2, of the secondary mirror. But first let us review the derivation of the law of reflection for mirrored surfaces.

6.2.7.1 Classical derivation of the Spherical Mirror Equation

We follow the classical derivation of Halliday and Resnick as presented in their chapter on "Geometrical Optics" [5].

Let's start by considering two rays emanating from an "object" at point A on the optical axis, being reflected at points B and P respectively on the concave mirror (as shown in Figure 6.8). The vertex of the mirror is at point B.

In Figure 6.8, Ray #1 from point A travels along the optical axis to point B (a distance = "o") where it is exactly reflected back along that axis. Ray #2 travels from point A to point P located on the spherical mirror above the optical axis. The slope of Ray #2 relative to the optical axis is angle α. The center of the spherical surface is at point C on the optical axis, located a distance R (the radius of the sphere) from the spherical vertex at point B. A dashed line from C to P is inclined at slope angle β relative to the optical axis, as shown in Figure 6.8. This line is, by definition of a radius, normal to the spherical surface. The angle between Ray #2 and this normal is the angle of incidence θ_i. Based on the law of reflection, the angle of reflection for Ray #2, θ_r, is equal to the angle of incidence: $\theta_r = \theta_i \equiv \theta$. Reflected

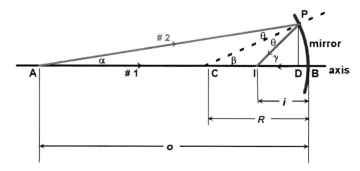

Figure 6.8. Reflection of rays from a spherical mirror (based on sketch in [5, Chapter 39-8]).

ray #2 crosses the optical axis at point I. The slope of this reflected ray, clockwise from the axis is γ. At point I, reflected ray #2 intersects axial ray #1 forming an "image" of the object at A. The distance from the vertex at B to I on the optical axis is defined to be the image distance = "i". For the triangle APC the exterior angle is the sum of the opposing two interior angles

$$\beta = \alpha + \theta \tag{6.28}$$

For the triangle API, the exterior angle γ is related to the interior angles by

$$\gamma = \alpha + 2\theta \tag{6.29}$$

Solving for θ from Equation 6.28 and substituting it into Equation 6.29 yields

$$\gamma + \alpha = 2\beta \tag{6.30}$$

We now write down an exact equation for β in terms of the ratio of the length of the circular arc (PB) to the radius R of the circle; i.e. $\beta = \dfrac{(PB)}{R}$. We define the height of the point P above the optical axis as $DP \equiv y_P$, where D is the point on the axis located near the vertex of the mirror (as in Figure 6.7). If the angle β is very small, then y_P is approximately equal to the arc length (PB), i.e. $y_P \approx (PB)$. Therefore for small angles, β is approximately given by

$$\beta \approx y_P/R \tag{6.31}$$

Let's develop a similar equation for the angle α. For right triangle APD, we write the trigonometric relation: $\tan \alpha = y_P/(AD)$. For small angles, the distance AD is quite close to the distance to the vertex of the mirror and $(AD) \approx (AB) = o$. For very small angles we can approximate the tangent by the angle itself, i.e. $\tan \alpha \approx \alpha$. Using these approximations we write the following equation for α

$$\alpha \approx y_P/o \tag{6.32}$$

Applying the same arguments to triangle IPD gives us the corresponding equation for γ

$$\gamma \approx y_P/i \tag{6.33}$$

Substituting Equations 6.31 to 6.33 into Equation 6.30 and canceling y_P, yields the *Spherical Mirror Equation*

$$\frac{1}{i} + \frac{1}{o} = \frac{2}{R} \tag{6.34}$$

We can also rewrite Equation 6.34 using the nomenclature adopted for the image and object distances, used in most elementary optical textbooks, as

$$\frac{1}{s'} + \frac{1}{s} = \frac{2}{R} = \frac{1}{f} \tag{6.35}$$

Note that if we set the object distance $s = \infty$ so that the incoming rays are parallel to the optical axis, then the image distance $s' = R/2 = f$. This is the definition of the focal length, "f", for the spherical mirror.

6.2.7.2 Sign conventions for optical distances and focal lengths

For concave and convex mirrors the usual sign convention for focal length is: if the mirror is concave $f_{\text{concave}} > 0$, but if it is convex, $f_{\text{convex}} < 0$. The signs for the distances s, s', and R in Figure 6.8 are the same as they would be in a *standard Cartesian x, y coordinate system* [7, Schroeder]. Thus, an object distance "s", or image distance "s'", or radius of curvature distance "R" is negative if it *lies to the left* of the given mirror's vertex. Conversely any one of these distances is positive if it *lies to the right* of the associated mirror's vertex. An example of this sign convention is shown in Figure 6.4, a schematic of the light rays for our two-mirror Cassegrain telescope model in which the radius of curvature of the concave primary, R_1, is to the *left* of the primary mirror's vertex. Thus the primary focal point is plotted to the left of the vertex, i.e. as a negative distance. Since by our convention we take the concave primary's focal length to be positive, the distance x to the primary focal point is

$$x_{\text{primary focal point}} = R_1/2 = -f_1$$

We will now use the *Spherical Mirror Equation* and the Cartesian coordinate sign conventions to calculate the radius of curvature R_2 for the HST secondary mirror.

6.2.7.3 Calculating the HST secondary radius of curvature from the Spherical Mirror Equation

We can use the *Spherical Mirror Equation*, Equation 6.35, to determine the radius of curvature of the secondary mirror based on the object and image distances for the HST secondary in Figure 6.4. The secondary mirror *object* distance s_2 is the distance from the vertex of the secondary mirror to the primary mirror focal point. The equation for the object distance is from Equation 6.5: $s_2 = -kf_1$ where $k = y_2/y_1$. Equation 6.25 provides the relation $k = \dfrac{L}{mf_1}$, in which the constructional length, L, is the secondary *image* distance s'_2 (> 0). Hence the secondary *object* distance s_2 is

$$s_2 = -\frac{L}{m} \tag{6.36}$$

As stated above, the secondary image distance s'_2 is defined to be

$$s'_2 = L \tag{6.37}$$

Substituting Equations 6.36 and 6.37 into Equation 6.35 gives an equation for the secondary mirror radius of curvature, R_2

$$\frac{1}{(-L/m)} + \frac{1}{L} = \frac{2}{R_2} \tag{6.38}$$

Solving Equation 6.38 for R_2

$$R_2 = -\frac{2L}{(m-1)} \qquad (6.39)$$

As we have previously calculated values for L and m, we could numerically calculate R_2 with Equation 6.39, but we can develop a better understanding of the relative size of the secondary mirror's radius of curvature by writing down the equation for the ratio of the secondary to the primary radius of curvature, R_2/R_1. Dividing both sides of Equation 6.39 by R_1, and recalling that $R_1 = -2f_1$ and that $L = \frac{m}{(m+1)}(f_1 + b)$ from Equation 6.23, we obtain the ratio

$$\frac{R_2}{R_1} = \frac{(L/f_1)}{(m-1)} = \left[\frac{m}{m^2-1}\right]\left(1 + \frac{b}{F_1 D}\right) \qquad (6.40)$$

Note that this ratio (as with that for y_2/y_1 given by Equation 6.21) is inversely proportional to the secondary magnification m, i.e. $R_2/R_1 \sim 1/m$ when m is large. Thus if our design calls for a secondary magnification m that is twice as large as that chosen for the HST design, then the magnitude of R_2 will be cut in half. So let us evaluate R_2/R_1 using $m = 10.435$, $b = 1.5$ m, $D = 2.4$ m, and $F_1 = 2.3$:

$$\frac{R_2}{R_1} = \left[\frac{10.435}{107.889}\right]\left(1 + \frac{1.5}{5.52}\right) = 0.123 \qquad (6.41)$$

Note that the secondary radius of curvature is a small fraction (about 1/8) of the primary mirror radius of curvature. For $R_1 = -2F_1 D = -11.04$ m, Equation 6.41 gives the radius of curvature R_2

$$R_2 = -1.358 \text{ m} \qquad (6.42)$$

Correspondingly, the convex secondary mirror focal length is

$$f_2 = R_2/2 = -0.69 \text{ m} \qquad (6.43)$$

For an ideal *object* very far in front of an isolated secondary mirror, the minus sign tells us that the image will be "virtual" with a focal point located less than one meter *behind* the convex secondary mirror.

6.3 MODELING THE HST LENGTH

In Section 6.1 we posed three HST design questions. The first question, pertaining to the basic optical telescope dimensions was addressed in Section 6.2 where we used our BotE techniques to estimate the key HST optics dimensions. The third question, the equivalent optical system focal length needed for the HST to satisfy the diffraction-limited resolution requirement, was answered in Section 6.2.2 where we estimated that the diffraction-limited resolution requirement would be satisfied by a system focal ratio F_{eq} of 24, or a system focal length f_{eq} of 57.6 m.

We shall now address the second question: *What HST design tradeoffs must be made to keep the Hubble payload within the total cargo bay length and cargo mass constraints imposed by the Shuttle design?* We first estimate the HST total system length, based on the lengths of the three major components identified in Figure 6.1.

In order to obtain a BotE estimate of the total HST length, an engineer must add to the optics package length L (calculated in Equation 6.24 as $L = 6.406$ m) the estimated lengths of the shield or baffle that protects the Cassegrain optics from stray light, ($L_{\text{light-shield}}$), and the scientific package located behind the optical focal plane and housed within the aft shroud, ($L_{\text{instrument}}$).

Let's first describe the light shield and then use a simple BotE argument to estimate its length.

6.3.1 The light-shield baffle extension

Stray light from different astronomical sources affects the quality of Cassegrain-focused images because it lowers the image contrast of some weak celestial source of interest. One of the primary means for reducing the impact of stray light is to add a "tube" or "baffle" in front of the optics. By incorporating various cylindrical or conical baffles in the optical system designers can suppress any stray light that impinges on a telescope's focal plane [6].

The HST light shield displayed in Figure 6.1 is basically a tube, 3 meters in diameter, designed to keep almost all of the light from the bright earth, moon, or sun from the primary mirror. Of course, the light shield must still allow light from the region of sky of astronomical interest to pass unimpeded through the HST optics to the focal plane. One of the primary functions of the light shield is to protect the instruments from direct or indirect solar illumination during the daytime portion of the HST's orbit around the earth. Preceding the secondary mirror, it is painted black on its interior walls and uses a number of internal concentric rings or vanes to absorb as much stray light as possible. Its exterior is coated with a solar-reflecting material [6].

There are two other important baffles installed on the HST telescope. One is a truncated cone-shaped baffle which extends outward from the upper edges of the secondary mirror, and it is designed to keep stray light away from that mirror. The other baffle is an almost cylindrical tube that extends forward from the hole in the primary mirror, through which the secondary beam passes on its way to the focal plane. This "central" baffle is present to keep stray light away from the primary mirror and the focal plane. We will not address here the design of these interior baffles, we will concentrate our analysis on determining the length of the exterior light shield.

Our BotE approach focuses on finding the tube length that keeps any off-axis stray light rays, above a certain incidence angle, from reaching the central optical axis.

6.3.2 Modeling the length of the light shield

Bely [13] states that any ray closer to the optical axis than 15° will illuminate the primary mirror directly, boosting stray light enormously (by scattering off any dust on the primary mirror). Our goal is to find the light shield length $L_{\text{light shield}}$ that prevents a 15° off-axis light ray from directly striking an outer edge of the primary mirror.

An off-axis light ray, inclined 15° to the optical axis, is shown in Figure 6.9. This hypothetical ray travels through an annular region radially beyond the outer edge of the secondary mirror and strikes the lower edge of the primary mirror. The vertical distance traveled by this stray light ray, measured from the point at which it passes the upper edge of the aperture door, is approximately equal to the primary diameter D. Any external ray encountering the light shield at an angle of incidence greater than 15° will be fully blocked by a light shield whose length, $L_{\text{light shield}}$, is long enough to block a 15° ray. For direct or indirect solar rays less than 15°, the aperture door at the entrance end of the light shield is closed, thus not only preventing stray light from entering but (of course) eliminating any HST observations as well. Normally the celestial objects to be viewed are selected to lie well outside this 15° incidence range based upon the principal solar or lunar angles encountered during a collection orbit. However, the earth (and often the moon) does prevent astronomical viewing (called occultation) about 50% of the time.

The magnitude of $L_{\text{light shield}}$ is determined using the following crude geometric approximation for the 15° incidence problem as sketched in Figure 6.9

$$\tan \theta \approx \frac{D}{L_{\text{light shield}} + d} \tag{6.44}$$

Substituting the key dimensions $D = 2.4$ m and $d =$ distance from the secondary to the primary mirrors $= 4.906$ m, we calculate the light shield length to be

$$L_{\text{light shield}} = \frac{D}{\tan 15°} - d = \frac{2.4 \text{ m}}{0.268} - 4.906 = 4.05 \text{ m} \tag{6.45}$$

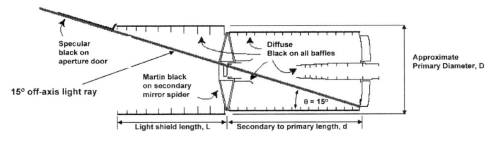

Figure 6.9. Fifteen degree off-axis stray light blocked by the forward light shield of length L. The secondary is not to scale, it is actually about 12% of the primary diameter D.

6.3.3 The length of the instrument section

The HST schematic of Figure 6.1 shows a large cylindrical shell at the rear of the HST. This is called the aft shroud. As stated in [14], "The aft shroud houses the box-shaped Focal Plane Structure (or FPS) that contains the axial science *instruments*. It is also the location of the Corrective Optics Space Telescope Axial Replacement unit (COSTAR)". COSTAR was added to the HST during the first servicing mission in order to correct the astigmatism created by an incorrectly ground primary mirror. When installed, COSTAR together with the primary and secondary mirrors, produced diffraction-limited resolutions for the HST images, as called for in the original design specifications. The COSTAR functionality was subsequently replaced with other optical devices during a later HST repair mission. NASA [14, Section 5] states that, "The aft shroud is made of aluminum, with a stiffened skin, internal panels, and reinforcing rings. It is 3.5m long and 4.3 m in diameter."

From the HST drawing, Figure 6.1, as well as a host of other schematics, it appears that the front edge of the full instrument section is located about 0.5 m ahead of the focal plane. Hence we estimate that the length of this "additional instrumental section" aft of the focal plane is approximately 3.5–0.5 m, or 3.0 m.

For our total HST length calculation this added instrument section length will therefore be set to: $L_{\text{instrument}} \approx 3.0 \, \text{m}$.

6.3.4 Calculating the total HST telescope length

To calculate the estimated HST total telescope length, we sum the lengths from the light shield, the distance from the primary to the secondary mirrors, the back-length, and the length of the aft instrument module, namely

$$\text{Total telescope length} = L_{\text{light shield}} + d_{\text{primary-secondary}} + b_{\text{back length}} + L_{\text{instrument}}$$
$$\approx 4.0 \, \text{m} + 4.9 \, \text{m} + 1.5 \, \text{m} + 3.0 \, \text{m} = 13.4 \, \text{m} \quad (6.46)$$

A typical value given in the literature for the total HST length [12] is

$$L_{\text{HST total system}} \approx 13.2 \, \text{m}$$

Although we didn't independently model the instrument section length of 3.0 m, we conclude that our BotE estimate for total HST length of 13.4 m is quite close to the reported total length of 13.2 m.

Based on our BotE estimate, it looks like the HST will indeed fit lengthwise into the 18.3 m Orbiter cargo bay with a little over 2 m to spare at each end! Allowing for the other equipment in the bay, the HST package should actually have a length no greater than about 16 m.

If the HST primary mirror diameter had been set at $D = 3.0 \, \text{m}$, as called for in some early designs, the total length (which scales with D) would be increased by 25% over the current 2.4 m design ($3.0/2.4 = 1.25$) and the total length would be scaled to

$$L_{\text{HST total system}} \approx 1.25(13.4 \, \text{m}) \approx 16.8 \, \text{m}$$

which would be too long to fit into the cargo bay along with the necessary supporting apparatus. This was one of the important reasons why the smaller 2.4 m mirror was chosen for the combined HST-Shuttle launch mission.

6.4 SUMMARY OF CALCULATED HST DIMENSIONS

Table 6.1 summarizes the calculations performed in the previous sections. All of the listed dimensions are within 1–2% of published values for HST [12] with the exception of the secondary mirror diameter D_2 estimated in Equation 6.27.

A more correct estimate for D_2 takes into account the finite "field of view" of the focal plane. (As related in Section 6.2.3, the radial field of view is about 820 arc-seconds or 4×10^{-3} radians in its angular radius). Schroeder, [7, Section 2.5.b] states that for the secondary mirror to cover a field on the sky of angular diameter 2θ without obstructing any light from the primary, its diameter must be larger by

$$\Delta D_2 = 2\theta \cdot \text{(primary to secondary length)} = 2\theta d \qquad (6.47)$$

Using our calculated value of "d" from Table 6.1, $d = 4.91$ m, and the required angular radius $\theta \approx 4 \times 10^{-3}$ radians, we estimate from Equation 6.47 a correction to the secondary mirror diameter

$$\Delta D_2 = 0.04 \text{ m} \qquad (6.48)$$

Table 6.1. Calculated dimensions/scales for HST telescope.

Given quantity	Given dimension	Derived quantity	Calculated dimension
Primary mirror aperture	2.4 m	Equivalent focal length, f_{eq}	57.6 m
Primary mirror focal ratio, F_1	$f/2.3$	Secondary mirror magnification, m	10.43
Equivalent focal ratio, F_{eq}	$f/24$	Secondary mirror to focal plane spacing, $s'_2 = L$	6.406 m
Back focal length, b	1.5 m	Primary to secondary spacing, d	4.906 m
Primary mirror focal ratio, $f_1 = F_1 D$	5.52 m	Secondary mirror diameter uncorrected for field size	0.267 m
Primary mirror radius of curvature, $R_1 = 2f_1$	11.04 m	Secondary mirror radius of curvature, R_2	1.358 m
		Light-shield length	4.05 m
		Total telescope length $d + f_1 \beta +$ light-shield + instrument package	13.2 m

Thus the total *corrected* secondary mirror diameter becomes

$$D_{2\text{final}} = 0.267 \text{ m} + 0.04 \text{ m} = 0.307 \text{ m} \tag{6.49}$$

This is very close to the actual HST secondary mirror diameter as quoted in [12].

6.5 ESTIMATING HST MASS

When we first examined the requirements for "fitting" the HST to the Orbiter (in Section 6.1.3) we made the obvious point that the mass of the entire HST system must not exceed the maximum cargo carrying capability of the Shuttle, which is dependent on the mission's orbital altitude. Our BotE estimate in Chapter 3 found that the maximum cargo mass is approximately 15,000 kg for the HST's operating orbit at 600 km.

In fact, NASA set the specific mission requirement for HST designers about 20% lower at 11,540 kg (25,500 lbm), a tight limit on the total HST engineering system mass. NASA probably built into this requirement a conservative "safety factor" which took into account the mass of the additional cargo bay equipment required to support the HST during launch and deploy it in space. The system engineers would have specified an overall mass budget (or in their parlance, a "weight statement") to keep track of the weights of all the HST elements from the initial design to the final component delivery. If a given component exceeded its budgeted mass allocation, then either its design would have to be revised or its manufacturing would be altered to keep its mass within the allocated limit.

Let's go back in time and assume that you are an engineer working the initial HST design phase and are given the job of quickly estimating the masses of the primary mirror, the full Optical Telescope Assembly (OTA), and the total system mass (including the Support Systems Module, the solar arrays, and the optical instruments). We will show how the masses of these three elements, can be crudely estimated using BotE "rules of thumb". These scaling rules are based on empirical information culled from analogous systems, including telescopes built in the world of amateur astronomy.

Our first task is to estimate the mass of the primary mirror. In his book on astronomical telescope design, J. Cheng [15] states, "The primary mirror is the most important component of an optical telescope ... The weight and cost of the primary mirror are determining facts for the telescope total weight and total cost."

6.5.1 Primary mirror design

Reflecting telescopes for astronomical observatories have traditionally been built around simple one-piece glass primary mirrors. Owing to the great weight of "all-glass" mirrors, modern mirrors are typically "light-weighted" by a *factor of five or more* whether for earth or space-based applications [16].

All of the early space-based telescope mirrors were "lightened" using a web-stiffened design and such a method was used for the primary mirror of the HST,

Sec. 6.5] **Estimating HST mass** 301

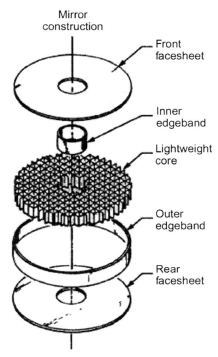

Figure 6.10. HST primary mirror construction with honeycomb lightweight core [14, pp. 5–22].

which had a *sandwich* type of construction design in which the central part (the meat) was cast with honeycomb web stiffeners and thin face-sheets (the bread) were affixed at either end. The primary mirror was made from honeycomb-shaped light-weighted Corning (ultra-low expansion) silica–titania glass built with a rectangular grid and then coated with a very thin layer of aluminum to reflect visible light after grinding and polishing (Figure 6.10).

6.5.2 Estimating primary mirror mass

Based on the comments in the preceding section, an approximate rule of thumb for estimating the mass of a honeycomb mirror is that it weighs only one-fifth as much as would a solid mirror of the same size [16]. We will first estimate the mass of a *solid* glass mirror made from Corning glass and then take 20% of that as our estimate of the mass of a lightened-honeycomb equivalent.

6.5.2.1 *Mass of a hypothetically solid glass mirror*

The density of Corning ultra-low expansion silica glass is approximately 2,200 kg/m^3 [16]. Let's model the HST 2.4 m mirror as a cylinder with

$$\text{Base area} = (\pi/4)(2.4)^2 = 4.52 \text{ m}^2 \qquad (6.50)$$

The height, or thickness, of the mirror is given by a rule of thumb adopted by amateur telescope makers that to produce a good "stiffness" the thickness of a mirror should be about 1/6 of the mirror's diameter; so let's use 1/6 as our rule of thumb. Hence

$$\text{Mirror thickness} \approx 1/6 D = 2.4\,\text{m}/6 = 0.4\,\text{m (about 1.3 feet thick)} \quad (6.51)$$

Then the volume of the solid cylindrical mirror = (Base area) × (thickness) is

$$\text{Volume} = 4.52 \times 0.4 = 1.80\,\text{m}^3 \quad (6.52)$$

Multiplying this by the density of the solid glass given above, we obtain the estimated mass of the solid form of the mirror

$$m_{\text{solid}} \approx 2{,}200 \times 1.80 = 3{,}960\,\text{kg (8,730 lbm)} \quad (6.53)$$

6.5.2.2 Estimated mass of honeycomb-structured mirror

Using our 20% rule for the lightening due to a honeycomb construction, we obtain

$$m_{\text{honeycomb}} \approx 3{,}960\,\text{kg}/5 = 792\,\text{kg} = 1{,}750\,\text{lb} \quad (6.54)$$

NASA documents list the HST primary mass as 828 kg or 1,825 lb, so our estimate is within 5% of the "true" primary mass.

Another way of expressing mirror mass that is scalable to mirrors of other diameters, is to calculate a metric for the mass per unit mirror cross-sectional area; this is called the *areal* density. Our estimate for the *areal* density of the primary mirror is

$$m_{\text{areal-primary}} = \frac{\text{mirror mass}}{\text{mirror Area}} = \frac{792\,\text{kg}}{4.52\,\text{m}^2} = 175\,\text{kg/m}^2 \quad (6.55)$$

6.5.2.3 The estimated mass of the entire Optical Telescope Assembly, m_{OTA}

Figure 6.1 shows that there are a significant number of components comprising the optical portion of the HST in addition to the primary mirror. We list the major OTA components:

- Secondary mirror assembly and baffle
- Central baffle
- Focal plane structure array
- OTA graphite epoxy metering truss
- OTA equipment section (e.g. the thermal system for regulating the mirror temperatures)
- Fine guidance sensors
- Main ring (integrates the telescope to the spacecraft)

It would be quite time-consuming and almost impossible to analyze all these components in order to arrive at a combined OTA mass. However we can obtain a crude estimate of the OTA mass by examining the ratio of OTA mass to the mass of the primary for significantly smaller Cassegrain telescope systems. As our reference, we calculate this ratio based on the weights given for a number of amateur telescope designs that are posted on the internet. Here are a few of the calculated values we obtained for (OTA mass)/(Primary mass):

1. A 1 m Newtonian (or Dobson) amateur telescope, primary focal ratio $F_1 = 3.0$; http://www.astrosurf.com/altaz/1000_e.htm, a website for those who wish to build a Dobsonian telescope.

   Primary mirror mass = 221 lbm
   Total Optical tube mass = 1,100 lbm
   OTA/Primary mass ratio = 1,100/221 ≈ 5.0

2. Ultra-light weight deployable 1-m-class optical telescope; www.amostech.com, a small government contractor

   Primary mirror mass = 12 kg (26.5 lbm)
   Complete OTA without instrument = 80 kg (176 lbm)
   OTA/Primary mass ratio = 176/26.5 ≈ 6.6

3. *Cruxis Telescope*. Large, but light-weight, amateur Newtonian telescope. A 1.1 m primary diameter with supporting truss, but no external tube. Built by Robert Houdart. http://www.cruxis.com/scope/scope1100.htm

   Primary mirror mass = 120 kg (260 lbm)
   Total weight of telescope = 270 kg (600 lbm)
   OTA/Primary mass ratio = 600/260 ≈ 2.3

For these three telescopes, the average OTA/Primary mass ratios is 4.6. We will use this ratio to estimate the mass of the Optical Telescope Assembly of the HST.

We estimated that the primary mirror mass ≈ 790 kg (Equation 6.54). Using the average empirical OTA/Primary mass ratio = 4.6, we estimate the OTA mass to be

$$\text{OTA mass} = 4.6(790 \text{ kg}) = 3{,}630 \text{ kg} = 8{,}000 \text{ lbm} \qquad (6.56)$$

The actual HST Optical Telescope Assembly mass is 4,125 kg or 9,120 lb according to reports on NASA documents [17], so our estimate is low by about 14%. Still our estimate is fairly respectable given the crudeness of our extrapolation method and the limited amateur telescope data from which we extrapolated.

We also estimate the *areal* density of the OTA to be

$$m_{\text{areal-primary}} = \frac{\text{OTA mass}}{\text{mirror Area}} = \frac{3{,}630 \text{ kg}}{4.52 \text{ m}^2} = 803 \text{ kg/m}^2 \qquad (6.57)$$

6.5.3 The estimated total HST system mass and areal density

Finally, we need to make a best guess at the ratio

$$\frac{\text{Total HST spacecraft mass}}{\text{OTA mass}}$$

in order to be able to estimate the overall HST spacecraft mass from our OTA mass estimate.

304 Estimating the dimensions and performance of the Hubble Space Telescope [Ch. 6]

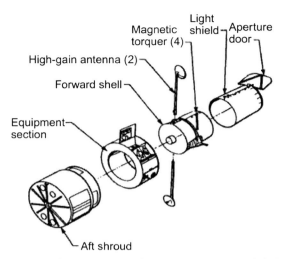

Figure 6.11. Structural components of support systems module [14, pp. 5-3].

In addition to the OTA, the HST has three other complex systems [14, p. 5-3]:

- The Support Systems Module (SSM), comprising the overall satellite structure, including the light shield tube (Figure 6.11), electrical power system, and pointing control system.
- The "solar cell array" (Figure 6.1), including the array deployment drive mechanisms.
- The "scientific instruments", comprising the cameras and spectrographs enclosed within the aft shroud shown (Figure 6.1).

Let's make some "very crude" guesses for the masses of each of the three systems listed above, as a fraction of the OTA mass. We include a measure of the range of our uncertainty.

Since the volumetric scale of the SSM components seems to be as large as that of the mirrors and the OTA structural support truss (Figure 6.11), our guess is that the SSM mass is comparable to or somewhat larger than that of the OTA, hence

$$\frac{\text{SSM mass}}{\text{OTA mass}} \approx 1.0 \text{ to } 2.0 \tag{6.58}$$

The solar cell panels are quite light but there is additional mass associated with the drive mechanism, so we guess

$$\frac{\text{Solar cell array mass}}{\text{OTA mass}} \approx 0.1 \text{ to } 0.2 \tag{6.59}$$

The instruments are complex optical systems whose entrance apertures are smaller than the main Cassegrain mirror optical system (the OTA), so we guess

$$\frac{\text{Instrument mass}}{\text{OTA mass}} \approx 0.2 \text{ to } 0.4 \tag{6.60}$$

Total HST mass is the sum of the masses of the OTA and these three other components

Total HST mass = OTA mass + SSM mass + Solar array mass + Instrument mass
(6.61)

Using Equations 6.58 to 6.61, the ratio of total HST mass to OTA mass is

$$\frac{\text{Total HST mass}}{\text{OTA mass}} = 2.3 \text{ to } 3.6 = 2.95 \pm 0.65 \quad (6.62)$$

Using our previous estimate of OTA mass = 3,630 kg, our overall total HST mass estimate

$$(\text{OTA mass}) \cdot \left(\frac{\text{Total HST mass}}{\text{OTA mass}}\right)$$

lies in the following range

$$\boxed{\text{Total HST mass} \approx 8,300 \text{ to } 13,100 \text{ kg} = 10,700 \pm 2,400 \text{ kg} \quad (6.63)}$$

where we have rounded to the nearest 100 kg.

According to NASA the "true" total HST mass is approximately 11,000 kg or about 24,000 lbm [17, HST Summary Weight Statement]. Thus our estimate might be low by as much as 24% or as high by 19%. Our mean or average mass estimate is, interestingly, 10,700 kg which is rather close to NASA's 11,000 kg.

The *areal* density for the HST (the mass to primary mirror area ratio) ranges from

$$\boxed{m_{\text{areal-HST total}} = \frac{\text{HST total mass}}{\text{Primary mirror Area}} = 1,800 \text{ to } 2,900 \text{ kg/m}^2 = 2,350 \pm 550 \text{ kg}}$$
(6.64)

A typical number cited in the literature for the overall areal density of the HST of $\approx 2,500$ kg/m^2 [18].

Engineers involved in designing the next-generation space telescope, called the James Webb Space Telescope (JWST), often quote areal densities for their designs that are more than one order of magnitude smaller than the 2,500 kg/m^2 quoted for the HST [18]. This is possible due to major improvements in the development of super-lightweight composite primary mirrors for both land-based and space-based telescopes that use active wavefront sensing and control. The JWST mirrors are not constructed from a single piece of glass, but are built up from a large number of semi-rigid hexagonal segments, each of which can be adjusted in real time.

6.5.4 Some final words on the HST mass estimation exercise

The goal of our HST mass estimation exercise was to show that its mass is less than the maximum Shuttle cargo mass that can be carried to a circular orbital at an altitude of 600 km.

Our estimate, calculated in Section 6.5.3, is $10,700 \pm 2,400$ kg. The estimated mean mass, 10,700 kg, satisfies the specified constraints for both total cargo of 15,000 kg (calculated in Chapter 3) and NASA's specific design goal for the HST of 11,500 kg. But the largest magnitude of our estimate (at the upper end of its range of uncertainty), 13,100 kg, would clearly exceed the design requirement.

This result shows that while BotE estimation methods might get us close to the correct mass in some average sense, a more detailed analysis is needed to carry forward an early design effort. (This additional analysis is called for as a result of modeling uncertainty, particularly at the upper end of the estimation range.) Nevertheless, we have demonstrated that our crude analysis of mass places us in the right "ball park" compared to the NASA cargo requirements in the wake of their decision to use a 2.4 m primary mirror aperture. For a 3.0 m primary, our estimated mass would have exceeded the cargo design limit by close to 50%. So we see that this kind of BotE analysis provides guidance to engineers in the early stages of design.

The reader should recognize that our last step in the estimation process, in which we scaled from OTA mass to HST mass, was performed with extremely crude information. A random group of engineers, addressing this problem for the first time, might well produce a considerably wider range of estimated masses. When developing our scaling procedures and final estimates, we were guided a good deal by NASA's detailed drawings and many reports on the design of the HST that are widely available in texts and on the internet.

It is clear that if we were to start a mass estimation problem like this without such background information, then the uncertainties in our predictions would be much greater.

6.5.5 Onward to an estimate of HST's sensitivity

We now turn our attention to the important problem of assessing the detection *sensitivity* of the HST, i.e. its ability to detect very faint stars or related astronomical targets.

A standard figure of merit for sensitivity is the estimated (or measured) focal plane signal to noise ratio (SNR) for a point source of light.

In the following section we develop a model for HST focal plane SNR as a function of apparent stellar magnitude, m, for a given observation time period.

We also invert this calculation in order to estimate the telescope exposure time required to achieve a given signal to noise ratio when imaging a star of a given magnitude.

6.6 BACK-OF-THE-ENVELOPE MODELING OF THE HST'S SENSITIVITY OR SIGNAL TO NOISE RATIO

The sensitivity of a telescope is the smallest signal that it can clearly measure from a source in space against the background. It is the minimum brightness that a telescope can detect. A telescope with high sensitivity can detect very dim objects. Simply stated, the more light a telescope can gather from faint objects (by using, for example, a very large primary mirror), the more sensitive it is [19].

To produce a sensitive image of a weak celestial target in a given exposure time, "t", the HST aperture must capture a sufficient number of signal photons, then convert them by focal plane CCD detectors into electronic signals, so that the star being photographed can readily be detected against competing random detector and signal-scaled Poisson electronic noise levels.

6.6.1 Defining signal to noise ratio

The principal measure of detectability is the signal to noise level, SNR, given by the following ratio

$$\text{SNR} = \text{mean signal/root mean square noise} \qquad (6.65)$$

which is also sometimes written as S/N.

This quantitatively measures how well an object captured by an image can be detected and characterized in the presence of noise. Here are some *practical* qualitative interpretations of the meaning of various SNR magnitudes [20]:

- SNR = 2 to 3: Object barely detectable
- SNR = 5: Object detected. One can start to believe what one sees, because the probability of a false alarm is quite low
- SNR = 10: One can start to do quantitative measurements. Also, at least 10 levels of intensity can be measured for an extended object [6, Section 1.5]
- SNR = 100: Excellent measurements.

In the analysis that follows, where we calculate the exposure (or integration) time required to achieve a given signal to noise ratio, the nominal value chosen for SNR is 10.

6.6.2 Modeling the mean signal, $\bar{S}$

The total number of photons, $\bar{S}$, from a star of interest that reaches the detectors is equal to the product of the mean *photon flux*, S (measured in units of photons per second) from a star, and the selected total exposure time, t, is

$$\bar{S}_{\text{total photons}} = St \qquad (6.66)$$

Following Schroeder [2, p. 435], the flux S collected from a point-like star of magnitude m, by a primary mirror of aperture diameter D, with small optical blockage fraction ε due to the secondary mirror, and transmitted to the detector through an optical system with transmittance τ and instrument bandwidth $\Delta\lambda$ is given by

$$S = (N_0 10^{-0.4\,m})(\pi D^2/4)(1 - \varepsilon^2)(\tau)(\Delta\lambda) \qquad (6.67)$$

For the star Vega, which is magnitude zero and of spectral type A0, $N_0 \approx 10^8$ photons/(sec-m^2-nm) centered at a wavelength of 550 nanometers in the visible.

The magnitude "m" can be thought of as a relative logarithmic scale for stellar photon flux. In Equation 6.67, the term $10^{-0.4m}$ is a factor that scales the reference spectral flux magnitude N_0 to the appropriate level for a star of magnitude m. If a star is of magnitude $m = 5$, its brightness is 1% of that for the reference star with magnitude $m = 0$. Similarly, if $m = 10$, the brightness is 0.01% of that reference star. Some of the brightest stars in the sky have negative magnitudes. Since the magnitude of the full moon is $m = -12.7$, this scaling tells us that the full moon is 120,000 times *brighter* than Vega.

In Equation 6.67, the term $\pi D^2/4$ is the area of the entrance aperture, A, of the telescope. The amount of light, or number of photons, collected by the telescope is clearly proportional to A.

The overall transmittance factor τ equals the product of three components

$$\tau = [\tau(\text{telescope})][\tau(\text{filter})][\tau(\text{relay optics})] \tag{6.68}$$

Nominal values given by Schroeder [7, p. 438, Table 17.3] are

$$\tau(\text{telescope}) = 0.81$$

$$\tau(\text{filter}) = 0.8 \text{ for a filter with bandwidth } \Delta\lambda = 100 \text{ nm}$$

$$\tau(\text{relay optics}) = 0.5$$

so the total transmittance for use in Equation 6.67 is $\tau = 0.324$.

Now let us model the number of electrons measured by the CCD detectors. When a certain number of photons hit a CCD detector, they are converted via the photoelectric effect into a comparable number of electrons. We won't get into the detailed physics here. Because the process is not 100% efficient, the number of detectable electrons is lower than the incident number of photons by a factor Q, called the quantum efficiency. Q depends on the design and materials used in fabricating the detector. Q also depends on the frequency of the incident light.

The average total number of signal-generated electrons, $\bar{S}_{\text{total electrons}}$, that are output after an exposure time, t, by a small group of focal plane detectors that capture the star's photons is

$$\bar{S}_{\text{total electrons}} = S\kappa Q t \tag{6.69}$$

where κ is the fraction of transmitted light that reaches the detector. Nominally, κ is 0.8 for a telescope such as the HST for visible wavelengths close to 550 nm [2, p. 438]. Q is the nominal quantum efficiency within the filter bandwidth, i.e. the fraction of input photons that are converted into detectable electrons by the CCD elements. Schroeder's value for Q is 80% or 0.8 [2, Tables 17.1 and 17.3].

Combining Equations 6.67 and 6.69, we obtain the expression for the output mean signal $\bar{S}$ from the focal plane detectors in equivalent number of photons

$$\bar{S}_{\text{total}} = S\kappa Q t = (N_0 10^{-0.4m})(\pi D^2/4)(1-\varepsilon^2)(\tau)(\Delta\lambda)(\kappa Q t) \tag{6.70}$$

6.6.2.1 Quantifying the number of photons collected by HST optics

Let's get a feel for the size of the collected signal $\bar{S}_{total}$ using Equation 6.70. We select the optical input constants corresponding to the HST's Advanced Camera for Surveys (ACS) in the Wide Field Camera (WFC) mode [7, Table 17.3].

We calculate the signal $\bar{S}_{total}$ based on a 40 minute collection of the light from a star of magnitude $m = 25$. $\bar{S}_{total}$ is measured in units of signal-generated CCD electrons, sometimes referred to as equivalent photons.

Input parameters

Star magnitude, m:	25
HST primary mirror diameter, D:	2.4 m
Instrument filter bandpass, $\Delta\lambda$:	100 nanometers
Secondary mirror optical blockage, fraction of primary diameter, ε:	0.33
System transmittance; telescope, filter, relay optics to detector, Equation 6.68, τ:	0.324
Fraction of 550 nm transmitted light that reaches the detector, κ:	0.8
Quantum efficiency within the filter bandwidth, Q:	0.8
Exposure time (40 minutes), t:	2,400 seconds

S, the magnitude of the photon flux (photons/sec) hitting the detector is from Equation 6.67

$$\left. \begin{array}{l} S = (N_0 10^{-0.4\,m})(\pi D^2/4)(1 - \varepsilon^2)(\tau)(\Delta\lambda) \\ S = (10^8 \cdot 10^{-10})(\pi(2.4)^2/4)(1 - 0.33^2)(0.324)(100) = 1.31 \text{ photons/s} \end{array} \right\} \quad (6.71)$$

This is quite a low photon rate, but it is consistent with the weak light output from a star of magnitude $m = 25$. Note that a larger 4.8 m primary mirror (double that of the HST) would quadruple this flux to 5.24 photons/s.

In 2,400 seconds, the total number of photons that pass the aperture of the HST and enter into the telescope's optical system is

$$\bar{S}_{total\ photons} = St = 3{,}144 \text{ photons} \quad (6.72)$$

6.6.2.2 Quantifying the number of electrons output by the HST detectors

If we now take into account both the fraction of transmitted light that reaches the detector through the telescope optical system, κ, and the probability that photons are converted into detectable electrons by the CCD elements, Q, then we obtain the average total number of signal-generated electrons, $\bar{S}_{\text{total electrons}}$, output by the focal plane detectors after an exposure time of 40 minutes

$$\bar{S}_{\text{total electrons}, m=25, t=40 \text{ min}} = \kappa QSt = (0.8)(0.8)(3{,}144 \text{ photons}) = 2{,}012 \text{ electrons} \quad (6.73)$$

Although less than the number of entrance aperture photons (Equation 6.72), the number of electrons (or "equivalent" photons) is easily detected and counted by any of the standard 15 micron CCD pixels. There will, of course, be detector and background noise to contend with when assessing how well a celestial object can be detected and characterized in the presence of noise.

Now that we have a "feel" for the magnitude of the average signal, $\bar{S}$, we can model the noise in the system. Estimating the noise (measured either in terms of electrons or equivalent photons) as well as the mean signal will enable us to calculate the signal to noise ratio, SNR.

6.6.3 Modeling the noise

The principal sources of electronic noise are: (1) photo-electron signal fluctuations, (2) photo-electron celestial background fluctuations, (3) extraneous detector dark current in the absence of light, and (4) read-out-noise associated with solid-state detectors. Since these noise sources are uncorrelated, the total mean square noise (or variance) is the sum of the variances of each component.

The total noise measure, N_t, in the denominator of the signal to noise ratio expression is simply the square root of the total variance

$$N_t = \sqrt{\text{var(signal noise)} + \text{var(sky background)} + \text{var(dark counts)} + \text{var(read noise)}} \quad (6.74)$$

Another way of writing the variance is in terms of the square of the standard deviation, i.e. variance $= N_t^2 = \sigma_t^2$.

We shall now develop detailed models for each of these terms, starting with variance (signal noise), or σ_{signal}^2.

6.6.3.1 Photo-electron signal fluctuations; the Poisson noise model

Since light is *quantum* in nature, there is a natural variation in how many photons will arrive in a specific time interval, t, even when the *average* flux S (photons/s) is fixed [20].

The Poisson probability distribution, $P(N)$, is an equation for estimating the *discrete probability* that N photons hit the detector pixel of a telescope in a fixed period of time t. For the Poisson distribution, the arrival of the photons is random and independent of time; and most importantly the photons impact the detectors with a known average rate, S (photons/s). The Poisson probability distribution for N photons to be counted in an observation time t, with average rate S, is

$$P(N) = \frac{(S \cdot t)^N e^{-St}}{N!} \quad (6.75)$$

This distribution is valid for any similar constant rate process such as when one wants to determine the probability that a given number of raindrops will hit the ground in a given area based on a known rainfall density, e.g. raindrops per unit area. A random process that has these characteristics is called a Poisson process.

Consider the numerical example given in Section 6.6.2.1 of photons collected by the HST entrance with light from a star of magnitude $m = 25$ passing through a 100 nm filter and hitting the CCD detector at a rate of approximately $S = 1.0$ photons/s. If the exposure time t is very low, say $t = 5$ seconds, then it is evident from the given rate that the average number of photons that impact the detector, $\bar{N} = S \cdot t = 5$ photons. Using Equation 6.75 we can calculate the probability that we will measure a smaller number, say 3 photons, i.e. $P(3)$, as

$$P(3) = \frac{(5)^3 e^{-5}}{3!} = 0.140 = 14\% \tag{6.76}$$

The complete Poisson distribution for N from 0 to 15 is shown in Figure 6.12 for the mean $\bar{N} = 5$. This figure also indicates the statistical spread about the mean, plus or minus the standard deviation σ. In this case we calculate the standard deviation to be

$$\sigma = \sqrt{\sum_{N=0}^{\infty}(N - \bar{N})^2 P(N)} \approx \sqrt{5.0} = 2.24$$

One can readily derive from Equation 6.75 the general mathematical result that the variance of a Poisson distribution is equal to its specified mean [21], i.e.

$$\text{Variance} = \sigma_{\text{signal}}^2 = \text{mean} = S \cdot t \tag{6.77}$$

Thus the noise, σ_{signal}, for our photon-count Poisson process is

$$\sigma_{\text{signal}} = \sqrt{\text{mean}} = \sqrt{S \cdot t} \tag{6.78}$$

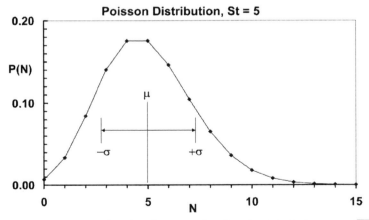

Figure 6.12. Poisson distribution for $\bar{N} = St = 5.0$. Standard deviation $\sigma = \sqrt{5.0} = 2.24$.

If the only source of noise was signal-generated Poisson noise, the signal to noise ratio, SNR, for this signal dominant case would be

$$\text{SNR} = \frac{\bar{S}}{\sigma_{\text{signal}}} = \frac{S \cdot t}{\sqrt{S \cdot t}} = \sqrt{S \cdot t} \quad (6.79)$$

This result, that SNR is proportional to the square root of "t", has important implications for the HST astronomical problem, because to increase SNR by a factor of 2 requires a factor of 4 increase in the observation or integration time.

The inverse of this calculation is to determine "t" for a given value of SNR

$$t = \frac{\text{SNR}^2}{S} \propto \frac{\text{SNR}^2}{D^2}(10^{0.4\text{m}}) \quad (6.80)$$

Equation 6.80 indicates that the required observation time "t" to obtain a given SNR increases as the square of the required SNR and decreases inversely with the square of the aperture diameter D. Thus if we double a telescope's primary mirror diameter, D, holding all other parameters the same, the observation time drops by a factor of 4. This allows for a greater number of observations during a given observation schedule. But the considerably higher telescope construction cost in dollars associated with a doubling of the primary mirror diameter is also a very important factor to consider when evaluating the design of the entire optical system [6]. We further note from Equation 6.80 that t increases exponentially with the star magnitude m. If the magnitude of the celestial source under study was 2 units "higher" than some reference star (e.g. $m = 25$ compared to $m = 23$) then the required integration time would be longer by a factor of $10^{0.8} = 6.3$.

To complete our analysis of the signal dominant case we need to take into consideration two telescope and detector parameters: the fraction of transmitted light that reaches the detector through the telescope optical system, κ, and the fraction of photons converted into detectable electrons by the CCD elements, Q. With these additional factors, the total number of electrons measured by the CCD detectors in an observation time t is given by

$$\bar{S}_{\text{total electrons}} = \kappa Q S t \quad (6.81)$$

From Equation 6.77 the Poisson variance is equal to the mean, so for the signal-dominant case

$$\text{Variance} = \sigma^2_{\text{signal}} = \text{mean} = \kappa Q S t \quad (6.82)$$

The signal to noise ratio for this signal-limited case becomes

$$\text{SNR}_{\text{signal-limited case}} = \sqrt{\kappa Q S t} = \sqrt{(N_0 10^{-0.4\text{m}})(\pi D^2/4)(1-\varepsilon^2)(\tau)(\Delta\lambda)(\kappa Q t)} \quad (6.83)$$

By solving Equation 6.83 for t, the corresponding observation time for a given SNR in the signal-limited case is

$$t_{\text{signal-limited}} = \frac{\text{SNR}^2}{\kappa Q S} = \frac{\text{SNR}^2}{(N_0 10^{-0.4\text{m}})(\pi D^2/4)(1-\varepsilon^2)(\tau)(\Delta\lambda)(\kappa Q)} \quad (6.84)$$

Next, we shall present the basic models for the other prominent sources of noise, starting with the model for the sky background noise.

6.6.3.2 Photo-electron sky background noise

The sky background (composed of the light from low intensity stars or celestial features near the star of interest) can contribute significant Poisson noise to the SNR calculations, particularly when the HST is imaging weak star sources with high star magnitudes, m.

Typical background sky flux rates quoted for HST's wide field camera fall in the range of 0.01 to 0.09 electrons/s per pixel [22]. The resolution of each pixel is about twice the nominal diffraction limit, or 0.1 arcsec. By way of comparison, our flux estimate for a single star with magnitude $m = 25$ is of order 1 electron/s for one pixel. However, for an $m = 28$ star the single star flux would be only about 0.06 electrons/s, which is comparable to the sky background in [22]. So for weak, more distant stars, the background sky Poisson noise is comparable to that of the Poisson noise from the source star. The background sky noise must therefore be taken into account in the SNR and exposure time calculations when the magnitude m of the star under observation is quite high.

6.6.3.2.1 Modeling the background sky flux and noise

The background flux, B, associated with sky light and other forms of stray light, is characterized by both its mean and its fluctuating intensity components. The sky mean signal is usually subtracted directly from a given focal plane image.

For Poisson noise, the background noise variance $\bar{B}$ is simply equal to the background mean number of electrons captured by n pixels after an exposure time, t. We shall adopt the background model presented by Schroeder [2, Section 17.3]

$$\text{Background variance} = \bar{B} = BQt = (N_0 10^{-0.4m'})(\phi\phi')(\pi D^2/4)(1 - \varepsilon^2)(\tau)(\Delta\lambda)Qt \quad (6.85)$$

which is similar to Equation 6.70.

In Equation 6.85, the units of N_0 in the sky flux B term are modified from the definition given in Section 6.6.2 for the signal flux. Here the units are $N_0 = 10^8$ photons/(sec-m^2-nm) per arcsec square. The sky brightness magnitude, m' in the term $10^{-0.4m'}$, has a nominal value of $m' = 23$. The angular area, $\phi\phi'$, of the detectors is given in units of arcsec squared projected on the sky.

6.6.3.2.2 Quantifying the sky photon flux and sky Poisson noise

For stellar photometric measurements with HST's wide field camera mode, the detector area projected on the sky is, $\phi\phi' = (0.1 \text{ arcsec}) \times (0.1 \text{ arcsec})$ [2, Table 17.3].

The magnitude of B, the sky photon flux (photons/sec) hitting the HST wide field camera detector with a 100 nm filter, is determined by Equation 6.85 to be

$$\left. \begin{array}{l} B = (N_0 10^{-0.4m'})(\phi\phi')(\pi D^2/4)(1-\varepsilon^2)(\tau)(\Delta\lambda) \\ B = (10^8 \cdot 10^{-9.2})(0.1)^2(\pi(2.4)^2/4)(1-0.33^2)(0.324)(100) = 0.082 \text{ photons/s} \end{array} \right\} \quad (6.86)$$

This very low photon rate (only 82 photons in 1,000 s), falls in the range of sky flux rates [0.01 to 0.09 photons/s] quoted above for the wide field camera.

The standard deviation of the sky background noise, $\sigma_{\text{sky background}}$, for Poisson noise with an exposure of 2,400 s is

$$\sigma_{\text{sky background}} = \sqrt{BQt} = \sqrt{(0.082)(0.82)(2,400 \text{ s})} = 12.7 \text{ electrons} \quad (6.87)$$

In contrast, for the Poisson noise for the single star signal, the standard deviation of that noise for an $m = 28$ star with the same exposure time is

$$\sigma_{\text{Poisson noise from signal}} = \sqrt{\kappa S Q t} = \sqrt{127} = 11.3 \text{ electrons} \quad (6.88)$$

It is evident that the signal-related and background-related noises are both comparable for a star magnitude $m = 28$.

For any value of m higher than 28, the major portion of the noise will be from the background. Thus the signal to noise ratio can be calculated using solely the background noise for $m > 28$. In that regime, the SNR equation is given by

$$\text{SNR}_{\text{background-limited}} = \frac{\kappa S Q t}{\sqrt{BQt}} = \sqrt{SQt} \cdot \sqrt{\frac{S}{B}} \cdot \kappa \quad \text{for } B \gg S \quad (6.89)$$

For a given SNR in the sky-limited case the corresponding observation time is

$$t_{\text{sky-limited}} = \frac{\text{SNR}^2}{S^2} \left(\frac{B}{Q\kappa^2} \right) \propto \frac{10^{0.8m}}{D^2} (\phi^2) \quad (6.90)$$

If the angular resolution of the focal plane cameras is diffraction-limited, as it is for the HST, then $\phi \approx \lambda/D$, and the proportionality relationship of Equation 6.90 can be re-written as

$$t_{\text{sky-limited}} \propto \frac{10^{0.8m}}{D^4} \quad (6.91)$$

This shows that the dependence of exposure time on aperture diameter D for the sky-background-limited case is even stronger than it is for the signal-limited case, with $t_{\text{sky-limited}}$ inversely proportional to D^4.

6.6.3.3 Photo-electron detector dark current and detector read-out noise

6.6.3.3.1 Detector dark current

The dark current mean from the CCD detectors, determined by measurements with zero light exposure, is equal to the dark current rate C_{dark} electrons/pixel/s multiplied by n_{pixels}, the number of pixels spanning the image, and the exposure time t. Since this is a Poisson process, the variance is equal to the mean

$$\text{Dark current variance} = C_{\text{dark}} n_{\text{pixel}} t \qquad (6.92)$$

Schroeder's suggested value for $C_{\text{dark}} = 0.003$ electrons/s/pixel for HST's detectors.

6.6.3.3.2 Readout noise

This describes the noise introduced by the camera electronics. Unlike the other noise sources, this noise is not Poisson distributed. It behaves rather like shot noise, and so must be squared to obtain the following variance

$$\text{Read noise variance} = n_{\text{pixel}} R^2 \qquad (6.93)$$

where the standard deviation R is the read-noise per pixel. The HST's wide field camera has an angular resolution of 0.1 arcsec. As shown earlier, each 15 micron pixel has an angular resolution of 0.05 arcsec. Because the angular resolution area of 0.1×0.1 arcsec is equivalent to 2×2 pixels, we can set $n_{\text{pixel}} = 4$ in Equation 6.93.

Schroeder quotes a typical value for R of 5 electrons-rms/pixel for the HST detectors [2, Section 17.4]. Hence, the standard deviation of the readout noise, $\sigma_{\text{read-noise}}$, is

$$\sigma_{\text{read-noise}} = \sqrt{5^2 \cdot 4} = 10 \text{ electrons} \qquad (6.94)$$

Since this value is of the order of the standard deviation of the signal-driven Poisson noise associated with a signal from a star of $m = 28$, the CCD readout noise cannot be ignored.

6.6.3.4 The standard deviation of all the total noise

From Equation 6.74, the standard deviation of the total noise is the square root of the total variance

$$\sigma_{\text{total}} = \sqrt{\text{var(signal noise)} + \text{var(sky background)} + \text{var(dark counts)} + \text{var(read noise)}}$$

and substituting the models for each of these variances yields σ_{total}.

The total noise measure, N_t, in the denominator of the signal to noise ratio expression is simply the square root of the total variance

$$\sigma_{\text{total}} = \sqrt{\kappa QSt + BQt + (C_{\text{dark}}t + R^2) \cdot n_{\text{pixel}}} \qquad (6.95)$$

Rearranging the terms, we obtain the following expression for the standard deviation

$$\sigma_{\text{total}} = \sqrt{(\kappa S + B)Qt + (C_{\text{dark}}t + R^2) \cdot n_{\text{pixel}}} \qquad (6.96)$$

6.6.4 Final equation for signal to noise ratio

Using Equation 6.96 we can write down the final equation for SNR as a function of time for a celestial source of magnitude m covering n_p pixels.

Substituting σ_{total} into Equation 6.65 for the signal to noise ratio, SNR is written in the form presented by Schroeder [2, Section 17.3]

$$\text{SNR} = \frac{\kappa S Q t}{\sqrt{(\kappa S + B)Qt + (C_{dark}t + R^2) \cdot n_{pixel}}} \qquad (6.97)$$

In Equation 6.97 the source flux S (collected from a point-like star of magnitude m, by a primary mirror of aperture diameter D) is given by Equation 6.67 based on the parameters in Section 6.6.2.1. The background sky photon flux rate B is given by Equation 6.86 with parameters suitable for the HST's wide field camera, which has an area projected on the sky of 0.1×0.1 arcsec square [2, Table 17.3].

The HST's detector dark current parameter is $C_{dark} = 0.003$ electrons/s/pixel. Also the readout noise per pixel R is equal to 5 electrons-rms/pixel for the HST detectors [2, Section 17.4].

We can now plot SNR as a function of star magnitude m using all the parameters in Equation 6.97.

6.6.4.1 SNR graph

Figure 6.13 plots, on a logarithmic scale, SNR as a function of magnitude (the measure of star signal strength) for long (3,000 s), intermediate (300 s), and short (30 s) exposure times.

According to Schroeder [2, Section 17.4] the maximum exposure time for any single 90 minute orbit is approximately 2,400 s (or 40 minutes, which is why we chose this duration for our calculations above). This is the approximate dark time of a single HST orbit. Also, multiple images can be collected and averaged together to increase overall SNR and improve image quality.

A dashed line in Figure 6.13 at SNR = 10 shows the maximum star magnitude that can achieve that high signal to noise ratio. Note that a weak $m = 27$ star can achieve an SNR > 10 with a 3,000 s exposure. But for only a 30 s exposure, the brightest star that can be imaged and still achieve an SNR of 10 is one that has $m < 23$.

6.6.4.2 Exposure time graph

We can invert the SNR problem presented above and determine the exposure time t required to achieve an SNR of a given level. To do this we must solve a quadratic

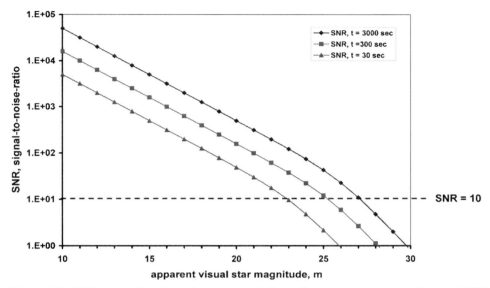

Figure 6.13. SNR as a function of star magnitude m for several exposure times t. HST parameters are used in Equation 6.97.

equation in t obtained directly from Equation 6.97, the algebraic solution of which is

$$t = \frac{1}{2}\left(\frac{(\text{SNR})^2}{\kappa SQ}\right)\left\{\left(1 + \frac{(BQ + C_{\text{dark}})n_{\text{pixel}}}{\kappa SQ}\right)\right.$$
$$\left. + \sqrt{\left(1 + \frac{(BQ + C_{\text{dark}})n_{\text{pixel}}}{\kappa SQ}\right)^2 + \left(\frac{2R}{\text{SNR}}\right)^2 n_{\text{pixel}}}\right\} \quad (6.98)$$

Figure 6.14 plots exposure time as a function of star magnitude m for two values of SNR (SNR = 10 a high quality image, SNR = 2 a low quality, barely detectable image).

Note the rapid increase in exposure time for the weak stars near $m = 30$, with t approaching 28 hours for SNR = 10. Such extremely long integrations require a number of sub-images of shorter duration to be summed.

6.6.4.3 Estimating exposure time in the sky-limited case

We have obtained simple limiting or asymptotic solutions for exposure time, t, for the signal-limited (Equation 6.84) and background-sky-limited noise (Equation 6.90) cases. The most important limit of interest for the HST system is when the signal is so weak that the background sky noise dominates the total noise variance. Yet it is the region where the HST attempts to image stars in the farthest reaches of the universe, and the farthest back in time.

318 Estimating the dimensions and performance of the Hubble Space Telescope [Ch. 6

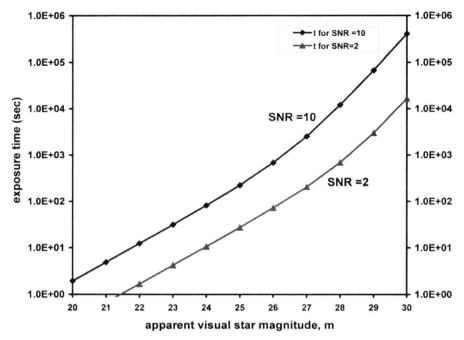

Figure 6.14. Exposure time t versus apparent star magnitude m for SNR values of 10 and 2. HST parameters are used in Equation 6.98.

For this sky limited regime, the appropriate model (Equation 6.90) is

$$t_{\text{sky-limited}} \approx \frac{\text{SNR}^2}{S^2}\left(\frac{B}{Q\kappa^2}\right) \propto \frac{10^{0.8m}}{D^2}(\phi^2)$$

when m is very large and the readout noise can be ignored (a valid assumption when SNR is large). Note that t grows exponentially with m as $10^{0.8m}$. This trend appears as a straight line in the following semi-log plot.

Figure 6.15 shows that the background-limited exposure time approaches the exact "all noise components" solution for very weak signals, $m > 27$.

In summary, we see that the background-limited solution yields reasonable exposure times only for very faint stars ($m > 27$). These exposure times where the solution is valid are quite large with $t > 160$ minutes. To achieve such long exposures the HST must collect and sum many sub-images of shorter duration.

6.6.5 Final thoughts on BotE estimates for HST sensitivity

In Section 6.6 we made a classic model for HST focal plane SNR as a function of apparent stellar magnitude m for a given observation time period, then inverted this calculation to estimate the exposure time required to achieve a given signal to noise

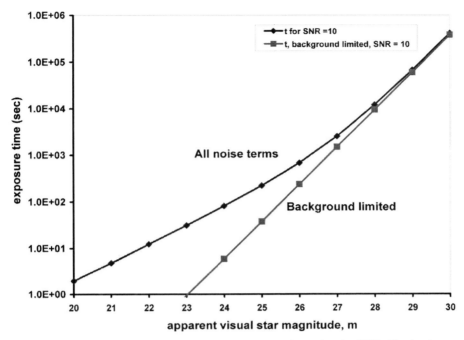

Figure 6.15. Exposure time as a function of star magnitude, m for the HST. The background sky limited approximation, Equation 6.90 is compared to the exact solution (Equation 6.98) for SNR = 10.

ratio when imaging a star of a given magnitude m, and finally showed that the sensitivity limits of the HST system could be estimated using a simple sky-limited equation for SNR and exposure time. Figure 6.13 revealed that with exposure times as long as 50 minutes, the HST can obtain high quality (SNR = 10) images of stars with magnitudes as high as $m = 28$. To sum up this chapter, using our BotE techniques for the HST we were able to obtain simple estimates of some of its important optical dimensions, its total mass in conjunction with the Space Shuttle, and its overall sensitivity to weak, very distant celestial sources of light.

6.7 REFERENCES

[1] Aurora, T.S. and M. Kirk, "Eye in the sky: Science with the Hubble Telescope", *Physics Education*, vol. 30, no. 1, pp. 163–170, Institute of Physics Publishing, January 1995.
[2] Schroeder, D., *Astronomical Optics*, 2nd edition, Chapter 11, Academic Press, 2000.
[3] Endelman, L., "HST mission, history, and systems", 19th International Congress on High Speed Photography and Photonics, International Society for Optical Engineering, SPIE 1358, 1990.

[4] "STS-31 Press Information", Rockwell International, Space Transportation Systems Division, Publication 3546-V, Rev. 4-90, see p. 13, April 1990.
[5] Halliday, D. and R. Resnick, *Fundamental of Physics*, 2nd edition, Chapter 39, John Wiley & Sons, 1981.
[6] Bely, Pierre, *The Design and Construction of Large Optical Telescopes*, Section 4.5.2 "Selection of f-ratio", Springer-Verlag New York, 2003.
[7] Schroeder, D., *Astronomical Optics*, 2nd edition, Chapter 2, Section 2.5a, Academic Press, 2000.
[8] Simpson, D.G., "Optics of the Hubble Space Telescope", note to Prince George's Community College students, May 12, 2010.
[9] Riedl, Max, *Optical Design Fundamentals for Infrared Systems*, 2nd edition, SPIE Press Book, 2001.
[10] Bely, Pierre, *The Design and Construction of Large Optical Telescopes*, "Angular Resolution", Section 4.1.6, Springer-Verlag New York, 2003.
[11] "Advanced Camera for Surveys Instrument Handbook for Cycle 15", Space Telescope Science Institute, October 2005.
[12] Soulihac, D. and D. Billeray, "Comparison of Angular Resolution Limit and SNR of HST and large ground based telescopes", (see Table 1: Characteristics of the Hubble Space Telescope, p. 507), published in "Space Astronomical Telescopes and Instruments; Proceedings of the Meeting, Orlando, FL, Apr. 1-4, 1991", edited by Bely, Pierre, and J. Breckinridge, SPIE Proceedings. Vol. 1494, 1991.
[13] Bely, Pierre, private communication, April 2006.
[14] Lockheed Martin Servicing mission 3A media reference guide, Section 5, "Hubble Space Telescope Systems", Goddard Space Flight Center, December 1999. http://sm3a.gsfc.nasa.gov/downloads/sm3a_media_guide/HST-systems.pdf
[15] Jingquan Cheng, *Principles of Astronomical Telescope Design*, p. 101 Springer.
[16] Carlin, Patrick, "Lightweight Mirror Systems for Spacecraft—An Overview of Materials & Manufacturing Needs", paper IEEE 8.0401, IEEE Aerospace Conference Proceedings, 18-25, 2000.
[17] Mattice, James J., "Hubble Space Telescope Systems Engineering Case Study." Center for Systems Engineering at the Air Force Institute of Technology, Wright Patterson Air Force Base, 2005.
[18] Bekey, Ivan, "Extremely large yet very low weight and low cost space-based telescopes," Bekey Designs, Inc., 1999.
[19] Cool Cosmos Website portal at the Infrared Processing and Analysis Center & the SIRTF Science Center, Caltech, Pasadena, California. http://coolcosmos.ipac.caltech.edu/cosmic_classroom/cosmic_reference/sensitivity.html
[20] Lawrence, Jon, "Telescope Sensitivity", Lecture in Advanced Astronomy course, ASTR278, Macquarie University. http://www.physics.mq.edu.au/current/undergraduate/units/ASTR278/10_ASTR278_JL_5_Sensitivity.pdf
[21] Brownlee, K.A., *Statistical Theory and Methodology in Science and Engineering*, 2nd edition, John Wiley & Sons, 1965.
[22] Heyer, Beretta et al., "Wide Field and Planetary Camera 2 Instrument Handbook, Version 9.0", Space Telescope Science Institute, STScI, Baltimore, 2004.

Index

Advanced Camera for Surveys (ACS), 309
aerodynamic drag, 116, 178, 237
aerodynamic model, simple, 37
aft shroud, 298, 304
air density
   100 s after launch estimate, 174
   747 cruise altitude estimate, 10
   foam impact velocity problem, 196
   at max dynamic pressure, 118
   rule of thumb, 174-5
aircraft mass, 12
Airy diffraction disk, 284–5
Allen, H.J., 117, 230, 263
amateur telescope makers, 302
Ampere's law, 52-3
angle of attack, effective, 11
angular momentum, 44, 47, 75, 136
angular resolution, optical, 284
apogee, 82, 88, 136–7
Ariane, 88–9, 113, 116
ascent-to-orbit period, 20
ATK Thiokol, 101

Back-of-the-Envelope
   engineering estimate, 4
   estimation techniques, 3
   model, 43, 140, 154
   model calculations, 35, 259
   quick-fire methodology, 31
   reasoning, 4, 6
baffle *see* light shield, baffle extension
Ballistic Number (BN), 183, 186
bank angle, 260–2
   effective, 261
banking, 256, 259–60, 262
Bely, Pierre, 281, 288–90, 297
BotE methods *see* Back-of-the-Envelope
Bravo Network, 14

CAIB *see* Columbia Accident Investigation Board and CAIB report
CAIB estimates, 185, 193
CAIB report, 178, 183, 207–8
   volume II, 204
cargo mass, 15–17, 20–3, 82, 84
   cargo bay, 20, 27, 29, 39, 95, 279
   predictions, 15, 25, 27
   rule of thumb, 17
   Shuttle, 19, 27
Cassegrain telescope, 30–1, 280
   primary mirror, 29
   secondary mirror, 30–1
CCD detectors, 287
   dark current, 314
   quantum efficiency, 308
   total electrons measured, 312
Chapman, Dean, 241, 275

I. E. Alber, *Aerospace Engineering on the Back of an Envelope*.
© Springer-Verlag Berlin Heidelberg 2012.

Columbia Accident Investigation Board, 169–70, 177
Conservation of angular momentum, 47, 136
Conservation of energy, 46, 136
COSTAR, 298
Cunningham, J., 267, 273, 276

DC-9, 95
Δv 1st stage with gravity and drag losses, 120
Δv budget, predicted, 126–7
Δv burnout at second stage MECO, 125
Δv gravity loss estimate
  Shuttle (first stage), 110
  Shuttle region 2a (second stage), 120
Δv ideal, Shuttle first stage, 106
density, material, 24, 98, 172
deorbit burn, 239
detection sensitivity, HST, 306
diffraction-limited *see* optics, angular resolution
direct insertion, 139–40, 144
drag force
  drag area, 118, 196, 230
  drag coefficient, 116, 173, 175, 195
  drag coefficient uncertainty, 196
  equation, 116
  lift to drag ratio, 249
dynamic pressure, 10, 116–18, 179
  q (max), 118
  vs time, 117–18
  units, 179
dynamics of rolling ball
  natural roll velocity, 48
  plunger velocity for given natural roll velocity, 48

earth rotation, 20, 126, 144
effective focal length *see* HST, equivalent focal length (diffraction limited)
Eggers, A.J., 117, 230, 232, 263
Eggers–Allen glide solution, 264
energy
  chemical, 18, 139
  gravitational potential, 225
  internal thermal, 223, 227–8
  magnetic field, 52
  stored electro-magnetic, 50
  total light intensity, 285
energy flux, incident aerodynamic, 228
engineering model
  guidelines for building, 12
equivalent focal ratio *see* HST, equivalent system focal ratio (diffraction limited)
equivalent nose radius for Orbiter heat transfer calculations, 266
estimation
  747 thrust estimation, 9
  by analogy, 24
  impact of knowledge and complexity, 25
  tradeoff between complexity and knowledge, 5
exit velocity
  effective, 83–4, 98–100
  OMS, 153
  rocket, 92, 98
external fuel tank
  diameter, 118
  estimated 1st stage propellant consumption, 102
  estimated volume, 94
  estimated wall thickness, 94
  foam separation, 37
  liquid propellant mass estimate, 96
  mass, 23
  soda can analogy, 24
  structural mass estimate, 94
  takeoff propellant mass, 103
  thin walled pressure vessel, 23

Faraday's law, 52
Fermi problems, 4, 7
  engineering, 8
Fleming, F. and Kemp, V., 115, 167
flight-path angle, 226, 238, 243–4
flight trajectory model *see* reentry trajectory
foam
  density, 202
  elastic wave speed, 202
focal lengths, equivalent system, 281
Fortescue, P. and Stark, J., 113, 116
Fraunhofer diffraction, 284

Goddard Space Flight Center, 39, 320
Griffel, W., 213, 220
Griffen, M. and French, J. "Space Vehicle Design", 115

Halliday, D. and Resnick, R., 57, 281, 292
Harwood, William, 128
heat load, 229, 271–3
  NASA data, 273
heat transfer
  blunt body, 231, 266–7
  max rate by Quick-Fire modeling, 228
  maximum rate in glide trajectory, 266
  maximum rate of, 223, 266
  model vs radiometer heat flux data, 270
  radiation, 267
Hirschel, E. and Weilend, C., 260, 276
Hohmann transfer, 135, 139, 151
  Δv, 139
hoop stress, 94
HST
  see also Hubble Space Telescope
  equivalent focal length (diffraction limited), 288
  equivalent system focal ratio (diffraction limited), 288, 299
  exact constructional length, 290
  plate scale, 288
  primary mirror focal ratio, 290
  secondary magnification, 290
  summary of calculated telescope dimensions, 299
Hubble Space Telescope, 277
  aft shroud, 298
  estimated total system mass, 303–5
  estimating total length, 30, 295
  instrument section, 298
  missions, 26, 132, 138, 140
  operating orbit, 300
  optical telescope assembly (OTA), 302
  in Orbiter Cargo Bay, 30
  schematic, 278
  servicing mission, 144, 147
  system support module (SSM), 304
  systems, 278
  total HST mass, 305
  total length, 30, 34, 296, 298

impact height on ball, 44, 65
impact load, 197, 208
  area, 209, 218
  estimated, 209
impact modeling, 203, 207, 219
  constant acceleration approximation, 181
  inverse problem (Osheroff), 194
  non-constant acceleration solution, 189–90
  parametric failure curve, 214
  RCC panel stress model, 213
  threshold load for failure onset, 214
impact speed see impact velocity
impact stress, 198–200, 202, 205–7, 218
  elastic, 200
  elastic–plastic model, 204–5, 209, 218
  full, 203
impact time, 188
  non-constant acceleration solution, 192
impact velocity, 172, 175, 177, 191
  exact solution, 191
  related to failure-level stress, 198, 207
  relative error, 187
International Space Station, 3, 23, 77, 154
Isp see Specific impulse

JWST (James Webb Space Telescope), 305

Kennedy Space Center, 125, 128, 141, 154, 247
Keplerian elliptical orbits, 135

launch trajectory
  modeling assumptions, 110
  NASA computer simulation for Shuttle, 128–9
Lees, Lester, 1, 231–2, 234, 245–6
lift, 10, 178, 239, 243–4, 250, 260
  747, 10–11
lift area
  effective, 11, 241
lift to drag ratio, 251, 254, 260–1, 264
light shield
  aperture door, 297
  baffle extension, 296
  incidence rays greater than 15 degrees, 297
  length, 30, 34

324 Index

light shield (*cont.*)
  protect instruments, 296, 304
  stray light, 30, 296
Linder, Benjamin, 8
liquid propellant mass, 91, 96, 103

Mach number, estimated, 118
magnetic field, 51–2
mass estimate
  HST, 305
  predicted structural, 89
  revised cargo, 25
mass fraction, rules of thumb, 87
mass fractions
  equivalent single stage structure, 27
  structural, 20, 22–3, 25
  STS exact, 88
McDonnell Douglas DC-9, 95
MECO
  main engine cutoff definition, 82
  time to MECO estimate, 125
Mueller, George, 78

NASA data, 112, 114–15, 259, 273
natural roll velocity, 42, 48, 56
  final, 45
Newton, 201, 250, 280
  law of motion for rocket, 98
  rotating rigid body, 45
  second law of motion, 10, 17, 192, 250
noise, shot, 315
nose radius, effective, 266
Nusselt number, 71–2
Nyquist, sampling theorem, 288

ohmic heating, 63, 70, 73
  Ohmic heating law, 68
OMS (orbital maneuvering systems), 82
  cargo mass reduction, 146
  specific impulse, 153
optics
  angular resolution, 279, 284, 288
  equivalent system focal length, 281–2
  focal ratio, 281
  mapping focal plane to the sky, 286
  resolution (angular), 284
  secondary magnification, 290

spherical mirror equation, 292
telescope plate scale, 288
orbit
  circular, 132, 134
  elliptical, 135, 139
  final altitude, 139
  geostationary transfer (GTO), 88
  period, 134
  transfer, 135
Orbital Maneuvering System, 36, 146, 168
orbital mechanics, laws of, 83, 132, 140–2
orbital velocity requirements, Shuttle
  missions, 132
Osheroff, David, 194–6
OTA (Optical Telescope Assembly),
  279–80, 300, 303–4

paraxial, ray approximation, 32
payload mass, 19, 21, 82
  full analytical model (Shuttle), 144
  vs orbital altitude, 145
perigee, 82, 136
  to apogee ratio, 137
  geostationary, 88
  STS-7, 82
photons, 308; *see also* sensitivity (HST),
  photon signal from a star
pitch angle
  definition for launch trajectory, 109
  time-dependent model, 110
pixel
  match to diffraction limited resolution,
    287
  number for Nyquist sampling, 287
plate scale, 288; *see also* HST, plate scale
Poisson noise, 313–15
  Poisson probability distribution, 310
Pool Player, Amateur, 75
Prandtl number, 232
primary mirror, 29–32, 280, 289, 300–2,
  309
  focal ratio, 290
  mass, 300
  sandwich type construction, 300–1
propellant
  average consumption rate, 92
  average SRB density, 92
  external fuel tank estimate, 91
  loaded in OMS, 149

mass consumed to time t, 99
mass fraction, 87
OMS, 82, 146
SRB mass, 92

q see dynamic pressure
Quick-Fire, 14
Quick-Fire modeling
  foam impact velocity, 173
  HST optics length, 30
  max nose heat transfer rate, 228–9
  reentry geometry, 226
  reentry heat load, 229
  reentry heating problem definition, 224
  Shuttle cargo mass (simple), 15
  Shuttle takeoff mass and thrust, 89, 95, 97
  solenoid temperature change, 69, 74
  two stage Shuttle cargo mass, 21

radiation temperature see temperature, nose radiation equilibrium
Rayleigh criterion, 284, 286
RCC failure
  model, 211
  RCC panel, 198, 200, 208, 211, 214
RCC material, 198, 200, 204, 213, 268
reentry trajectory, 37, 250
  distance traveled estimate, 256
  effect of banking see banking, bank angle
  equilibrium glide flight-path angle, 252
  equilibrium glide velocity vs time, 253, 259
  glide model vs STS-5 data, 256–7
  initial entry period, 239
  max flight time estimate, 254
  Orbiter, 246
  simple flight model, 237
Reynolds number, 71, 232
Ritchey–Chretien see Cassegrain telescope
rule of thumb
  air density, 174
  angular diameter of moon and sun, 289
  cargo mass, 17, 91
  first stage burnout velocity, 115
  payload mass vs orbital altitude, 149
  rocket structural mass, 93
  SRB structural mass, 91
  thickness of primary mirror, 302

scale height (atmosphere), 159, 225
Schroeder, D., 281, 308, 315–16, 319
secondary mirror, 28, 31–2, 279, 290
  corrected diameter, 299
  magnification, 30, 282
  radius of curvature, 292
sensitivity (HST)
  background sky flux and variance, 313
  detector dark current and readout noise, 314
  exposure time equation, 313
  photon signal from a star, 308
  signal noise model, 310
  SNR equation, 316
  SNR vs star magnitude given exposure time, 316–7
  system requirements, 29
Shepard, R., 48, 75
Sibulkin, M.J., 232, 275
signal to noise ratio see SNR
SNR, 306–7, 310, 312–13, 316–19
solenoid actuator, 53
  manufacturer, 58, 74
  physical description, 51
  temperature change, 73
solenoid kicking device
  circuit schematic and control board, 63
  maximum plunger force, 55
  plunger travel time estimate, 57
  required ampere-turns, 61
  required soccer ball velocity, 42
  voltage estimate, 61
  wire size and resistance, 59
Southwest Research Institute, 170, 206–7, 210
Space Mission Analysis and Design (SMAD), 16, 38, 88, 114
Space Shuttle
  cargo bay requirement (HST), 279
  estimating first stage drag loss, 117
  gravity loss modeling (second stage), 121
  main components, 80
  mass of Orbiter, 18–9
  model altitude history vs NASA data (first stage), 112
  model altitude history vs NASA prediction (to orbit), 130–1
  model velocity history vs NASA data (first stage), 114

Space Shuttle (*cont.*)
  model velocity history vs NASA prediction (to orbit), 129
  preliminary design phase, 14
  sketch, 16
  takeoff mass estimate, 90–5
  takeoff thrust estimate, 89, 97
  testing philosophy, 79
Sparrow criterion, 284, 286
Specific impulse, 18
  see also exit velocity
  sea level and vacuum, 87, 105
  Shuttle solid rocket motors, 18
  single stage average Isp, 20
  SSME liquid hydrogen/oxygen, 18
spherical mirror equation, 292
SRB
  burnout time, 101
  cylindrical volume, 92
  diameter, 92
  propellant density, 93
  propellant mass, 93
  solid rocket boosters, 80–1, 87, 92–3, 96, 100–2, 105
  specific impulse, 87
  thrust, 81, 97, 100–1
SSME engines, 101, 103, 115, 139
  vacuum thrust, 87, 99
Stanton number, 228, 230, 232, 236–7
  St, 228
star magnitude (m), 4, 309, 312, 314, 317–18
Stefan–Boltzmann constant, 268
STS (Space Transportation System Mission #)
  STS-1, 113, 167, 237, 247, 256, 273
  STS-5, 226, 244, 247, 249, 257–9
  STS-7, 82
  STS-31, 29–30
  STS-41, 148
  STS-61, 148
  STS-107, 169, 183, 221
  STS-119, 129, 167
Sutton–Graves equation, 264, 271
Sutton–Graves constant, 235–6

telescope
  amateur Newtonian, 303
  Hubble space (HST), 277
temperature
  free stream, 233
  nose radiation equilibrium, 267–8
  peak nose-tip, 223, 227
thermal protection system, TPS, 82, 221
thrust, 8, 12, 20, 82, 84, 87, 97, 105, 222
  engine thrust problem, 36
  inventory, 84
  per jet engine, 12
  thrust bucket, 115
Timoshenko, S., 213, 220
Top Chef, 14, 38
Tsiolkovsky rocket equation, 17, 36, 99, 141

Walker, J.D., 204–6, 209, 211, 214–15
wide field camera (HST, WFC), 289

yield stress (Y), 202–5; *see also* impact stress, elastic

Zoby, E. V., 266–7, 273, 276
Zukas, J. A., 200–2, 219

Lightning Source UK Ltd.
Milton Keynes UK
UKOW011030070812

197174UK00009B/1/P

9 783642 225369